AUSTRALIA'S LOST WORLD

PREHISTORIC ANIMALS OF RIVERSLEIGH

MICHAEL ARCHER
SUZANNE J HAND
HENK GODTHELP

INDIANA UNIVERSITY PRESS
BLOOMINGTON & INDIANAPOLIS

Contents

This book is a publication of

Indiana University Press
601 North Morton Street
Bloomington, Indiana 47404-3797 USA

www.indiana.edu/~iupress
Telephone orders 800-842-6796
Fax orders 812-855-7931
Orders by email iuporder@indiana.edu

First US edition 2000

Cataloging-in-publication data is available from the Library of Congress

1 2 3 4 5 05 04 03 02 01 00

ISBN 0-253-33914-6 (cloth)

Produced in Australia
Edited by Louise Egerton
Designed by Robert Taylor
Art Assistant Linda Maclean
Cover design by Nanette Backhouse
Proofreeding by Dawn Hope
Principal fossil photography by Ross Arnett

Dioramas by Dorothy Dunphy
Additional illustrations by Alistair Barnard

Typeset in Australia by
Deblaere Typesetting Pty Ltd
Printed in Singapore through
Imago Productions

The Riversleigh Project is primarily supported by Environment Australia, the Australian Research Council, the University of New South Wales and the Australian Museum.

Foreword

by Sir David Attenborough

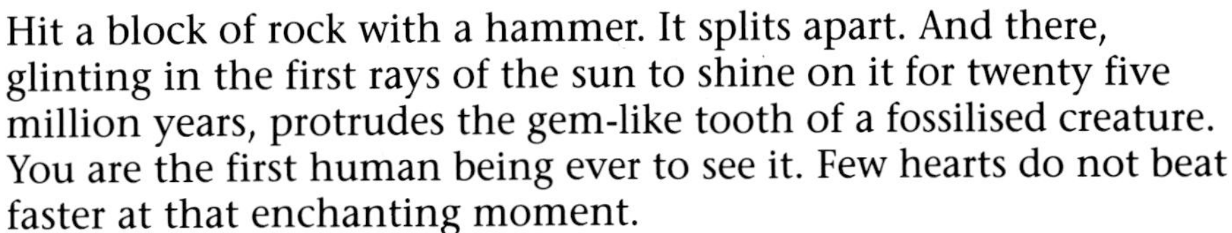

Hit a block of rock with a hammer. It splits apart. And there, glinting in the first rays of the sun to shine on it for twenty five million years, protrudes the gem-like tooth of a fossilised creature. You are the first human being ever to see it. Few hearts do not beat faster at that enchanting moment.

But imagine the excitement if the fossil you see is not a well-known kind but that of a creature so curious that few suspected its existence—a kangaroo that ate not grass but flesh, a toothed platypus with a bill far larger than that of the living species, a marsupial lion, a wombat-like marsupial the size of a rhinoceros and other animals so extraordinary that you are driven to invent bizarre names for them, such as Thingodonta and Weirdodonta.

Such were the discoveries made at Riversleigh in the remote bush country of north-western Queensland by Mike Archer, Suzanne Hand, Henk Godthelp and their team. Their finds were so rich that on one occasion, they picked up bones and teeth from thirty different unknown species in as many minutes. So far, their work there has uncovered the remains of over 150 hitherto unknown mammals—more than twice as many as were discovered in the whole of Australia during the previous century. And in addition there are exquisitely preserved reptiles and fish, amphibians and birds, even insects and millipedes, the vast majority entirely new to science.

The preservation of animal remains over millions of years is a very chancy affair. Only an infinitesimally small proportion of even the most abundant animal that lived in previous geological periods have left any trace of their passing. Nearly all that have done so lived in water or were washed into it after death where deposits of sand, mud or lime entombed them. The remains of *land*-living animals are nearly always totally destroyed within a short period of their death. So vast numbers of species must have existed of which we have no trace or relic. Only in one or two places on the surface of our planet, in the course of the last three thousand million years, have conditions been just right to preserve anything like a representative sample of the species living at any particular time. Those places are the rare treasure houses of palaeontology. Riversleigh is one of them.

What those conditions and remarkable creatures were in this one small area of prehistoric Queensland, you must read this book to discover. You will also find out how the work there has charted the climatic changes that have overtaken the Australian continent as it drifted northwards into the tropics, how it has added whole new chapters to the history of Australian animals, and how this knowledge gives us a new and quite possible critical appreciation of the conservation status of the living survivors.

Read this book for all those reasons. But read it mostly to experience the spine-tingling excitement that comes from exploring an extraordinary lost world.

London, 1991

PREFACE

The Big View of Life

A Palaeontological Perspective

The palaeontologist's view of the world is often very different from that of the biologist's. The latter sometimes presumes that the significance of living creatures is best determined from study of all that breathes—a view which is, perhaps, a bit myopic. Australian creatures, like those of every other area of the world, are the outcome of millions of years of unique interaction between hundreds of thousands of different natural selection pressures and at least as many features inherited from their ancestors.

The Numbat, the Echidna and the Elephant Snail, creatures that we can see and touch, are really ephemeral bundles of interacting cells unintentionally racing through time and space, unable to stop themselves from becoming something else. These unique Australian creatures and their environments are, in fact, illusions. Humans only see life in three dimensions. We are like a camera focused along the finishing line of a horse race, blind to everything but the moment life appears within our limited view. Surely to fully understand life's 'race', we also must know *how* each creature has run its race, how it interacted with others as it ran and the range of its potential for response to variations or changes in its environment.

Questions about our animals, such as 'why do Koalas lack tails?', 'why does the Mountain Pygmy-possum live only in alpine regions of south-eastern Australia?' and 'is the Platypus as secure in its environment as the Brushtail Possum is in its?', require information that simply cannot be provided from the study of living animals or their environments alone. Palaeontology, the study of ancient creatures, *can* provide a more accurate vision of life's parade through time.

Despite the extraordinary variety of previously unrecognised species documented in Riversleigh's fossil record, we have not in fact discovered a single new, distinct creature! This seeming paradox stems from an awareness that LIFE—the whole colossal thing that is, was and will be—quite literally is one gigantic time-travelling creature, without a single physical or temporal break in its four-dimensional being. All of it, every squishy piece from Amoebas, Apples and *Acrobates* to Zebras, Zygomycetes and Zen Buddists, is in reality a single shape-changing mass of trillions of pulsating cells that first evolved on Earth more than

Looking backwards along the axis of time, the whole of life is seen as a single, time-travelling, shape-changing creature—the dynamic, four-dimensional 'Bioblob'. Back through time, every creature that has ever lived is connected by deathless cytoplasm to every other. None of us nor any blade of grass is alone; death and distinction are illusions of limited vision.

3.5 billion years ago. This is the concept of the four-dimensional Bioblob.

Each of our bodies is a mass of replicating cells, every one of which is the ultimate descendant of our parents' sperm and egg cells that united when we were conceived. If we could see these same cells within the fourth dimension of time and look back along time's axis, we would see that the cytoplasm of each cell in our bodies is quite literally an amoeboid extension from those first two cells. In the fourth dimension, every cell is in tangible contact with the one from which it arose. Looking further back, those sex cells of our parents, to which we are physically connected by living cytoplasm, are themselves living extensions through space and time of the original cells that united to become our parents' bodies, and so forth back to the first cell that gave rise to all subsequent life on Earth.

Further, and perhaps most difficult to accept, at every point along that time dimension axis, *every* one of those dividing cells remains alive. Death is an illusion. This must be so if what we call individuals contribute cell lines that extend into the future. And even if they do not, as with childlessness, the cytoplasm will still be physically connected *back* through time to living tissue that *does* pass on into the future—perhaps through the parents' other offspring or the offspring of the aunts, uncles or great-great-grandmother. At some point back in time, the lineage must connect to another piece of the time-travelling 4D Bioblob that continues to shape-change and move through time.

This palaeontological vision challenges common understanding about the biological world. For example, what we perceive as 'species extinction' should in reality be understood as changes in the relative growth of parts of the Bioblob. Similarly, the search for real species *boundaries* is futile because in four dimensions they are simply not there, species being once again figments of limited vision. This is why we can say, paradoxically, that although we have discovered hundreds of new species at Riversleigh, in fact we have discovered none! Rather we have found hundreds of previously unseen protoplasmic strands of the 4D Bioblob back along the time axis.

But we have also discovered two other attributes about the 4D Bioblob from studying the Riversleigh fossils: new branching points in space and time for the 4D Bioblob's protoplasmic tendrils; and a great deal of information about the length of many of Australia's tendrils back through time. Branching points are in fact one of the few real boundaries on the Bioblob—points in time where newly distinguished tendrils of the creature peel away from the others to begin their own history of subsequent branching. For example, in the extinct giant flightless bird from Australia *Dromaius gidju*, we appear to be very close to the branching point between emus and cassowaries.

The length of the tendrils of the Bioblob back through time, from the present to the date of the Riversleigh fossils, defines a minimal period since those tendrils separated from all others. These lineages might be considerably older but until a rich record preceding that of

Riversleigh's is found, we cannot determine how much longer. For example, the Riversleigh records for many possums, kangaroos and snakes define minimal lengths of time since they separated from their cousins.

Riversleigh's fossil record is one of the world's great refuse piles! Here discarded fragments from the Bioblob, like dandruff, were cast off during the creature's travels through space and time. From study of these discarded fragments, we can obtain an, albeit imperfect, vision of the true shape of the Bioblob.

The fossil record is also the most powerful tool humans have for understanding the nature of change in natural environments. This last opportunity is shaping up as one of the most important to come out of the Riversleigh Project. With conservation of the world's ecosystems and biodiversity an urgent priority facing every nation on earth, we *must* discover, and quickly, how best to comprehend what is happening to the remnants of unique life on our continent. As the Riversleigh Project took shape and the extraordinary richness of the resource became evident, it became apparent that this resource *must* be of vital concern to conservationists.

It is impossible to determine from a decade or two of modern field studies the conservation status of any living species. It simply can't be done—not with any confidence. Short-term perturbations in populations may alert us to short-term problems but we really cannot rely on these to define the long-term 'health' of a lineage. Rarity cannot be interpreted as anything other than rarity. To tell if a lineage is in the process of decline, stability or rise, we must have an historical perspective—and not just that of the last few centuries.

For some groups we can already provide this information, this fourth-dimensional perspective, for the vast majority of Australian terrestrial vertebrates over an interval of at least the last 25 million years—and for some over the last 55 million years (with the newly discovered early Tertiary Murgon deposits which contain, among other things, Australia's oldest-known marsupials). With more than 20 000 really superb specimens (increasing at about 4000 per year) from Riversleigh's deposits, representing more than 150 faunal assemblages spanning the last 25 million years, we are beginning to pull together one of the world's best opportunities to study evolving species-rich communities.

Ecologists working as co-investigators recognise the Riversleigh Project's potential to establish trends in Australian ecosystems as 'Greenhouse' conditions loom on the horizon. Without this information, human-induced changes cannot be distinguished from pre-human trends. Without understanding the links between past and present, well-meaning but short-sighted conservationists could risk serious mismanagement of the precious remnants of Australia's unique biota.

CHAPTER 1

Riversleigh

A Search for Beginnings

Our exhausted vehicle death-rattled its way over the corrugations of the north Queensland 'road' that led, we hoped—signs being a luxury last seen in Mount Isa 250 kilometres to the south—to Riversleigh Station. The noise of four numbed shock absorbers, an irreparably-holed muffler and camp gear doing the rounds of a never-ending cycle between suspension in mid-air and deafening crashes as the floor of the vehicle lurched upwards again, made conversation impossible. Outside, whenever the explosions of bulldust subsided, the countryside, rock and vegetation alike, was a kaleidoscope of yellows and browns. Volcano-like termite mounds erupted between scrubby trees and needly cushions of spinifex. Too frequently, the overturned hulk of a rusting car declared the folly of trying to make anything mechanical survive a trip like this.

When at last a dilapidated road sign with the roughly scrawled word 'Riversleigh' loomed out of the swirling dust, we headed off onto an even less distinct track, doomed to spend still another hour imitating the contents of a cement mixer. It was getting late and there was a growing certainty that we would have to set up camp in the dark. As the vehicle dropped into darkening gullies and hurtled over still-glowing rises, Agile Wallabies seemed to spring out of nowhere, leaping across the road like deranged rabbits. Suddenly the track swerved left and dropped out of sight. The vehicle came crashing down onto a rutted slope as we hit the brakes, swerving to a stop. Pans, tents, tins, gear, all momentarily possessed by poltergeists, dropped out of the air onto heads and laps. Thus, ingloriously, we

Site D (centre) from the air. The road north to Lawn Hill Station, visible on the right, was an old Cobb & Co. Stagecoach track.

Riversleigh Station is a rugged cattle property of just over 3000 square kilometres in far north-western Queensland. Most of the Tertiary fossil-bearing rocks on this Station have now been included in an extension to the Lawn Hill National Park.

arrived at the O'Shanassy River Crossing and, soon after, Riversleigh Station itself, a vast privately-owned cattle station in Queensland's Gulf country.

The then owner Ted Naughton had given us permission to search for fossils on the Station so we were warmly welcomed by his manager, John Nelson. He suggested

we camp on an island between two branches of the Gregory River so we wasted no time in rattle-clanking our way to the river crossing. After becoming used to the dust-covered, semi-arid vegetation that led us this far, we were utterly delighted to find the Gregory River itself a tropical paradise. Its banks were lined by huge trees including palms and giant cluster figs, remnants of lusher times when vine forests were spread far out across what is now scrubby woodland and drying spinifex. Golden insects, purple-crowned wrens and darting things half unseen wove arcs through the last rays of sunshine streaming in across the surface of the darkling river. We stared in delighted silence for a while as the ringing in our

The tree-lined, spring-fed Gregory River on its way to the Gulf of Carpentaria nourishes Riversleigh.

In the cool of an early June morning, mist rises off the warm waters of the Gregory River.

S. WILLIAMS

Because the waters of the region are supercharged with dissolved limestone, they often build tufa dams in places where the passage of water is obstructed.

R. ARNETT

ears from the vehicle subsided, gradually replaced by the sounds of rushing water, frogs, crickets and myriad birds. A large, seemingly unperturbed egret wandered into our unfolding camp, curiously peering into tents and boxes.

By night, our island refuge seemed fraught with mystery. Half-heard screams cut across the sounds of running water, animals of unknown dimensions and food

R. ARNETT

The clear, freshwaters of the Gregory River mirror the remnants of a once lush vegetation that dominated the whole region.

preferences hungrily sniffed the sides of our tents before disappearing into the darkness, four-metre pythons searched the area for midnight snacks. Ants by the millions, unseen and unheard, explored every ruptured box of food and giant spiders found strange new places in which to hide—one causing severe heart palpitations when it plopped out into a bowl of muesli the next morning. Wide-eyed and willing, we were being overwhelmed by a tropical paradise.

Next morning the real reason for the trip to Riversleigh took hold—an obsession with this Station's fossil deposits that has lasted more than 15 years. We drove across the deeper, second crossing of the Gregory towards the fossil deposits that had intrigued palaeontologists before us. At a place known as 'Site D', we found what we had come for. In the greying limestone blocks that tumbled down from the cliffs above there were fossil bones, bits of life's fabric torn from a world otherwise lost to prehistory. Here were all manner of unfamiliar creatures, such as the jaw and gem-like teeth of a tiny species of primitive kangaroo, leg bones of enormous flightless birds some of which may have weighed 400 kilograms in life, and the skull of a bizarre crocodile in whose mouth, as a remnant of its last supper enjoyed 25 million years ago, was the jaw of a tiny marsupial lion—all creatures that lived and died on the edge of a north Queensland rainforest millions of years before anything remotely human gazed across the plains of Africa. Riversleigh was a gateway, a beckoning time portal through which we could pull into the present things long since lost to the past, unfamiliar beginnings and much stranger ends.

Today, Riversleigh is a cattle station on the Gregory River, about 200 kilometres north of Mount Isa and about 200 kilometres south of the Gulf of Carpentaria, suspended as it were between the blue waters to the north and the waterless tracts of Australia's red centre. So, too, is the biota of this northern area suspended between worlds for along the Gregory River the vestiges of a tropical past, such as palms and cluster figs, cling precariously to the banks of the deep, spring-fed river while the surrounding sea of yellow grass and Red Kangaroos threatens to wash over the oasis.

From the gravel track that winds through the Station, threading together the Gregory Developmental Road to northern points such as Lawn Hill Gorge National Park and Burketown, the hills of Riversleigh may at first look monotonous: amorphous, jumbled piles of stone, spotted with scraggy trees, covered in a patchy cloak of hummocky spinifex grass and the whole peppered with red termite mounds of wildly varying shape.

But these hills are not all the same. When the visitor pauses long enough to gather an entourage of flies, it soon becomes clear that some of the hills are the off-white colour of massive quartzite. These are the upthrust and then eroded remnants of a once vast sea floor that spread across this area during Precambriam time perhaps 1600 million years ago. Life in this ancient sea consisted of little more than bacteria and mats of algae-like stromatolites. Other hills are dark grey and made of marine limestone rocks stacked like Sao biscuits. These are the remains of a shallow sea that spread across the region 530 million years ago. Its waves would have lapped against the flanks of the Precambrian mountains, slowly tearing down their substance to bury it once again with the shells and spicules of trilobites and sponges, creatures that thrived in its younger waters. Nothing as 'advanced' as even a fish darted in this sea.

Still other, rarer, hills are of a much younger age. These are the paler grey younger limestones that range in age between 25 and 15 million years. They often appear like mounds of muffins, carved into this shape by millions of years of naturally acid rain. They were formed in freshwater pools, lakes and perhaps streams that nourished the floor of a vast area of rainforest.

Palms, cluster figs and other luxuriant plants persist as an oasis along the Gregory River which winds through an ocean of dry spinifex grass and dust.

Rearing up through the forest, like arcane giants from long-dead worlds, would have been the Precambrian and Cambrian hills, both interminably gouged by the searching roots of hungry vines and towering trees.

Our work has revealed that the younger rocks of the Riversleigh landscape are more complex than they seemed at first. Some, the oldest, appear to have formed in large lakes where bottom sediments entombed the bones of strange crocodiles and giant flightless birds. These in turn were buried by at least two sequences of limestones that formed in shallower pools whose margins swarmed with a staggering variety of forest creatures. These layers were then dissolved by groundwater to form caves in which even younger bones of bats and other forest creatures accumulated. The cave deposits span at least the last 20 million years and provide vital clues as to how

Australia's continental rainforests gave way to the dry habitats that now blanket 44% of the continent.

The Riversleigh fossil vertebrate deposits are among the richest and most extensive in the world. For this reason, in 1992 these deposits were included in an extension to the Lawn Hill National Park. The more than 150 fossil deposits now recognised (at least 100 of which represent extinct rainforest communities) span a period of some 25 million years and are found over an area of more than 40 square kilometres. The fossilised remains are also exceptionally well preserved. The reasons why the fossil record of this particular region is so spectacular are complex, but perhaps basic to them all is the fact that for millions of years the groundwater of the region has been saturated with dissolved calcium carbonate. For this reason, wherever it appears at the surface, as springs, pools or rivers, it cannot help but encourage the development of fossils. Even today in the waters of the Gregory River cow bones can be found partially coated by precipitated limestone. Within perhaps months, these coatings can cement the bone to the floor of the pool or river and then, over years,weld it rock hard into the constantly accumulating limestones. Thus trapped, the processes of fossilisation proceed and the bone is gradually mineralised into an enduring piece of biological history. Because the coating process is so rapid, it has been common for whole skulls and other bones of ancient animals to be entombed before they could be broken by the feet or jaws of other animals. For this reason, Riversleigh's fossils, despite their often considerable age, include some of the most perfect recovered from the continent. Fortunately, the process will continue. This extraordinary area cannot help but accumulate a rich record of all life that has made Riversleigh its home.

Among the most common rocks in the region are Cambrian limestones that formed in a shallow inland sea 530 million years ago.

Why Riversleigh is important

Modern mammals in Australia are significantly distinct from those in the rest of the world. During the last 50 million years, most continents at one time or another have swapped many of their creatures which is why, for example, there are tapirs in South America as well as south-eastern Asia, lions in Africa and (until recently) North America and Eurasia, rhinoceroses in Africa, Asia and (until recently) North America, bears in all northern continents and even members of the camel family in Africa (camels) and South America (e.g. llamas). But *only* Australia, which lacks all of these, has kangaroos, koalas and numbats.

Because Australia was relatively isolated from the other continents for at least the last 45 million years, it became a second, highly distinctive laboratory for the production of biological diversity. Its fossil record provides, in addition to a lithic notebook of natural experiments that led to kangaroos, koalas and platypuses, an army of even more bizarre creatures that left no living descendants. The compulsion of palaeontologists, students of the history of life, is to find that long lost notebook or whatever remains of it.

Australia, again unlike all other continents, seems intent on making that search difficult. With few mountain ranges or canyons to expose the older record, each rare glimpse we have of Australia's biotic past is treated with unashamed reverence. In a recent book on the fossil mammals of the world, Australia rated an embarrassing 5 out of 432 pages—and, to add insult to injury, the map of Australia was printed upside down! The records for other Australian vertebrate groups, all just as fascinating in a world context, are unfortunately not much better than that for mammals. But that is about to change. With the growing additions from Riversleigh's record as well as from other newly discovered sites around Australia, the next edition of that book cannot help but devote a vastly larger section to the unique and intriguing biological history of this continent.

To those of us consumed with the need to explore 'lost' worlds and understand how

Discoverers Sue and Jim Lavarack inspect the fossil bones of a large extinct bird at a place they named Sticky Beak Site.

Australia's modern biota developed, Riversleigh has been a vast step forward. By the late 1980s, discoveries of, for example, Tertiary mammals from Riversleigh had nearly trebled the record previously known for the whole continent. While Australia formerly provided about 40 Tertiary mammal assemblages, Riversleigh has now added an additional 150 such assemblages. Most of these are far more diverse than any previously known. In some cases the fossilised jaws and skulls of more than 64 distinct kinds of mammals have been recovered from as little as a cubic metre of ancient pond floor.

In addition to new and diverse mammals, Riversleigh's record is giving up the bones of hundreds of other kinds of vertebrates such as: masses of frogs; lungfish; giant horned turtles; snakes and lizards, including some groups that occur in South America; a wide range of birds, including parrots, song-birds, giant predatory raptors and a host of giant flightless 'mihirungs'; beetles; ants; millipedes; and now, as a result of the 1990 Riversleigh Expedition, even leaves, seeds and wood. In short, the biological paraphernalia from almost all of the occupants of what were once complex rainforest ecosystems are tumbling out of these ancient limestones.

Perhaps most importantly, we have in Riversleigh an opportunity to study how prehistoric ecosystems gradually developed into those that persist today—and to use this

GLADA

As the Riversleigh fossil-rich limestones are dissolved in acetic acid, their surfaces begin to bristle with the bones and teeth of long extinct creatures.

understanding of change through time to make informed predictions about probable trends into the future.

While it is certainly true at this stage that more is missing from Riversleigh's fossil record than is represented, the significance of its resources stand out for many reasons (see Chapter 5). With each successive year of field collecting, preparation and research, more of significance about the Riversleigh resource becomes apparent. While what we are already aware of is enough to keep us busy for years to come, each year of research leads to new questions and new ways of looking at old problems. The challenge of understanding the past will take a great deal of the future.

SOME BASIC CONCEPTS

In talking about what has been discovered at Riversleigh it is useful to acquire a bit of palaeontological jargon. Technical terminology is too often used by specialists to impress the curious. It is quite human for someone to feel 'superior' when in a group they seem to be the only one who knows what the word 'obfuscate' means! But unless use of the term helps to communicate an idea or contributes to an economy of words, it may only confuse. Our aim here has been to use as few technical terms as possible. Where we have had to, commonly to avoid repeating a longer-winded alternative expression, we have explained the term the first time it has been used and, in most cases, also added it to the Glossary.

Here we mention only a few of the more frequently used terms in this book. The Glossary should be consulted for more detailed or additional explanations.

The Oligocene (38 to 24.6 million years ago), **Miocene** (24.6 to 4.5 million years ago), **Pliocene** (4.5 to 2 million years ago), **Pleistocene** (2 million to 10 000 years) and **Holocene** (last 10 000 years) Epochs are names of intervals of Earth's history (see geological time scale, page 177). They are parts of **Tertiary** (65 to 2 million years ago) and **Quaternary** (last 2 million years) Periods which are in turn the two parts of the **Cainozoic Era** (65 million years to the present).

Sites are specific places where we have sampled the rocks and their contained fossils. Most often, the amount sampled will have been taken from no more than two square metres of area. Often site names commemorate some physical feature such as Upper Site or Outasite, the discoverers such as V.I.P. Site or Syp's Siberia, some conspicuous aspect of the site such as Gotham City Site (which was loaded with fossil bats) or an event connected with the discovery such as Gag Site.

Local Faunas (abbreviated LF) are the names given to the total assemblage of kinds of fossil animals recovered from a particular site. Commonly these take the site name as their base such as Upper Site LF. Sometimes they are based on local Aboriginal words such as Nooraleeba LF which means 'stuck in the mud' in reference to the way in which that particular assemblage was presumed to have formed.

Systems A-C refer to our current understanding of the relationships of the older Riversleigh sites to each other. System A is a collection of sites that appears to be among the oldest in the area, probably late Oligocene to early Miocene in age. System B is a collection of sites, most of which occur in an isolated area of the Station called Godthelp Hill, that appear to be younger, probably early to middle Miocene, than the System A sites but older than the System C sites. The latter also occur in a discrete region of the Station, known as the Gag Plateau, and contain what are presumed to be middle to early late Miocene sites.

CHAPTER 2

Extraordinary Discoveries

A history of understanding

In a strange way, the current Riversleigh research program really began in the howling sands of central Australia. Here thousands of years of dry winds have sand-blasted through the crust to reveal, like an old painting, a previous world of fossil-bearing clays and sands that accumulated in vast inland freshwater lakes perhaps 25 million years ago. Exploration of this green and wet heart of the continent was carried out most aggressively in the 1950s and 60s by Reuben Stirton with students and colleagues from the University of California at Berkeley. At Lake Palankarinna on Etadunna Station in the Tirari Desert (the south-eastern corner of the Simpson Desert), they found a series of mainly flat-lying sediments which they interpreted to be Oligocene in age. In fact, they later realised that several different ages were represented by these sediments and their fossils.

It was the fossils in the lower, older units that attracted the most attention for they included the remains of creatures unlike any that had been seen before. They found many previously unknown kinds of marsupials and birds hiding like rare gems among masses of catfish and turtle bones. Among the mammals they found koalas (*Perikoala*) and wombat-sized palorchestids (*Ngapakaldia*). As it turned out, this was just an appetite whetter and in the years that followed, Stirton's group continued successful exploration and excavation of the region.

Of the subsequent discoveries they made, the most important to Riversleigh turned out to be that of the 'Leaf Locality' on the eastern edge of Lake Ngapakaldi, between the channels of the vast Cooper and Warburton Rivers. Here, the brightly coloured bedload of an ancient river arched steeply up from its burial ground under the massive dunes of the Tirari Desert. As the desert's sand and the lake's salt tore the ancient clays apart, they left behind a sprinkling of bright fossilised bones spread over the surface of the dry salt lake. These baubles were first spotted by Paul Lawson, from the South Australian Museum, and excavation of the Leaf Locality was smartly under way.

From the trenches of the Leaf Locality came the remains of many new animals, including marsupials, birds, reptiles and fish, the whole being known as the Kutjamarpu Local Fauna. The instantaneous appeal of this limited but fossil-rich deposit was such that it became the focus of a major expedition from the USA in 1971.

Although Stirton's pioneering work certainly got the ball rolling, the 'success' rate of work in the central Australian fossil deposits was not all that high. For example, a month-long expedition for four people to Lake Palankarinna has rarely resulted in the recovery of more than half a dozen incomplete jaws and

Palaeontologists at work on 20-25 million-year-old fossil deposits at Lake Palankarinna in the Tirari Desert, one of the most important sites in central Australia.

Brushes, picks and other collecting equipment are used to recover 20-15 million-year-old bones from the Leaf Locality at Lake Ngapakaldi in the Tirari Desert.

Richard Tedford, one of the pioneers of fossil work in the Tirari Desert and Riversleigh, inspects a bat bone at Microsite, Riversleigh, during a return visit in 1987.

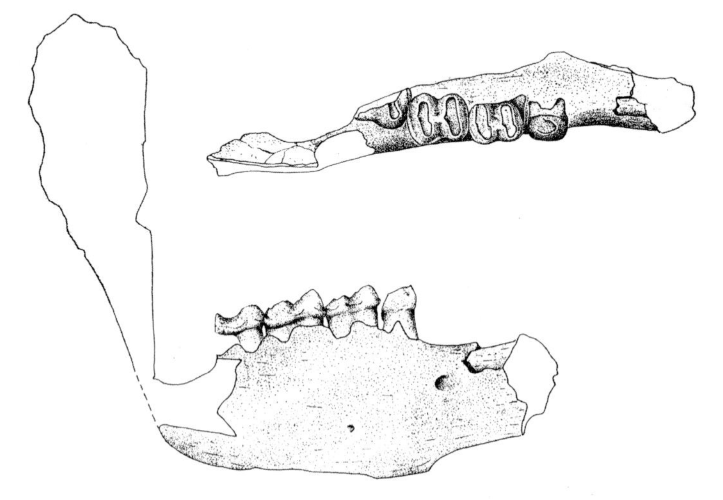

One of the first mammals found at Riversleigh and published in 1967 by Dick Tedford was *Bematherium angulum*, a large extinct plant-eating marsupial.

perhaps a hundred isolated teeth representing up to ten different kinds of mammals, of which perhaps three would be new. However that is not a *bad* recovery rate for fossil vertebrates on a world scale—good fossil vertebrates not being that easy to find, and at that time there was nowhere else in Australia where fossil mammals of this age could be found in any greater abundance. So, until 1983 the three of us migrated back to the centre like hypnotised lemmings for yet another bout of dunes, staked tyres and sandy sandwiches, ever hopeful and sometimes successful at finding yet another mind-boggler.

In July, 1963, Richard Tedford (then one of Stirton's students, now at the American Museum of Natural History) took the opportunity while being in Australia to join Alan Lloyd of Australia's Bureau of Mineral Resources for a brief exploration of fossiliferous deposits in Queensland's Gulf country. They spent five days on Riversleigh Station which led to their discovery of six sites (Tedford's Sites A to F), including the now famous Site D, the starting point for most subsequent palaeontological expeditions to Riversleigh.

Tedford's investigations at Riversleigh led to a 1967 publication in which he described the then new diprotodontoid marsupial *Bematherium angulum* and illustrated but did not name other fragmentary specimens, most of which were found at Sites D and C.

Like Tedford, Mike's early visits to Riversleigh, starting in 1976, followed five years of work in central Australia. The success of the first Riversleigh trip, although less than that of the years following 1983, was as great as any trip to central Australia. It was a different area, the prehistoric beasts were different and it was just a bit easier to discover better specimens. So, we persisted.

Early discoveries at Riversleigh

As far as we are aware, the first European to note that there were fossil bones on Riversleigh was W.E. Cameron. In 1900 he published a short paper about fossil bones he had collected and which were determined by the then Director of the Queensland Museum in Brisbane, the Rev. C.W. DeVis, to represent the 'Nototheriidae' (marsupial cow-sized diprotodontoids).

If such a discovery in Tertiary rocks were made now, it is almost certain that a palaeontologist would promptly organise an expedition to check it out. This is the way that most of the important palaeontological discoveries have been made in Australia. Unfortunately, two things mitigated against DeVis following up Cameron's 1900 lead. First, Cameron was convinced that the rocks that produced these fossils were no older than Pleistocene in age, i.e. less than 2 million years old, and many deposits of this age, even then, were known from Australia. DeVis himself had pioneered the study of many of these, particularly those from the richly fossiliferous eastern Darling Downs of Queensland with its giant lizards and rhinoceros-sized marsupial *Diprotodon*. Second, the Gulf of Carpentaria was a long way from Brisbane, perhaps weeks by train and stagecoach. Although a Cobb & Co. stagecoach track evidently went right past the fossil site (which we suspect was Site D), DeVis was 72 at the time and probably disinclined to organise a major 'stagecoach expedition' simply to investigate another Pleistocene site.

Over the next 50 years, several geologists and invertebrate palaeontologists debated the age of the bone-bearing Riversleigh limestones, variously estimating them to be Cretaceous, Miocene, Pliocene or Pleistocene in age because fossils then known were too few to provide an unambiguous age determination. By the time we began our work, Tedford had concluded that Riversleigh fossils were early Miocene in age on the basis of comparisons of the fossil mammals he and Lloyd had collected with those known from South Australia. Although this estimate was subsequently changed to middle Miocene, it was generally concluded that it was about as old or just a bit younger than the oldest mammal-bearing central Australian deposits from the Tirari Desert.

Our first trip to Riversleigh—1976

Our first trip to Riversleigh was motivated by a conversation with the Director of the Queensland Museum, Dr Alan Bartholomai. When previously visiting Riversleigh, he had seen the fossil jaw of an evidently new diprotodontid marsupial in a massive block of limestone at Site D but had been unable to retrieve the coveted object; so there it evidently still was. This was all the incentive we needed.

Site D

Site D, the Riversleigh locality where Dick Tedford collected most of his Riversleigh specimens and the starting point for our work in 1976.

With the permission of the then station owner, Ted Naughton, and managers John and Jan Nelson, we began work at Site D in 1976 and, following Alan's directions, soon found the abandoned jaw, projecting upside down from the cement-hard roof of a limestone overhang. The surface of the dome transected the specimen, revealing half a molar suspended over an oval section of the dentary. Our hammers and chisels, chosen for their hardness and carefully sharpened before the trip, proved useless. Barely hours after starting, the chisels were all blunt, the hammers chipped and our throbbing thumbs unrecognisable amalgams of dried blood and bandages. Far worse, there were distinct sounds of prehistoric laughter echoing from the limestone dome over our heads. So we, too, soon gave up on the perverse jaw, although we managed to get the fiendish thing out in 1978—only to find that it consisted of a single molar, the vast bulk of what should have been there having been eroded away by naturally acid rain thousands of years before.

Despite that initial failure, the 1976 trip was very important to us. In the first place, we managed to recover other palaeontological treasures such as giant bird bones, a crocodile skull and even a few mammal jaws from rocks on the surrounding hillside. One, in particular, was the dentary of a small kangaroo. Recovery of this jaw, first noticed as the tip of a caramel-coloured tooth projecting from a small boulder on the lower slope, took many hours of work. By pounding a blunt chisel round and round at a safe distance from the object, a deep channel finally separated the block containing what we hoped would be a complete jaw from the parent rock. The small, blood-stained crater from which we removed that block (and recovered a nearly complete jaw, published in 1979 as *Wabularoo naughtoni*) is still as fresh and hard as it was when we left in 1976.

When work first began at Riversleigh, the tools used proved hopelessly inadequate for the task and, as a result, very few, albeit intriguing, specimens were obtained.

Site D, seen here from the air, contains the bones of many kinds of large, extinct marsupials, birds, crocodiles, turtles and other creatures.

Discovery of Microsite

It was during that trip that Henk, after climbing over the top of Site D and walking back to the saddle west of the crest, made a momentous discovery. The others heard his excited shout a moment later. He had found a rock that was bristling with tiny bones—millions of mouse-sized limb bones! The initial significance of his discovery, which became known as Microsite, did not take long to comprehend. These were Miocene bats, fascinating and very important animals that, before this, had had a Tertiary fossil record in Australia consisting of exactly two teeth. Henk had found literally millions of bones, jaws and teeth and, as things turned out, re-routed the entire course of the Riversleigh Project. The *nature* of those bats was even more surprising than the discovery itself.

Microsite, a pale patch of limestone in the centre right, was discovered in 1976, and was the first locality to produce fossil bats at Riversleigh.

When the Microsite limestone was dissolved in acetic acid, masses of perfect tiny jaws, each little more than a centimetre long, rained out of the dissolving surfaces.

THE MICROSITE BATS

Flying in the Face of Ignorance

When Henk first discovered bat bones at Microsite in 1976, he set in motion a chain of events that is still snowballing.

In 1971 a single bat tooth had been recovered (by Mike and Mike Woodburne) from an Oligo-Miocene site in central Australia. The procedure of classifying this bat was fraught with difficulty. It convinced us that bats were too diverse and too frequently convergent in molar morphology to enable a single fossil tooth to be confidently placed in any particular family-level group. Eventually, it was published as 'a possible rhinolophid'—the first paper devoted to an Australian fossil bat.

When *hundreds* of jaws and *thousands* of isolated teeth began to tumble out of the Microsite blocks, the opportunity to more precisely determine relationships came in a rush. This time the jaws and teeth were sent to Bernard Sigé in Montpellier, France, a world expert on fossil bats. We were not prepared for the response. Bernard instantly recognised the Riversleigh bat as a member of a group he had spent years studying, one he previously thought confined to the Oligo-Miocene deposits of France. In 1982 the new bat was named *Brachipposideros* (the same genus as many other Oligo-Miocene fossil bats from France) *nooraleebus* (in reference to the Nooraleeba Local Fauna, the assemblage from Microsite).

The palaeobiogeographic significance of the Microsite bat overwhelmed us. Not only did we suddenly have a mammalian overlap with another continent, we had a potential tool to help date the Riversleigh deposits. Its relationships to the French species suggested it was most closely related to early Miocene taxa. That date has now been corroborated by other means.

Since Henk's discovery, many more fossil bats from Riversleigh have been recovered and some sites, such as Upper Site, have as many as 10 different species, some of which have relatives in European fossil deposits. There is therefore a high probability of being able to cross-check this age interpretation.

The Microsite bat will, however, remain important because it is the only bat of its kind in that deposit. This means that it can be used to estimate how much variation one might expect for a species of *Brachipposideros*, something that is difficult to determine with confidence for bats from other sites where many often similar types occur together.

All in all, Henk's 1976 discovery of Microsite continues to open many doors.

UPPER SITE

About 20 million years ago, in the early Miocene, the lowland rainforests of Riversleigh teemed with life. In the crowns of the trees were possums of every imaginable kind, including strange types long since lost to the world. Leaves on the forest floor were stirred by the fleeting feet of herbivores ranging in size from tiny wallabies to cow-sized diprotodontids. Creatures as strange as 'Thingodonta' and as familiar as musky rat-kangaroos jostled each other for space. The flesh-eaters were only just less diverse. Marsupial lions licked their rapier-like incisors; flesh-eating kangaroos day-dreamed of reddening kills and deep-headed crocodiles prowled the forest margins. Even within the forest floor, ancestral marsupial moles and blind burrowing snakes pursued their prey and sometimes each other. For marsupials, this was a time of enormous variety. Bats, too, were here in diversity and abundance, streaming out of their diurnal hiding places to consume the myriad insects of the night. In the forest floors lime-rich, often carbonate-encrusted pools developed acting as perfect pit traps, cementing together the bones of unwary passers-by with those of creatures that made the pools their home.

1 Giant flightless dromornithid bird (*Bullockornis* sp.).
2 Small carnivorous thylacine (*Nimbacinus dicksoni*).
3 Dragonfly.
4 Browsing wynyardiid marsupial (*Namilamadeta* sp.).
5 Deep-headed, semi-terrestrial, quinkanine crocodile.
6 Chelid turtle.
7 Frog (*Lechriodus intergerivus*).
8 Snail.
9 Leaf-nosed bat (*Brachipposideros* sp. 1).
10 Ancestral musky rat-kangaroo (*Hypsiprymnodon bartholomaii*).
11 Marsh frog (*Lymnodynastes* sp.).
12 Water dragon (*Physignathus* sp. similar to *P. leseur ii*).
13 Skink (scincid lizard).
14 Long-toothed bettong (*Wakiewakie lawsoni*).
15 Burrowing marsupial mole (notoryctid).
16 Forest bandicoot marsupial (peroryctid).
17 Carnivorous dasyurid marsupial.
18 Carnivorous kangaroo (*Ekaltadeta* sp.) eating a bird.
19 Large browsing marsupial diprotodontid (*Neohelos* sp.).
20 Free-tailed bat (molossid).
21 Dragon lizard (varanid).
22 'Thingodontan' marsupial (*Yalkaparidon coheni*).
23 Bower bird (paradisaeid).
24 Ancestral emu (*Dromaius gidju*).
25 Leaf-nosed bat (*Brachipposideros* sp. 2).
26 Rainforest koala (*Litokoala* sp.).
27 Gecko (gekkonid lizard).
28 Honeyeater (meliphagid).
29 Cicada (cicadid).
30 Ants (fomicids).
31 Feather-tailed possum (acrobatid).
32 Marsupial lion (*Priscileo* sp.).
33 Tree frog (*Litoria* sp.).
34 Python (*Montypythonoides* sp.).
35 Pilkipildrid possum (*Djilgaringa* sp.).
36 Tiny, very primitive ringtail possum (*Paljara* sp.).
37 Ground cuscus (*Strigocuscus* sp.).
38 Rainforest mountain pygmy-possum (*Burramys* sp.).
39 Woolly rainforest ringtail possum (*Pseudochirops* sp.).
40 Millipede.

Dunphy

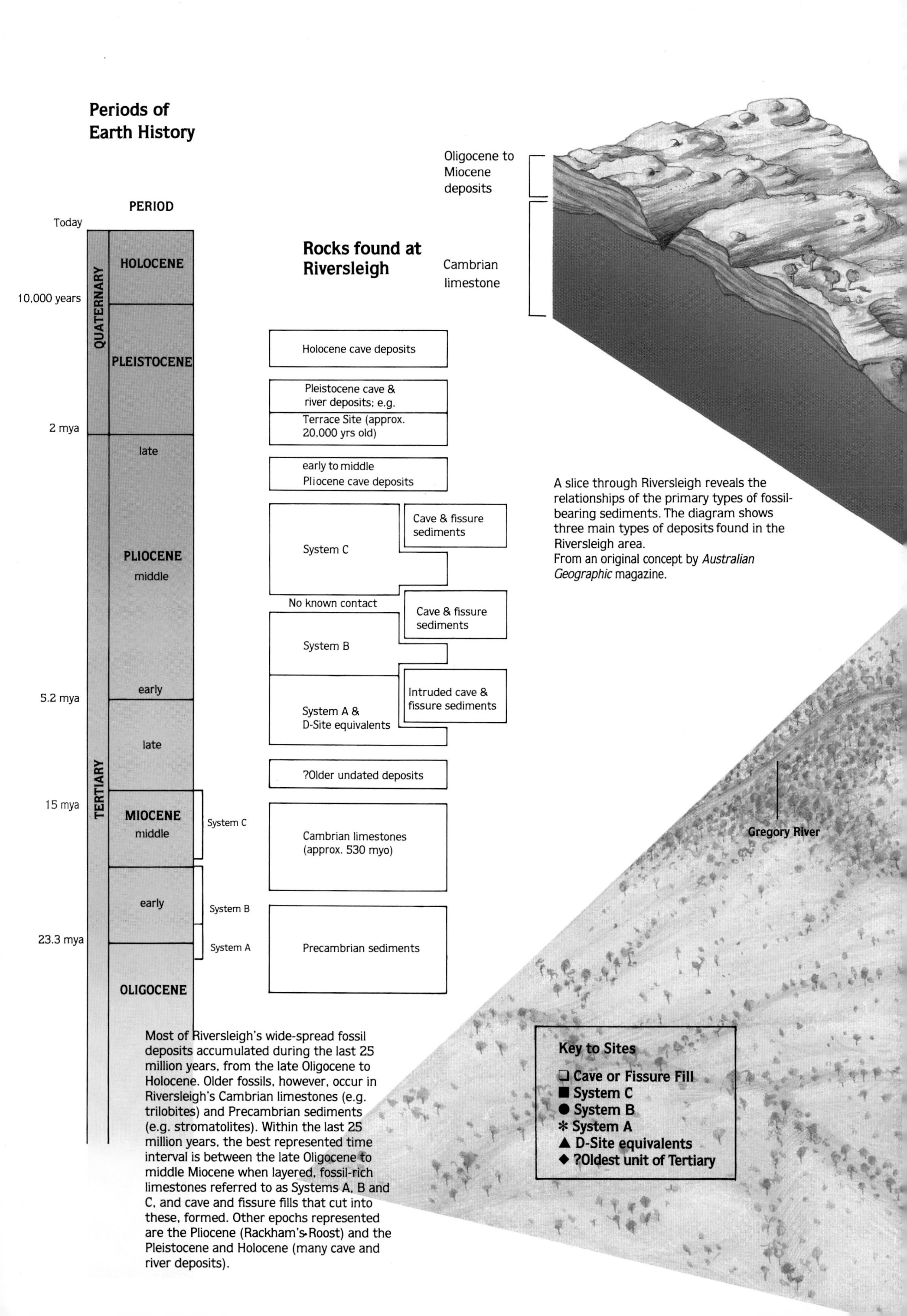

A slice through Riversleigh reveals the relationships of the primary types of fossil-bearing sediments. The diagram shows three main types of deposits found in the Riversleigh area.
From an original concept by *Australian Geographic* magazine.

Most of Riversleigh's wide-spread fossil deposits accumulated during the last 25 million years, from the late Oligocene to Holocene. Older fossils, however, occur in Riversleigh's Cambrian limestones (e.g. trilobites) and Precambrian sediments (e.g. stromatolites). Within the last 25 million years, the best represented time interval is between the late Oligocene to middle Miocene when layered, fossil-rich limestones referred to as Systems A, B and C, and cave and fissure fills that cut into these, formed. Other epochs represented are the Pliocene (Rackham's Roost) and the Pleistocene and Holocene (many cave and river deposits).

Riversleigh Sites

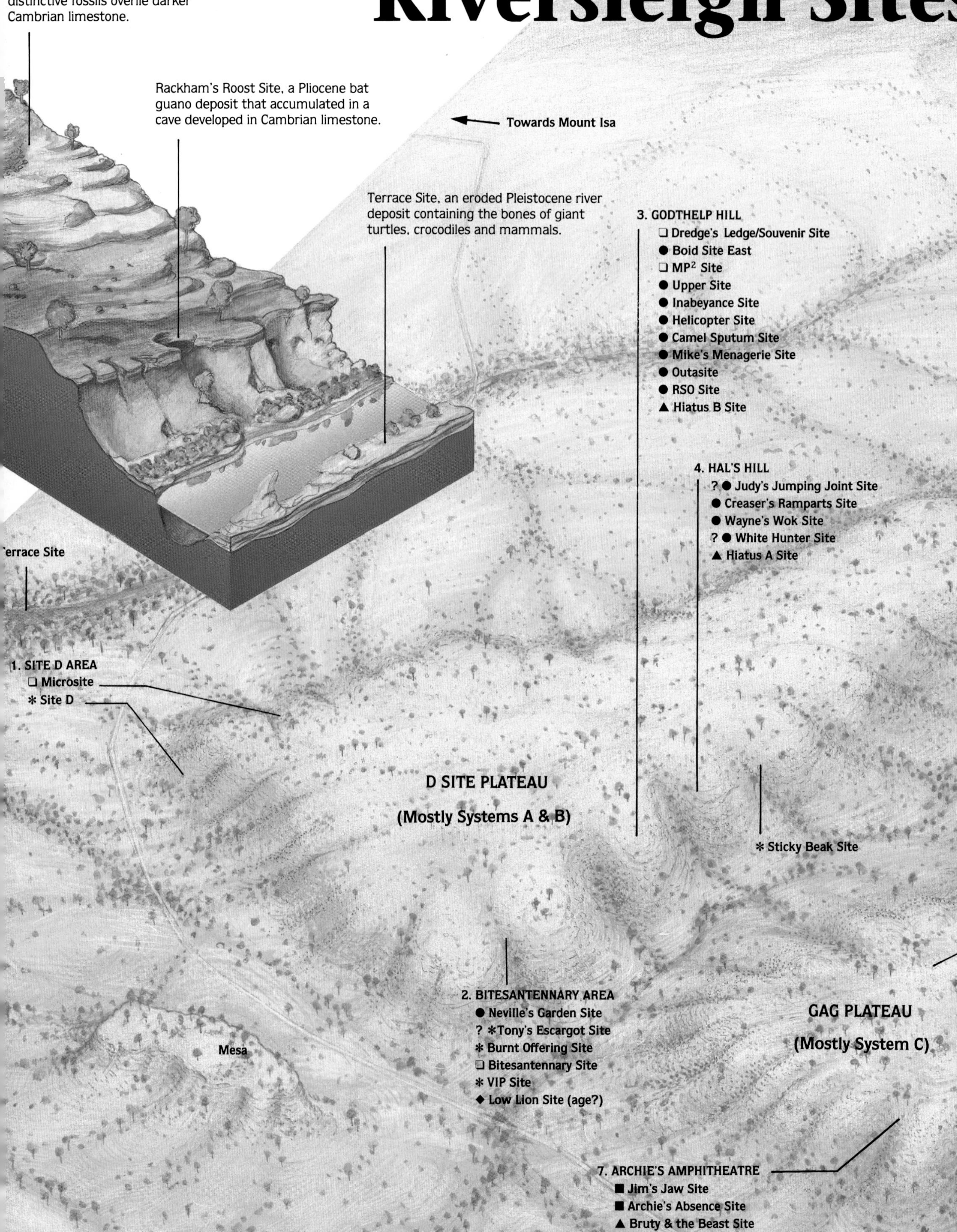

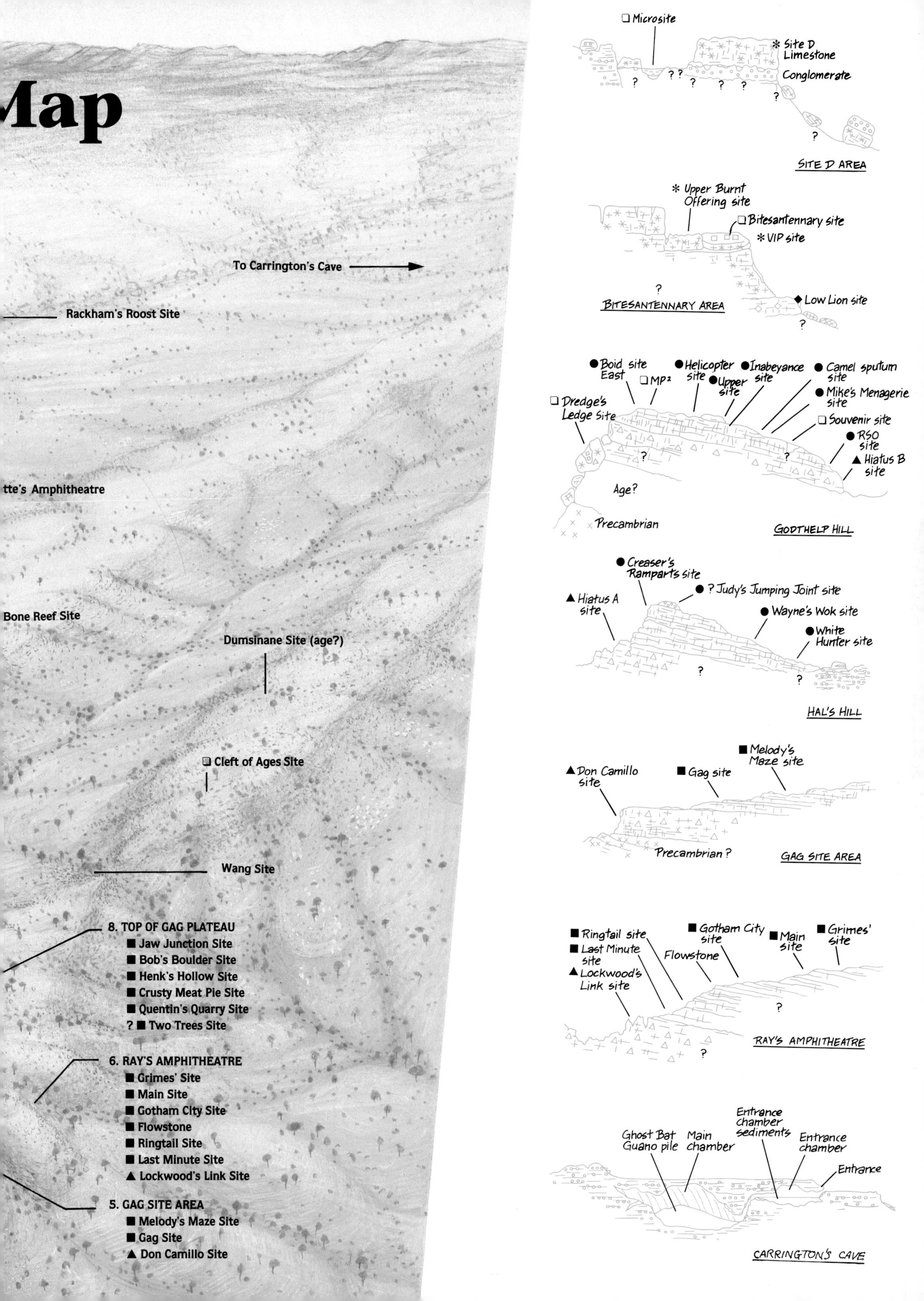
ap
To Carrington's Cave
Rackham's Roost Site
tte's Amphitheatre
Bone Reef Site
Dumsinane Site (age?)
Cleft of Ages Site
Wang Site
8. TOP OF GAG PLATEAU
Jaw Junction Site
Bob's Boulder Site
Henk's Hollow Site
Crusty Meat Pie Site
Quentin's Quarry Site
? Two Trees Site
6. RAY'S AMPHITHEATRE
Grimes' Site
Main Site
Gotham City Site
Flowstone
Ringtail Site
Last Minute Site
Lockwood's Link Site
5. GAG SITE AREA
Melody's Maze Site
Gag Site
Don Camillo Site
Microsite
Site D Limestone
Conglomerate
SITE D AREA
Upper Burnt Offering site
Bitesantennary site
VIP site
Low Lion site
BITESANTENNARY AREA
Boid site East
MP²
Helicopter site
Upper site
Inabeyance site
Camel sputum site
Mike's Menagerie site
Dredge's Ledge Site
Souvenir site
RSO site
Hiatus B site
Age?
Precambrian
GODTHELP HILL
Creaser's Ramparts site
? Judy's Jumping Joint site
Hiatus A site
Wayne's Wok site
White Hunter site
HAL'S HILL
Melody's Maze site
Don Camillo site
Gag site
Precambrian ?
GAG SITE AREA
Ringtail site
Last Minute site
Lockwood's Link site
Gotham City site
Main site
Grimes' site
Flowstone
RAY'S AMPHITHEATRE
Entrance chamber sediments
Ghost Bat Guano pile
Main chamber
Entrance chamber
Entrance
CARRINGTON'S CAVE

The gently dipping layers of limestone on Godthelp Hill are crammed with Miocene fossils.

GLADA

Henk is studying the most common animals found in the Rackham's Roost deposit: some of Australia's oldest rats and mice.

GLADA

The early Miocene fossils at Wayne's Wok Site occur in a 20-metre long horizontal layer (about knee-level) sandwiched between less fossiliferous limestones above and below.

A piece of Rackham's Roost limestone preserves tiny fossilised bone fragments and teeth—the remains of prey eaten 3-5 million years ago by cave-dwelling carnivorous bats.

GAG SITE

If any single discovery at Riversleigh stands out as most important, it must be that of Gag Site. Discovered, almost serendipitously, at the very end of the 1983 expedition, it was the first discovery of a really rich Tertiary site in the Riversleigh area. Almost hidden from site by a thick carpet of prickly spinifex, the limestone pavements at this site were covered with the jaws, teeth and bones of more than 30 new kinds of Oligo-Miocene mammals—nearly half again as many as had been previously known for the whole continent.

The single-most significant breakthrough for our work at Riversleigh came with the discovery of Gag Site in 1983.

WANJI LANGUAGE

That first year, we also met Ivy Stinkin, a very old woman of the Wanji Tribe who had lived for many years as a station hand and friend on Riversleigh Station. She spent evenings, with John and Jan Nelson, teaching us the Wanji names for the animals and plants of the Riversleigh region. We were interested in this information partly because whenever possible we try to enshrine local dialects in the names of new fossil animals (e.g. the name of the first-discovered Riversleigh kangaroo became *Wabularoo* after we learned that *Wabula* means 'long forgotten') and partly because Ivy was, she declared, one of the few surviving people who could fluently speak the Wanji language. She was concerned that if it wasn't written down, it would vanish as quickly and thoroughly as had her tribe.

Ivy Stinkin with Jan Nelson (left) and son Christopher: 1976.

The 1977 Expedition

Before the 1977 Riversleigh Expedition, we did our homework. First, we sent one of the Queensland Museum preparators, David Joffe, to 'explosion school' to learn how to use all manner of volatile substances. Second, we scoured the heavy hardware stores for gear that could divide limestone and came back with an electric jack-hammer (with which we finally secured that troublesome Site D jaw). Armed with a battery of new and somewhat indelicate gadgets and with conspicuous red 'EXPLOSIVES' signs hung on all sides of our Queensland Museum trailer, we headed up the track, mission-bent and feeling just a bit like the Dirty Half-dozen.

At the Station, we asked John Nelson's permission to use explosives. There was no problem but he asked if he could come out to the site when we began—he was curious about how it would work. So were we. The cry 'Shot!' went up as required by law. With hearts pounding, we crouched behind a massive boulder and put our fingers in our ears. When the half stick of gelignite went off, there was a colossal 'Whumph!', a fountain of dust and the barrel-sized boulder in which it had been placed simply vanished. All we ever found was a head-sized piece that came down a few seconds later about two metres from the vehicles. We could not have anticipated what would happen before the trial but clearly this was not the way to do it. The limestone, although hard, was evidently far too brittle for gelignite. Eventually we discovered that we needed no more than a few folds of 'Red Cord', an ICI light explosive normally used in quarrying operations to link widely-spaced charges to ensure their simultaneous detonation. This technique has now been perfected and, happily, we sent no more boulders to the Great Rock Pile in the Sky.

In 1978 Mike moved to the University of New South Wales where Sue was finishing a Zoology degree. Henk stayed on at the Queensland Museum until he joined the team at the University of New South Wales in 1982. We spent the 1978-80 field seasons working deposits in other areas of Queensland and South Australia, being (as it turned out, naively) in doubt that Riversleigh had much more to offer than we had seen in 1977.

The 1981 Expedition

In 1981, we were joined in the field by Tim Flannery and Georgina Hickey, as well as Cindy Hann and Rosendo Pascual (from Argentina). This time we collected about a tonne of Microsite material, as well as more sacks of bone-rich rock from Site D. This was the start of a major growth phase in the Riversleigh work but, exciting as they were, none of the discoveries at Microsite prepared us for the breakthroughs of the 1983 Expedition.

Discoveries of 1983

We discussed the discovery of Microsite with Mike Plane of the Bureau of Mineral Resources in Canberra and Ken Grimes of the Geological Survey of Queensland and in 1982 both told us that they could distinguish Microsite on the basis of its colour on air photos (things which we could not afford at the time) because the limestone of this locality appeared paler than that of the surrounding rocks. Both also noted from the air photos that similarly pale limestone hills appeared much further to the west of the areas we had so far explored. This raised the exciting possibility that there might be other small mammal-rich 'microsites' in those pale-coloured hills.

So, in 1983, we were joined by Ken Grimes and Debbie Wood (both from the Geological Survey of Queensland) and Ken Aplin and Klim Gollan (University of New South Wales). On the first day, Ken Grimes and Mike walked into a corner of the previously unknown area but found little more than a few Site D-type localities. On the second day, Ken returned to the campfire with a small bone in a rock which he passed around for comment. We recognised it immediately—it was unmistakably the limb bone of a bat. Excitement grew because it could mean another, perhaps slightly different, Microsite-type locality had just turned up, one that clearly contained small mammals in contrast to localities like Site D which at that stage had only produced middle-sized to large mammals.

Discovery of Gag Site

The following morning, we all headed towards Ken's new site—but some of us never got there. While most of the team followed Ken, Sue returned to the vehicle for the maps and Mike waited on the edge of the plateau near the vehicle. In the few moments while he waited, he looked down at his feet—and nearly passed out. There, bristling out from the stone he was standing on, etched by hundreds of years of naturally occurring acid rains, were the beautifully preserved jaws of bandicoots, dasyurids, kangaroos, bats and many more. He shrieked. Soon the rest of the team had gathered together at what was to be called Gag Site. In a mere hour's-worth of deranged oggling and crawling around on knees immuned by excitement to the pain of prickly spinifex and limestone, we spotted the remains of dozens of new kinds of animals, many more than we had found on all of the previous field trips to Riversleigh combined. By the time the excitement settled down to hoarse shouts, we had seen or collected jaws and skulls representing at least 30 mammals never before seen by anyone. When the material was later prepared in Sydney, this number was to rise to a staggering fifty-eight. As well, there were vast arrays of frogs, lizards, birds and even lungfish. This was paradise!

From that one site alone, we nearly doubled the sum of previous knowledge about the diversity of Australia's Tertiary mammals. Every palaeontologist dreams of moments like this as others dream of fishing a fist-sized ruby out of dirt. Yet you don't *really* expect it to happen, so much so that you even doubt your senses

From the air, the younger, paler Tertiary limestones of Riversleigh can be distinguished from the more common, older Cambrian limestones. This paved the way for the discoveries of 1983.

when it does. But, as it turned out, this discovery was only one of a vast number that followed hard on its heels.

When we returned to the camp that evening, we were ecstatic. The only 'champagne' in the camp was a $1.00 Don Camillo sparkling wine, so the plastic 'corks' were popped. Ken Aplin and Klim Gollan had just returned from a two-day trip to Mount Isa to put sacks of Microsite on the rail to Sydney. We wallowed in the looks of utter amazement on their faces as they listened to the tales of discovery. But with the following morning, came mischievousness; we apologised to Ken and Klim for the 'pretense' of the previous evening but declared that *today* someone would find a site as incredible as the 'tale' of the evening before. By the time we climbed into the vehicle for the day's explorations, Ken's face couldn't get much longer. We drove to the site of the previous day's discovery and suggested that he head *that* way, towards where we knew the site to be. Five minutes later, when Ken stubbed his toe on a jaw-filled block, we heard the excited shriek 'You bastards!' This tomfoolery earned for the locality the name 'Gag Site'.

The Gag Site discovery occurred the day before we were due to leave but there was just enough time to collect about 2 tonnes of material from the Site and for Henk to find another rich locality—Henk's Hollow Site—near the top of the hill which has come to be known as the Gag Plateau. All in all, it was a quite extraordinary year.

A 1985 bushfire which swept the limestones clean (below) of obscuring spinifex grass (above) revealed vertical sequences of horizontally bedded limestones.

Improving on Paradise: 1984 to the Present

The 1983 discovery of Gag Site and the realisation that it was the real start rather than the culmination of work in the area, pretty much guaranteed that Riversleigh would be an annual pilgrimage for years to come. Each year brought far more than simply discoveries of new, fossil-rich sites. With every trip we understood a bit more about the nature of the Riversleigh resource; ideas were generated both during and after the field trips about how the rocks formed, how they relate to each other in time and space and how the prehistoric animals came to be fossilised in the areas where we found them. As a consequence, each subsequent trip was an opportunity to test current ideas.

The 1984 Expedition, however, was one of pure discovery, unfettered by attempts to comprehend grand generative schemes. For us then, it was Christmas in Bone Land. We had been joined by a few newcomers,

including John Courtenay and Melody Nixon of the Gulf Local Authorities Development Association, and Ray Langford, a Queensland National Parks and Wildlife. Ranger resident on Lawn Hill National Park. Ray's first move after being shown the boundaries of what we knew at the time was to scramble over the edge of the Gag Plateau on its north-western side into an elongated, shallowly eroded depression which became known as 'Ray's Amphitheatre'. On its eastern flank, it oozed fossil deposits of all kinds. There were so many sites that we could almost guarantee to find another by picking any pile of spinifex and pushing it aside. That richness was the first clue that our initial ideas about the nature of the Riversleigh resource were hopelessly simplistic.

Also joining us for the 1984 field season were journalist Bob Beale and photographer Quentin Jones, both then from the *Newcastle Herald*. After putting in a week of hard yakka—sledging boulders, bagging rocks and lifting heavy sacks—Quentin decided to take a well-deserved day off. He chose for his soujourn the Gregory River crossing, over which the water flows with soporific gurgles, fast and clear. This, decided Quentin, was the place to be, and he climbed into the camp's 'rubber duckie', a small, plastic inflatable dinghy. He anchored the dinghy with a sturdy rope to a tree in the middle of the main flow of water, positioning the craft so that it remained in the shade of some fragrant *Melaleuca* trees and not too far from the bank. It wasn't long before the pleasant conditions got the better of Quentin and he drifted into a light sleep.

Quentin says that it was the shadow that passed over him that woke him. Snapping open his eyes, he

Journalist Bob Beale (left) and photographer Quentin Jones (centre) joined us on the 1984 Expedition and brought the significance of the Riversleigh deposits to the general public.

In the limestone layers exposed at Ray's Amphitheatre, many different-aged fossil sites were discovered.

instinctively threw himself from the dinghy into the river, even before realising what it was that had disturbed him. He tumbled around in the fast-flowing water, unable to get a footing on the smooth, jumbled stones of the crossing and gripped by the sort of panic only experienced in the worst of nightmares. Finally, after crashing into a large, partially submerged log, Quentin, gasping for air, looked up. There on the river bank was the cause of his terror—two local Aborigines in traditional dress stood over him. Leaning on their spears for support, they were doubled over with laughter—they had certainly 'caught' more than they expected on that fishing trip!

Stratigraphic sequences

When we returned to Sydney after the 1984 trip, material processed from the Ray's Amphitheatre locality called Last Minute Site produced a beautiful jaw of a small but very strange possum now described as *Djilgaringa gillespieae.* As curious to us was the fact that only one other tooth of this animal was recovered from another Riversleigh deposit—Gag Site. While there was a pattern beginning to form here, we did not see it. Because we had begun with Tedford's (1967) understanding about what was happening at Riversleigh, we had accepted that the bone-bearing units (except perhaps Microsite whose relationships to surrounding sediments were very unclear) were all approximately the same age from Site D to Gag Site. So why did this strange possum occur only in those two sites, despite the fact that by 1984 we had many more sites with an abundance of small mammals?

The unexpected answer came in 1985. Sometime in mid-summer, the volatile and prickly spinifex grass that had previously covered the area to almost hip height, was devasted by a widespread bushfire. When we arrived in 1985, months after the fire had vanished, it was a different Riversleigh that greeted us. Suddenly, there were rocks *visible* everywhere. New fossil sites turned up at a rate of almost six a day. But most importantly, while we used the opportunity to stand back and photograph the north-eastern flank of Ray's Amphitheatre, we were dumbstruck by what we had completely failed to notice before: the fossil sites occurred in a thick sequence of gently inclined layers. When tall spinifex cloaked the hills, there simply were not enough visual cues for the eye to perceive the pattern that now confronted us. Suddenly, the otherwise anomolous distribution of *Djilgaringa gillespieae* began to make sense. Gag Site was near the base of the Gag Plateau on its eastern edge; Last Minute Site was at the base of the Plateau on its northern flank—they both contained the same possum probably because both sites were about the same age and older than most of the other topographically higher sites we found in 1983 and 1984.

That simple observation was the key to a whole new way of perceiving the Riversleigh palaeontological resource. The incredibly high diversity of prehistoric animals found by that time at Riversleigh made more sense if the assemblages from which they came (the various local faunas) spanned time as well as space. Since that realisation, much of our effort has been devoted to working out the vertical sequences and/or horizontal equivalents among the more than 150 discrete fossil sites.

Volunteers, trucks and helicopters

From 1985 to 1987, we joined forces with John Courtenay and his organisation PROBE. They arranged for a team of self-supporting volunteers to help us with a range of tasks that had to be accomplished during each expedition.

This community involvement marked a major turning point in our 'normal' procedure. Before this, we were a small band of zealots in a tropical wilderness of palaeontological potential. As a consequence, most of us soon began to suffer

GLADA

Between 1985 and 1987 volunteers, gathered here at the base camp, were organised by PROBE to help with all aspects of the Riversleigh expeditions.

In 1985 and 1986, the Australian Army provided an Iroquois helicopter and crew from the Townsville-based Wallaby Squadron to help survey in and collect from remote areas.

infuriating lower back problems from toting tonnes on our backs over too great a distance. With the volunteer army, we evened the odds in our favour.

Looking back, there was more to this than simply acquiring a lot of extra hands. Until the arrival of the first teams of volunteers in 1985, Riversleigh expeditions had been a kind of 'private' experience, reserved for the palaeontological cognoscenti. Each year we arrived with a small band of committed enthusiasts (the number that could fit into one normally dilapidated four-wheel drive) and camped (with permission) on the edge of a large island in the Gregory River. Each day's activities were discussed in the morning over breakfast. The days were spent struggling with spinifex, flies and too many overweight boulders; nights were filled with the sound of rippling water, frogs and the snapping of wood on the fire. Dinners were whatever tin your hand landed on in the hessian bag. The deep river, with its overhanging wall of pandanus and paperbarks, was a thing of awe to us, in part because we had been told by locals that they had, from time to time, heard the bellowing of large saltwater crocodiles. Consequently, we always swam in the shallows where vehicles could cross without being swallowed whole by giant green saurians. In all, the living Riversleigh of those first years was as mysterious to us as the fossils.

Once the volunteers arrived, most of this changed. Nights filled with the sounds of rational voices, pots and pans in the kitchen tent, two-way radios and discussions about how the volunteers were to be deployed the following day. The days involved coordination of trucks and teams doing many different things simultaneously. Media groups turned up at a regular pace, attracted by the sudden public interest in what was happening in this tiny corner of the Gulf. Before we came to terms with how best to use this new resource, day's end would find us exhausted—not from hauling rocks but from organising people. And somewhere along the line, the giant crocodiles which we had never actually seen must have vanished because at the end of each day we dropped with weary abandon into the deepest pools—and were never eaten.

Once we found how best to use all of this help, the routine became less exhausting and the scientific output from the trips increased by many orders of magnitude. Whereas before to sample a site took three or four people a week of literally back-breaking and relatively inefficient work, we now needed just a day. While two or three individuals worked out the best way to loosen large rocks, others sledge-hammered, others helped to assess the richness of what was being uncovered, others sacked and labelled material for transport and still others ferried the bags—at a safe pace—as they were filled. It was marvellous!

What's more, we met a whole range of interesting people all of whom shared our enthusiasm for palaeontology. From their ranks came many 'regulars', like the Lavaracks and the Scott-Orrs, who became as hooked on Riversleigh as we were. They came back each year and eventually were regarded as part of our core team of research workers. Several went on to help form the Riversleigh Society Inc. and found ways in which the community at large could support the field and laboratory work. Some of the enthusiastic media personnel, like Geoff Burchfield (the ABC's Quantum program—'The Ghosts of Riversleigh') and Ian Anderson (writer for *New Scientist*), became similarly valuable in bringing to the public's attention, here and overseas, the extraordinary importance of the Riversleigh resource.

But there were some problems in the field that armies of enthusiastic volunteers simply could *not* solve. What did we do when rich fossil deposits were discovered in really remote areas? We had to find some way to carry out the masses of limestone blocks necessary to adequately sample these new sites. We contemplated hot air balloons led along by ropes (but the rocks were too sharp, the wind gusts unpredictable and daytime temperatures much too high), the purchase of a Unimog (too many thousands of dollars) and even camels (but the jagged, irregular limestone terrain would tear their feet and legs apart).

Finally the Army solved the problem, at least for a while. The Townsville-based Wallaby Squadron provided an Iroquois helicopter and crew to help with the 1985 and 1986 Expeditions and a Unimog truck in 1987. This life- and back-saving helicopter could carry almost a tonne a trip, using its underslung cargo net. Without this help, it would have been practically impossible to back-pack more than a few sacks of rocks from the very distant areas.

The Iroquois was also our first opportunity to explore well beyond our normal working range. On the first trip into the remote reaches of the Gag Plateau, we started approximately 11 kilometres to the west of Gag Site and worked back towards the north-eastern edge of the Plateau where we had first found Gag Site. Touching down every 3 kilometres or so, we found fossil deposits nearly everywhere we looked. The enormity of Riversleigh's fossil fields began to sink in. Clearly, we were working just the edge of a vast, richly-fossiliferous region whose boundaries were delightfully unknown.

Discovery of Godthelp Hill and systematic surveys

As if to emphasise these first impressions, soon after the Iroquois survey in 1985, Henk discovered what came to be known as Godthelp Hill, well distant from the Gag Plateau. The whole surface of this enormous inclined hill was riddled with complex fossil deposits, including the particularly rich Camel Sputum and Upper Sites. So rich is this area that every year we return to collect a bit more from one site or another, we find yet more new sites. For example, in 1989 two

Situated high on a cliff above the Gregory River, Rackham's Roost, Riversleigh's only Pliocene deposit (about 3-5 million years old), was discovered in 1985 by 14-year-old, Mt Isa resident, Alan Rackham Jnr.

M. SANDILANDS

distinctively different types of sites (including Dredge's Ledge Site) were found on the eastern flank of this hill.

The isolated nature of many of the fossiliferous hills in the Riversleigh area, such as Godthelp Hill, led us to organise systematic surveys in 1986 and 1987. Small teams of volunteers, led by members of our core research team, spent each successive day prospecting a section of unexplored Tertiary rocks. Although some survey days were better than others, the net effect was a mountain of new sites to be checked.

Another richly fossiliferous region found in this way was Hal's Hill, first explored by Hal Lescinsky. On its rear slopes were several important and very rich sites including Wayne's Wok and White Hunter. During the exploration of these sites, volunteers Jim and Sue Lavarack wandered across the valley to find yet another rich and distinctive site, Sticky Beak Site, from which bulged the bones of giant birds and cow-sized diprotodontid marsupials.

And so it went. Each new find soon resulted in the subsequent discovery of yet more new finds nearby. With an unlimited exploration team, the process of discovery of new sites at Riversleigh would be logarithmic.

Discovery of Rackham's Roost

Among the many other discoveries of the 1985 Expedition, two in particular stand out. Both were made by an extraordinary man and his son on an albeit ordinary holiday. When we arrived at our usual campsite on the Gregory River Water Reserve, Alan Rackham and his family were already enjoying the quiet beauty of the place as they had done on their holidays for many years. We invited them to join us in our field work. After learning about the type of rocks we were looking for and how to recognise fossil bones, they set off to explore the bluffs up and down the Gregory River. Alan senior was skilled in the knowledge of edible wild foods and a master bushman so we had no worries about him taking a small band of our volunteers with him. By the time the rest of us returned to the camp at the end of the day, the two Alans had grins so broad their heads were at risk of falling off. Alan Sr thrust at us a lump of relatively non-descript pinkish limestone.

At first we saw nothing. Then, as we looked closer, we saw dozens of minute teeth and thousands of tiny bone fragments bristling from its surface like coarse grains of salt. That would have been a sufficiently beautiful sight for us to share their barely controlled enthusiasm, but there was something else—most of the teeth visible were those of rodents! Previously this major group of Australian mammals (comprising 25% of Australia's modern native mammal diversity) was represented by no more than a few isolated teeth from early to middle Pliocene sites in eastern Queensland. Here were masses of rather primitive-looking rodents mixed in with a plethora of tiny bats and dasyurids in a deposit we soon came to realise was approximately 3-5 million years in age. This important site came to be known as Rackham's Roost in honour of its discoverer (Alan Jr) and its position on the top of a riverside cliff.

To get to Rackham's Roost, we had to go by boat—but boats seem to be a constant source of chagrin on our field trips. Between Alan Rackham's 'first start-no start' boat and Neville Whitworth's (a skilled mineralogist and chemical engineer from Mount Isa Mines) 'self-motivated gear-lever', many of us have had cause to curse. The first trip to Rackham's Roost was one such occasion. Alan successfully boated us to the area of the deposit where we spent the day collecting. At day's end, having carried our treasures down the cliff, a trip that took longer than expected, we reluctantly climbed into the boat for the return trip to camp. The sun was already well below the horizon and the short sunset of the tropics was rapidly fading. By the time we actually cast off, it was dark.

Good friend and companion that he is, Alan kept us amused on the way back to camp with an assortment of crocodile stories, assuring us of course there were none in the Gregory; none that were really big anyway. Leastways *he'd* never seen a really big one, but of course some locals claimed *they'd* seen them here. The stories became increasingly terrifying as Alan peered into the darkness ahead, anticipating and skilfully negotiating each obstacle. Soon our disbelieving titters transformed to swallow hard 'Yeah?'s. A lull in the gore while Alan concentrated his attention on a particularly narrow stretch of the river encouraged each of us to contemplate the likelihood of a horrific ending to what had until now been a great day.

Bringing us ever nearer to our camp and the waiting meal, the little motor was purring softly, reassuringly, occasionally surging as Alan increased the throttle to get around logs. The moon had begun to rise above the trees but visibility was still poor. 'The river widens around this bend,' said Alan and he confidently brought the little motor up to full power. The surge brought the front of the boat up with an unexpected lunge. Still higher and higher went the bow of the boat. Screams cut the air. A sudden heave and tilt and we all clung to the boat, which now seemed to have assumed the character of a bucking bronco. White knuckles were clearly visible in the dim light as we were thrown from side to side. Fossils forgotten, someone yelled out that we were all going to die. And still the boat rocked. Suddenly Alan leapt from the boat into the black waters ... no doubt to wrestle the monster of the deep. Brave beyond words we thought but what would we tell his wife? Moments later, from the inky blackness below came a booming voice, 'Would someone else care to jump in and give me a hand here or we'll never get off this snag.' A volunteer was *eventually* found.

Discovery of Terrace Site

Not content with the discovery of Rackham's Roost Site, Alan senior promptly set off downstream and returned with the news that he had found another bone-rich site but this time not in limestone. When he took us to the spot, it was clear that

GLADA

Under a sun-shielding tarp, volunteers sift through the gravels and sands of Terrace Site, a late Pleistocene river deposit containing many bones of giant extinct animals.

Bats are common inhabitants of the caves in the Riversleigh region, as they have been for millions of years.

GLADA

R. ARNETT

Half hidden by a jumble of tumbled-down blocks of Cambrian limestone is the entrance to Carrington Cave, which contains two very important fossil deposits.

this was an ancient perched terrace of an ancestral Gregory River. As soon as excavation work began bones and teeth of many animals were recovered, including those of the giant marsupial *Diprotodon*. This discovery made it clear that the site which became known as Terrace Site was Pleistocene in age, somewhere between 35 000 and two million years old.

Other similar deposits have since been found (and one of them mysteriously never relocated!) but not yet excavated. For example, when Alex Baynes of the Western Australian Museum wandered from the Terrace excavation site to the bank of the Gregory River, he spotted the end of a limb bone of a giant extinct dromornithid bird in a previously unexamined stretch of ancient river conglomerates. Very recently, during the 1990 Expedition, a number of promising deposits were found, including Dentists' Delight (discovered by two dentists and containing, among other goodies, an isolated kangaroo tooth) and Turtles in Trouble Site (which, not surprisingly, contained much freshwater turtle shell). We expect these sites to eventually add a great deal to our growing picture of Riversleigh's more recent history.

Discovery of Carrington's Cave

It was not just the volunteers who made major discoveries at Riversleigh—even helicopter pilots made their mark. After the Army decided that use of an Iroquois at Riversleigh was too expensive, Wang Australia enabled us to hire a lighter three-seater helicopter for several years. The pilots were experienced at the tasks of aerial mustering and able to dart around the jagged hillsides with consummate skill. One of these pilots, Russell Carrington, was bringing us back from a day of exploration in a remote area when he noted a strange hole in the flank of a low hill of Cambrian limestone some distance to the north-east of our camp. We

landed to investigate and found it was the entrance to a huge cave. After returning for torches, we clambered down the rubble pile at the entrance and pushed aside the curtain of hanging roots that draped the opening. The large entrance chamber led into a long tunnel that finally emerged into a vast domed chamber whose ceiling was approximately 25 metres high. As far as we could tell, this was the first time any European had set foot in this incredibly huge cave.

Palaeontologists inspect one of the most remote chambers in Carrington's Cave where, for thousands of years, ancient Ghost Bats dropped the remains of their prey.

From our point of view, the most important aspect of Russell's discovery was the fact that the largest chamber contained a colossal pile of bat guano, at least some of which had been produced relatively recently (probably within the last 50 000 years) by carnivorous Ghost Bats. The central floor of the chamber is a lightly cemented layer cake of millions of tiny teeth and bones, most of which have washed or have drifted downslope from the guano pile. The fine structure and composition of this bedded deposit is very similar to the rock-hard Rackham's Roost deposit. Here is a relatively modern cave situation that will hopefully help to explain the nature and origin of the unique Pliocene Rackham's Roost site.

In addition, this inner deposit in Carrington's Cave should record, upwards from its depths, progressive changes in the nature of the prey species available to the populations of Ghost Bats that have foraged in this area over thousands of years. This in turn will reflect changes over time in the region's climates, vegetation and perhaps even the prey species themselves. Palaeobotanist Helene Martin from the University of New South Wales expects to recover from this and other Carrington's Cave deposits a record of fossil pollens which should give an even better indication of recent changes in the vegetation of the area. Such records are extremely scarce for the northern half of the continent and particularly the now semi-arid to arid regions in which Carrington's Cave occurs.

Finally, Carrington's Cave will be of keen interest to the many palaeoanthropologists interested in the history of interactions between the Aborigines and the Australian biota. The entrance chamber, which is easy to get into and dimly lit by daylight, retains clear indications that humans occupied this cave, at least intermittently.

All in all, Russell Carrington's discovery was one of the most important of recent years. Inevitably, as excavations and other researches focusing on the Cave get underway, other caves and shelters in the area will be discovered, some of which may eclipse Carrington's Cave in importance. It's the nature of the game.

Discovery of Low Lion Site, Dredge's Ledge Site and VIP Site

Important discoveries at Riversleigh are frequent events. As noted above, each time a discovery is made, it provides a springboard for many more to be made in the same area. For example, in 1989, Alan Rackham was walking downslope from Bitesantennary Site, a bat-rich site discovered in 1988 by Syp Praeseuthsouk and Margaret Beavis, when he spotted the partial skull of a marsupial lion. It was emerging from a slab of limestone far below any level previously known to produce bone. Now that the lion's snout is out, it is clearly more primitive than any we have previously seen at Riversleigh and confirms that this locality, unavoidably named Rackham's Low Lion Site, was indeed a low-lying site, seemingly older than any previously encountered.

Similarly, while we excavated materials from the previously discovered VD Site (which contains a rare bat) in 1989, Martin Dredge spotted a strange sheet of bones and teeth cemented into a vertical crevasse further down slope. Now known as Dredge's Ledge, it may prove to represent a younger deposit, of an age not otherwise represented at Riversleigh, that filled a crevice eroded into the older limestones of the area.

In the same way, while most of our crew excavated a large block of limestone bristling with bat skulls, snails and the mysterious braincase of an as yet

Dredge's Ledge Site, discovered in 1989 by Martin Dredge, is a fissure containing fossils possibly younger in age than others on Godthelp Hill.

Fragment of a fossilised leaf from Dunsinane Site found alongside pieces of wood, fragments of insects and the teeth and bones of animals.

R. ARNETT

undetermined animal, Alan Rackham, Jenny and Michael Birt (Vice Chancellor of the University of New South Wales), Tony Wicken (Pro-Vice-Chancellor) and others explored a fossiliferous ledge about ten metres away, now known as VIP Site. The first swing of the sledge-hammer revealed a complete skull and partial skeleton of a new kind of diprotodontid.

The follow-ups of VIP, Dredge's Ledge and Low Lion Sites, in the 1990 Expedition, proved just as interesting. At VIP Site, a semi-articulated foot, various limb bones and the skull of a pouch young diprotodontid caused great excitement, while Low Lion Site relinquished a marsupial lion jaw and other goodies. Dredge's Ledge was attacked with light explosives and sledge-hammers and a large sample of limestone now awaits processing at the University of New South Wales.

Discovery of Cleft of Ages Site and Dunsinane Site

These two exciting, potentially very important sites were found within a few days of each other towards the end of the 1990 field expedition. Both were discovered while we were surveying remote areas of the Gag Plateau. On each occasion a

After a few hours, palaeontologists can barely contain their delight with their discoveries at Dunsinane Site.

Discovery of Dunsinane Site during a helicopter survey in a remote area in 1990 was extremely exciting—it was the first Oligo-Miocene deposit at Riversleigh to produce plants.

team of eight to ten people had been dropped by helicopter into that particular area to spend the day looking for new fossil deposits. On both days, the explorers were richly rewarded.

The first to turn up was Cleft of Ages Site. The general area in which the team was dropped was a good one. The morning had been very productive, with every member of the exploration team having found at least one site, many of which appeared to have good prospects. In the afternoon, Henk was following a dry, shallow creek bed when an area away from the creek bed caught his eye. Here, the limestone changed colour abruptly, as did its erosional features: the rock was mostly a rich chocolate brown colour rather than a cheesy yellow and the area was flat without limestone pinnacles. Henk investigated and in the first lump of limestone he picked up he found a partially exposed wombat tooth. In Riversleigh's other Oligo-Miocene deposits we had found only primitive pre-wombats and even they were quite rare. More wombat teeth were quickly found at the Site, as well as those of kangaroos and diprotodontoids. This was a very rich site. In fact, there were even teeth lying on the ground that had eroded out of the limestone. Causing most excitement, however, was the distinctness of the fossils at this site. Not only were there ancestral wombats but a whole suite of animals that had a more modern appearance than those in Riversleigh's other Oligo-Miocene deposits. It seems that we may have found in Cleft of Ages a slightly younger deposit, one that will hopefully add appreciably to our growing understanding of Riversleigh's history.

Dunsinane Site was found on the last day we had the services of our pilot Peter Forbes and his helicopter. On this day we chose to examine a region to the south of Cleft of Ages. When we flew over a ridge just south of Pilot Basin, we were disappointed to find that we had reached the end of the Tertiary limestone in that particular area. A small flat outcrop on the very edge of the Tertiary limestone was chosen from the air as the drop point for the survey. Still a little discouraged at the limited amount of Tertiary limestone available for searching in the area, Henk and Sue made a start on surveying the immediate vicinity of the drop point while waiting ten minutes for the next two members of the team to be dropped in.

It wasn't long before Sue had found fossil bone, at Rocky Road Site, so-named because, although the limestone was bone-rich, the bone itself was fragmented and poorly preserved. Within minutes, Sue found another site in which the bone was better preserved and a partial *Neohelos* (diprotodontid) jaw was exposed. Meanwhile the next group of surveyors had arrived and joined the search.

At about this time, Henk found himself staring at a most surprising object: a small but unmistakable piece of fossilised wood lying in the dirt beside the *Neohelos* jaw. This was something we never expected to see at Riversleigh. Previously, we had blamed the lack of fossil plant deposits at Riversleigh on the chemistry of the ancient environments, which because of all of the limestone, was almost certainly basic with a pH somewhere between 7 and 10. While these were great conditions for preserving bones, they were not suitable for preserving plants, a process that normally requires relatively acid conditions (i.e. a pH of less than 7). That is why it is unusual to find plant and animal fossils together in the same deposit. Ever hopeful, however, we had continued the search for a legacy of an acid palaeoenvironment in some out-of-the-way corner of Riversleigh. Now we had found it! And, as predicted, the associated fossil bone was, for Riversleigh, uncharacteristically very poorly preserved, many of the remains appearing in the limestone as ghostly outlines with little substance.

GLADA

Alan Rackham Jnr and Snr with a small Freshwater Crocodile. In 1985, father and son discovered two of Riversleigh's most important fossil deposits: the Rackham's Roost and Terrace Sites.

Fossil bone now all but forgotten, the now complete team of ten spread out to search for fossil wood. As steadily as we spread out, the pieces appeared. Among them were lumps of limonite that contained woody material. Then a layer of limestone containing both wood and bone was found. Discovered next were rounded nodules of limonite that when broken open revealed clear impressions of fossil leaves. In another nodule, an insect carapace was found. In the space of what seemed like minutes but was actually hours, and in an area of less than 50 square metres, a wealth of fossil material, both plant and animal, was found that morning. Our working hypothesis is that the fossilised plants were contemporaries of the Oligo-Miocene vertebrates with which they are bedded, although it is also possible that the plants date from a different period, have eroded out of their original beds and been redeposited with younger vertebrates. Until the plant material is identified and a more complete study of the area of deposition can be made we can say little except that the information about palaeoenvironments retrieved from the material should be very interesting indeed. With bated breath, we await the pronouncements of the palaeobotanists in whose care the fossil plants now rest.

We will be returning to Cleft of Ages and Dunsinane Sites again in the near future and expect many more new sites to turn up.

The never-ending search for new sites

Because many new discoveries are made each year, not all are examined in the year they are found. As an example, in 1990, volunteer Alan Crickman spotted a cluster of limb bones projecting from the corner of a low limestone ridge in a remote region of the Gag Plateau. It was logged as Site AL90 ('Alan's Ledge 1990') but we were too busy with other richer spots that year to examine his find. It wasn't until 1993 that three volunteers in the process of sampling AL90 found, besides the skulls and jaws of new diprotodontids and kangaroos, the first articulated skeletons from Riversleigh. This find may be one of the richest so far discovered. The UNSW Vice-Chancellor, John Niland, staring in disbelief at a magnificent skull jutting out from the end of an AL90 block he crow-barred up was heard to whisper 'I wonder how long it took them to bury this one!'

CHAPTER 3

Passage to Immortality

Four Serendipitous Ways to Turn to Stone

Two of the first questions people commonly ask us are: 'How did the Riversleigh fossils form?' and 'What has made Riversleigh so special in this respect?' In this chapter we attempt to answer the 'hows' and 'whys' of fossilisation at Riversleigh, though we hasten to add that these are questions for which we are often still searching for answers. Each of Riversleigh's 150 treasure troves has its own depositional history — its own story of life, death, burial and resurrection. Room does not permit an exploration of each of these histories so here we have attempted only to describe what we believe to have been four of the most common scenes enacted during Riversleigh's very complex history.

Fossils: where to look and what to expect

Before embarking on our date with destiny, a few words about fossilisation in general. Normally, fossilisation is a rare event. Typically, a plant or animal dies and quickly decays, leaving little trace of its existence. In situations in which decomposition has been slowed or prevented, however, fossilisation may occur. For example, this may happen if the plant or animal is rapidly buried under sediment such as sand, silt, mud, peat, ash and so on. Over time, sometimes over tens of years, sometimes millions of years, the sediment may be altered (e.g. by heat or pressure) and the mud may turn to shale, the sand to sandstone, the peat to coal, and oozy, limey mud to rock-hard limestone, preserving traces of the organism in the rock often for millions of years.

Some parts of plants and animals are more resistant to decay than others and these are relatively abundant as fossils. They include animal shells, bones and teeth and some parts of plants, for example thick-walled lignified or cutinised tissues. Conversely, soft tissues, such as the leaves of plants, animal skin and muscle; soft-bodied organisms, such as worms, jellyfish and many insects; and footprints, tracks and burrows are less commonly fossilised.

Rarely, whole plants and animals are preserved. Some of the most spectacular include 20 000 year-old mammoths of the last great Ice Age that have been frozen in the Siberian permafrost and the 4-5000 year-old remains of a Thylacine (or Tasmanian Tiger) found in a cave, now known as Thylacine Hole, on the very arid Nullarbor Plain of Western Australia. Some environments favour fossilisation over others. Situations in which sediments are accumulating, such as in caves, lakes, swamps, floodplains and on the sea-floor, are more likely to preserve fossils than eroding land surfaces, such as mountain peaks and river beds.

During fossilisation, the original composition of the bone, shell or cuticle of an organism is sometimes altered or petrified through the infiltration of groundwater, so that the original mineral composition changes but the structure is preserved. For example, at Lightning Ridge in northern New South Wales, the opalised remains of prehistoric animals and plants (including dinosaurs and pine cones, as well as Australia's oldest monotreme, the 110 million-year-old platypus-like *Steropodon galmani*) have been found. Long after the remains of the organisms were buried and firmly encased in rock, silica-rich groundwater seeped through the sediments replacing the original remains with brilliant, multicoloured opal.

Riversleigh's fossil treasure trove is mainly comprised of the mineralised bones and teeth of animals, although in some deposits the shells of snails and cuticle of insects and other arthropods have also been preserved. During our studies of the Riversleigh fossil deposits it has become clear to us that many different mechanisms were responsible for producing the variety of deposits at Riversleigh. Basically, however, the Riversleigh creatures have become fossilised in one of four situations: 1. at the bottom of Riversleigh's Oligo-Miocene lime-rich lakes and pools; 2. as drop-ins to Oligo-Miocene, Pliocene and Pleistocene limestone caves and fissures; 3. as prey remains accumulated by Oligo-Miocene and Pliocene flesh-eating bats; 4. by being preserved as part of Pleistocene river deposits.

Here, we follow the fate of four of Riversleigh's ancient inhabitants as they cross the threshold of eternity.

L. HALL

When cave-dwellers like this Ghost Bat die, their remains accumulate and, if circumstances are right, become fossilised as part of the cemented cave floor.

The many bones and skulls of ancient bats in this piece of limestone from Bitesantennary Site accumulated on the floor of a small, dome-like cave approximately 20 million years ago.

Life and death in the catacombs: Riversleigh's ancient cave deposits

Night falls and with it a tropical downpour of rain that temporarily drowns out a cacophony of buzzing cicadas, burring crickets and chorusing frogs. A multistoreyed, broad-leafed canopy, laden with epiphytes and dripping with lianes, acts like an umbrella for the forest's night creatures, many of whom are already out and about foraging on the forest floor or in the trees.

Carpeting the uneven limestone floor of the forest are mosses, ferns and eerily glowing toadstools, underlain by a thin layer of humus and leaf litter. The green carpet partially conceals a gaping hole into which curtains of vines and dripping vegetation hang. The hole is a gateway to the world under the forest —an underground cave system riddling the limestone bedrock like a cancer.

Inside the cave, restless twittering, stretching and the fluttering of wings disturbs the still, damp, acrid-smelling air. The rain eases and, as bat sentinels return from short reconnaissance flights, a thousand others stretch, yawn and, in waves, drop from the rocky ceiling to expertly manoeuvre their way through barely wing-wide passages towards the cave's entrances and the rich feeding grounds of the rainforest.

Ducking under an overhang on the way out, some bats with bizarre, fleshy, flower-like faces pause briefly at entrances to drink from pools stained brown by rotting vegetation before dispersing into the forest to glean mosquitoes and midges from the surfaces of bushes, branches, vines and fallen logs. Other bats with faces more like mastiff dogs fly fast and high above the canopy and, in amazing displays of aerial acrobatics, pluck evading moths from the blood-warm air, still tinged purply-pink by the sun's last rays.

A small, apricot-coloured leaf-nosed bat, suckling her single, tiny young, stays behind. One diaphanous wing is badly torn and the bone shattered — the result of an encounter with a hungry python during last night's foraging. Although able to limp home, she will never leave the cave again, infection speeding her demise and that of her young.

At Bitesantennary Site, in the centre of this photograph, the fossil deposit represents an old cave floor whose roof and walls have long since disappeared.

In death, the fragile bones of mother and young join those of generations of bats that have lived and died here. Some of these bones are scattered over the cave floor but many more have accumulated in a deep pool that extends from beneath the collapsed roof into the dark, bat-occupied recesses of the cave. In the pool, guano- and lime-rich sediment gradually covers the remains of the dead. Alongside the bat bones are the skeletal remains of small bat-eating pythons and less frequent visitors to the cave, such as beetle-eating bandicoots, scavenging dasyurids and a careless kangaroo that fell into the hole and drowned. Everywhere are freshwater snails that have taken up residence in the daylit area of the cave pool.

Over thousands of years, or perhaps much less, the watertable changes and the pool disappears, but a solid limestone floor, rich in fossilised bat bones and snail shells, remains. Over time, rainwater erodes away the roof and most of the walls of the old cave, leaving the cave floor deposit for scientists to find millions of years later.

Bitesantennary Site on the D-Site Plateau is typical of these kinds of sites. Discovered during the 1988 Bicentennial Riversleigh Expedition, Bitesantennary was the first of Riversleigh's Oligo-Miocene fossil deposits to be unambiguously recognised as an ancient cave-fill. The deposit covers an area of approximately 150 square metres. Bat bones, jaws, skulls, snail shells, the odd marsupial jaw, bird bones and frog skulls compete for space in the matrix. Almost all the bat bones and snail shells are complete, suggesting fossilisation occurred at or very near the point of accumulation.

The deposit is cut into an older, relatively non-fossiliferous limestone. (The latter may be contiguous with limestones underlying or containing two adjacent Oligo-Miocene fossil deposits, Neville's Garden and Burnt Offering.) At Bitesantennary Site, the boundary or contact between the fossiliferous fill and what is interpreted to be remnants of the older, non-fossiliferous cave wall has been identified at several points around the perimeter of the deposit.

Represented in the cave deposit are at least five species of leaf-nosed bats (hipposiderids), each by many near-perfect skulls as well as complete limb bones. Less well represented in the Bitesantennary deposit are fish, frogs, lizards, a python, birds, peramelid bandicoots, a dasyurid and a kangaroo. The deposit's many snails and what seem to be extensive, algal-like mats indicate that during its history, the depositional area was open to light and under water.

While there is little doubt that bat-filled caves once riddled the Riversleigh limestone, and a number of undoubted cave-fill deposits have been recognised (e.g. at Bitesantennary and Gotham City Sites), the majority of Riversleigh's fossil-rich deposits appear to have accumulated, not in caves, but in pools of

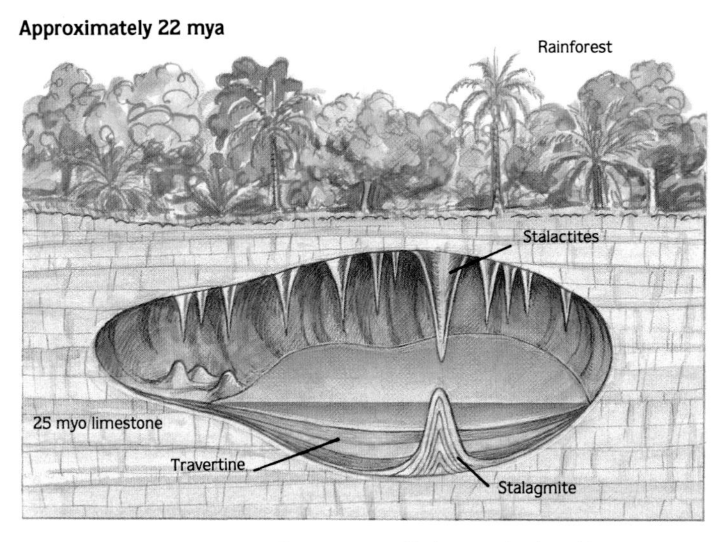

In some areas (such as Bitesantennary Site), caves developed in the older Oligo-Miocene limestone, perhaps not long after the limestone itself had formed.

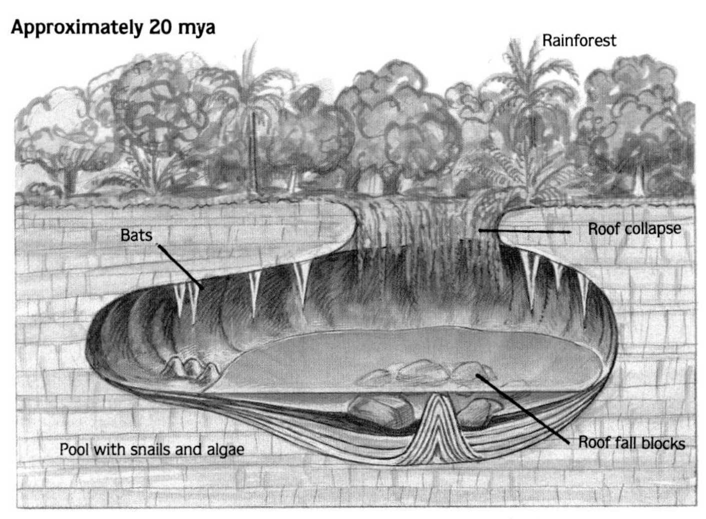

Eventually, erosion from within would have caused roof collapse enabling water, snails, algae, bats and accidentally trapped animals to accumulate. After thousands of years, the natural trap would fill with the detritus of the forest.

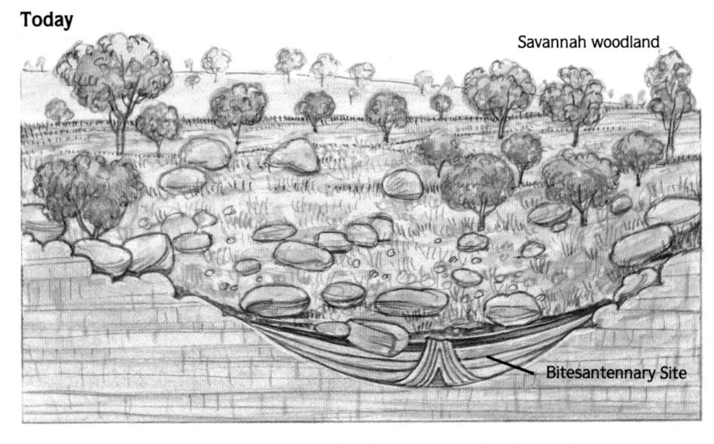

Millions of years later, the roof and walls would finally erode away exposing, to the seeking eye of the palaeontologist, pieces of the thick, fossil-rich cave floor.

various sizes and depths within the Riversleigh rainforest. The following recreates what may have been a common scene in the great forest.

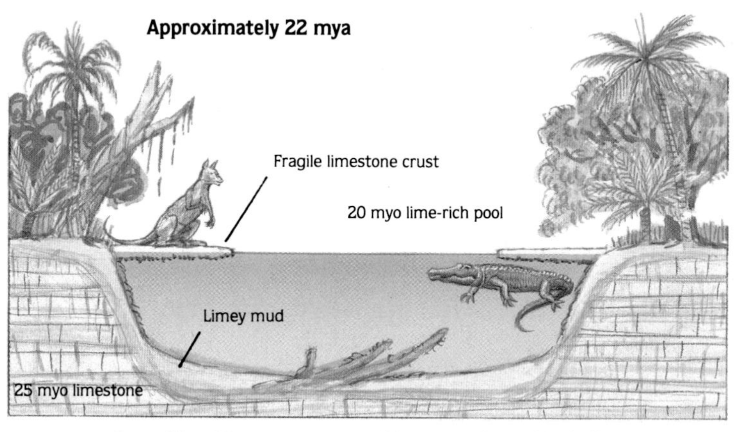

Some Oligo-Miocene deposits at Riversleigh formed in shallow pools where creatures may have ventured too far out onto crusts or floating debris.

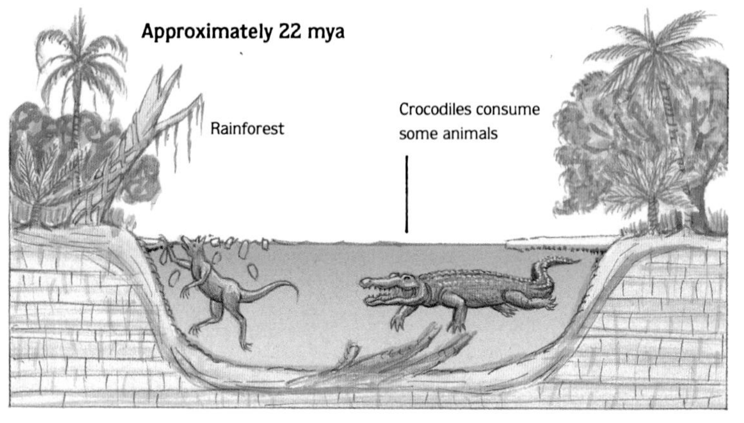

After falling in, the animals would either drown or be killed by crocodiles.

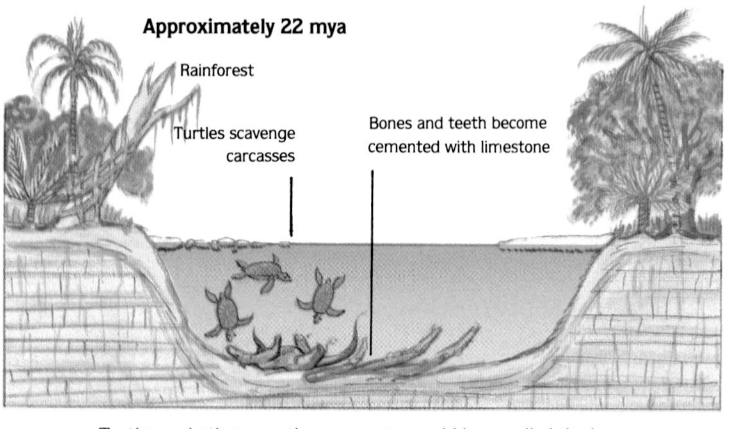

Turtles and other aquatic scavengers would have pulled the bones of the skeletons apart before they were finally buried in limey mud.

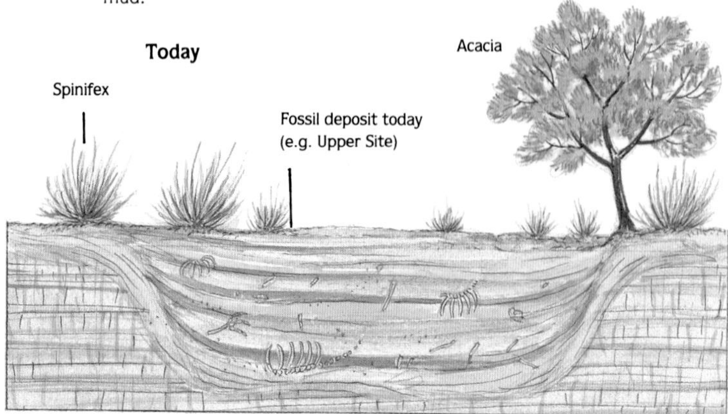

Eventually, the pools filled from the bottom up and the accumulating limy, bone-rich mud slowly turned to stone.

From an original concept by *Australian Geographic* magazine.

Gateway to eternity: Riversleigh's Oligo-Miocene lake and pool deposits

On a breathless night illuminated by a full moon, a small wallaby browses quietly on shrubs growing near the edge of the still, deep pool that is surrounded and overhung by warm, tropical rainforest with its many storeys, emergent trees, rich ground and litter layer and diverse liane and epiphyte flora. The forest is alive with the creatures of the night. In the ferny undergrowth along the pool edge as many as nine different bandicoots—ranging from mouse size to cat size—forage in the gloom for insects, fallen blossoms and soft figs, while above an incredible variety of possums—ringtails, brushtails, cuscuses, feather-tailed possums and pygmy-possums—concentrate on leaves, blossom and fruit of many kinds.

The wallaby positions itself beneath a branch bearing particularly appetising leaves, cranes its neck to reach the leaves, then, using its forepaws, it pulls the branch closer. Startled, a copper-coloured brushtail possum, previously occupied with the fruit of the same huge tree, runs nimbly across the branch and up a thin, swaying vine towards its siblings 20 metres up in the flower-strewn crowns of the ancient forest. In its haste, the possum neglects to see a 3-metre green python draped in a tree fork. As the possum passes it, the python strikes, hitting the possum hard in the head and quickly throwing a coil of its green length around the stunned animal. But the grip is bad and the

At this point, fossilisation begins. As the limey muds, with their entombed bones and teeth, steadily accumulate, chemicals in groundwater flowing through the sediments gradually alter those of which the teeth and bones were originally built. The alteration or replacement occurs molecule by molecule, so that the fossil

possum has a moment to sink its dagger-like incisors into the snake. The reptile recoils. Shaken but unhurt, the possum lives to see another day.

Not so the wallaby. Oblivious to the life and death struggle far above its head, the wallaby abandons the branch and moves towards the shrubs growing closer to the water's edge—and unknowingly onto a thin, ice-like crust of limestone on the pool's surface. Spring-fed, the pool is supersaturated with dissolved limestone and evaporation at its surface has left a limestone crust cementing together a fragile surface of leaves and other debris that looks, in the moonlight, like a solid part of the forest floor.

Too late, the wallaby realises its mistake as the bottom falls out of its world. It struggles desperately but vainly to scramble back onto the steep-sided bank. Its efforts attract the attention of a prowling family of small, ancestral thylacines, who allow their intended prey, a large, lumbering, wombat-like creature, to beat a hasty retreat into the forest. The python, wounded but hungry, glides silently down the moss-covered tree trunk to size up its competition for the wallaby meal.

But the wallaby cannot get back onto the bank. Exhausted, it finally drowns and sinks to the bottom, taking with it sheets of the pool's limestone crust. Scavenging turtles and lungfish eat their fill of wallaby flesh, eventually leaving the disarticulated skeleton to rest in the limey bottom muds with those of other incautious creatures of the forest.

bones and teeth are visually indistinguishable in structure from modern bones and teeth, although chemically they are quite different.

Twenty million years later, the story of that night in the rainforest, and hundreds like it, is being read in the rocks from Upper Site. Today, chains of python vertebrae snake their way over and around wallaby bones, thylacine teeth, bandicoot bits and crocodile crania in the 1.5 tonnes of limestone recovered from Upper Site, the richest of all fossil-bearing sites at Riversleigh. From a volume of limestone less than 2 cubic metres, the remains of at least 64 mammal species have been recovered. By any modern standards, this 20 million-year-old assemblage is very diverse with almost twice the number of mammal families and twice the number of marsupial species of any surviving Australian or New Guinean ecosystem. High diversity is the hallmark of Upper Site, the fossil bird, reptile, frog and arthropod faunas also being extremely rich.

The high mammal diversity and the presence of many differently sized herbivores, from mouse to cow size, suggests a comparably high diversity of local plant species. Further, the kinds of herbivores represented, from ringtail possums to diprotodontoids, suggest a forest in which the resources were stratified.

Preserved alongside Upper Site's terrestrial and arboreal animals, are freshwater fish, lungfish, turtles and crocodiles, testimony to the fact that the bones and sediment accumulated in water. The depositional environment for the Upper Site Local Fauna is interpreted to have been a lime-rich pool in the rainforest, the depth of the pool (as indicated by the size of the aquatic animals it contained) being probably between 0.5 and a few metres deep. The Upper Site limestone is characterised by black, iron-enriched bands that may reflect periods (or areas) in which anaerobic conditions existed in the pool. Wrinkled sheets of what we interpret to be algal mats also tell us something about the pool. Many of the invertebrate remains (in particular, dozens of beetle larvae) occurred in vertical tube-like extensions of these mats that may have grown down into the bottom mud of the pond or hung below the algal mats if they floated.

Although more diverse than any other from Riversleigh, the Upper Site Local Fauna, we feel, probably best reflects the way in which many of Riversleigh's Oligo-Miocene fossil deposits were formed. Each site has its own complex depositional history, however, and we fully expect that it will be many more years before we are able to come to grips with all of the mechanisms involved. In the remainder of the

Of the many Oligo-Miocene fossil deposits on Godthelp Hill, the relatively tiny Upper Site (at the far left, beneath the copse of trees) produced one of the most diverse assemblages.

chapter we will examine two younger and quite different kinds of fossil deposits, those of the Rackham's Roost and Terrace Sites.

Among Riversleigh's Tertiary fossil deposits, Rackham's Roost is unique. Estimated to be between 3-5 million years old, it is Riversleigh's only recognised Pliocene deposit. However, its depositional history, and in particular the method in which the bones were accumulated, by carnivorous ghost bats, appears to have been common throughout Riversleigh's long history. The early to middle Miocene Gotham City deposit (on the Gag Plateau) appears to represent such an accumulation, as does the huge bone and guano deposit in the Great Chamber of Riversleigh's Carrington's Cave. It is possible, too, that at least some of the small, fragmented bones and teeth found in a number of Oligo-Miocene deposits in which ghost bats are common may have also accumulated in this way.

A bat's eye view of the Rackham's Roost cave deposit

The low, smooth branch of a white-barked eucalypt tree provides the perfect camouflage and vantage point for the hungry, silver-grey ghost bat as it waits in ambush for its evening meal. In the moonlight, a chestnut mouse forages for seeds in the undergrowth while a small wallaby browses on a low shrubby bush nearby. Unsuspecting, the mouse leaves the shelter of leafy cover to explore the edges of the adjacent grassy plain. The ghost bat sweeps, its shadow reaching the victim just before a fatal, bone-crunching bite to the back of the hapless mouse's head.

Mouse in mouth, the ghost bat flies back to its roost in the cave-riddled cliff 60 metres above the ancestral Gregory River. There, hanging upside down in the warm blackness, it eats the mouse and drops the remains onto the cave floor. These join the discarded wings of birds, accidentally dropped pieces of lizards, dasyurids, bandicoots, possums, rodents and a steady rain of tooth-filled dung. Adding to the debris pile are the bones of pythons and even an occasional kangaroo that died in the cool depths of the cave, as well as the bones of the ghost bats themselves. Small leaf-nosed and plain-faced bats and larger sheath-tailed bats are also tenants of the cave, some destined to become convenient 'fast food' meals for their ghost bat neighbours.

Approximately 3-5 mya

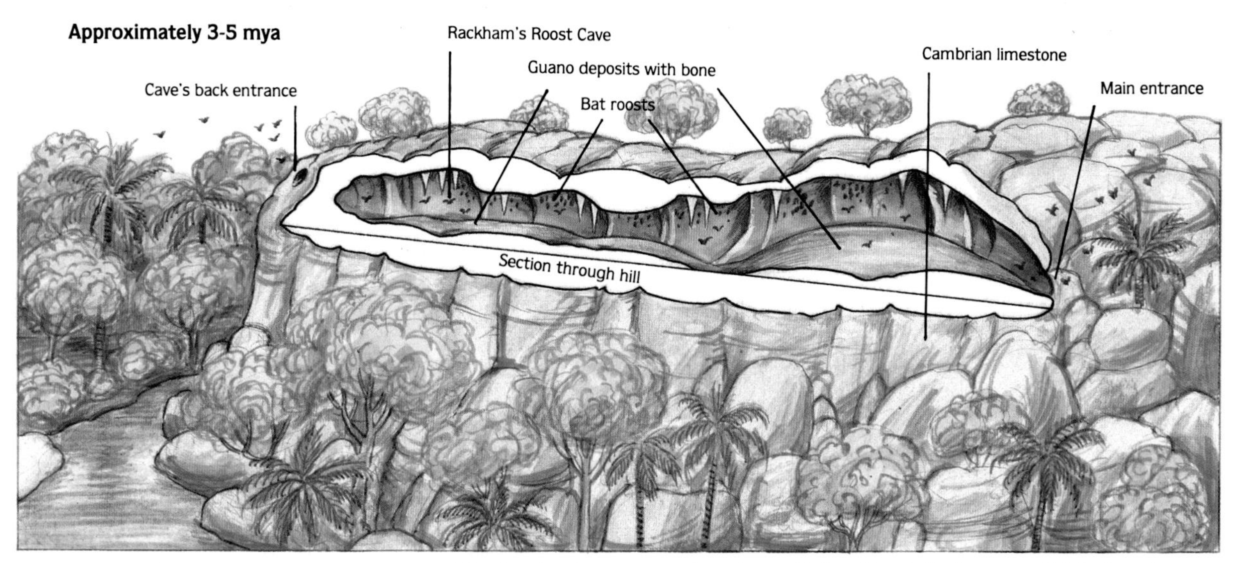

Perhaps 5 million years ago, the long, tunnel-like Rackham's Roost cave (seen here with the side artistically removed) linked small entrances above the ancestral Gregory River to a larger entrance to the east. Colonies of carnivorous Ghost Bats brought small prey back to the cave including, eventually, millions of rodents and bats. Bone-rich guano piles accumulated on the floor.

Approximately 2 mya

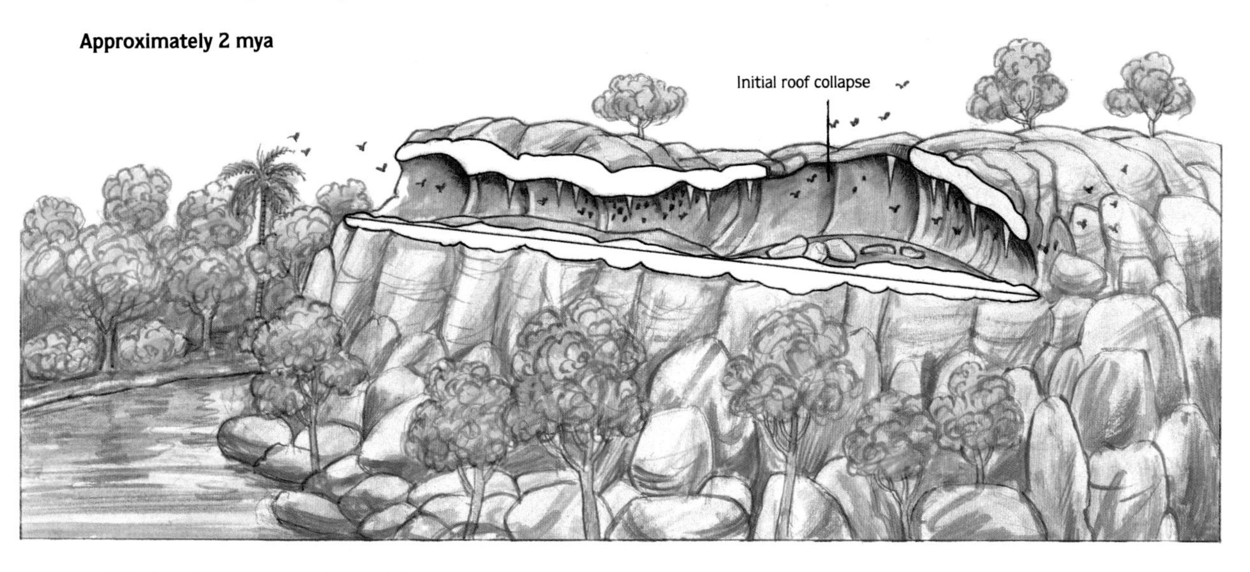

With time the cave roof began to collapse.

Today

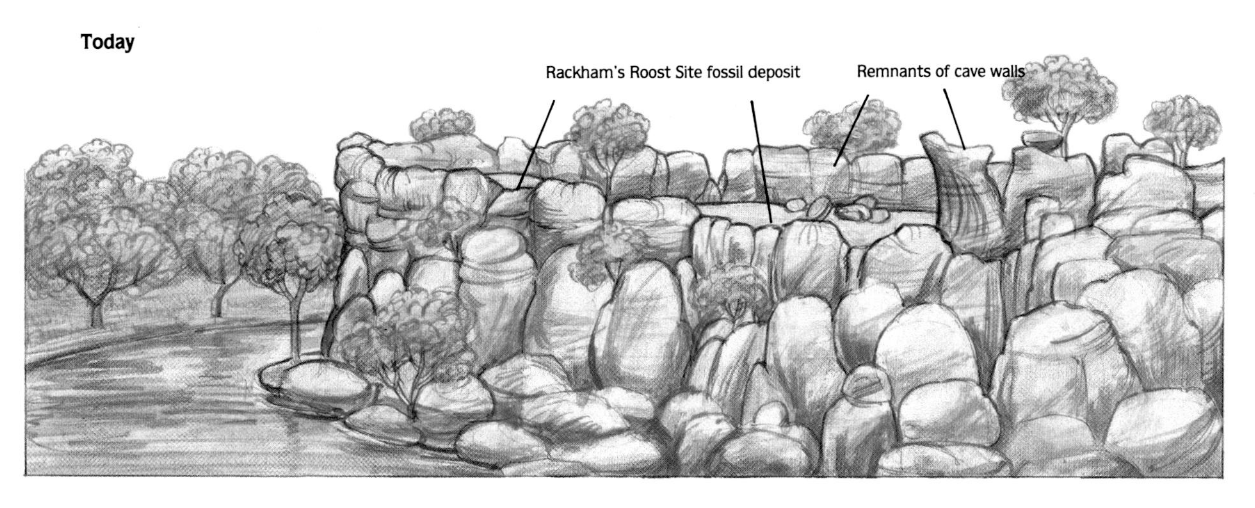

Eventually, the roof and most of the walls of the cave eroded away, leaving the pinkish fossilised cave floor exposed to the light of day and the sharp eyes of Alan Rackham Jnr.

Over thousands of years, water dripping on and running through the debris pile gradually cemented into a rock hard limestone. Eventually the softer roof of the cave collapsed, and it and the walls were dissolved away by rainwater, leaving exposed to the sun and palaeontological discovery a 3 to 5 million-year-old solid pavement of pink-coloured limestone, studded with bones and teeth.

The Rackham's Roost Site is providing invaluable information about life in the Riversleigh area at a time when the ubiquitous rainforests of the early and middle Tertiary (15-65 million years ago) had already retreated to the edges of the continent, and woodlands (rather than today's grasslands) covered great tracts of northern Australia.

In Rackham's Roost, Riversleigh's prehistoric ghost bats have conveniently collected together a precious sample of life from these ancient woodlands. Remains of animals recovered from the Rackham's Roost deposit include ghost bats, 13 species of rodents, one sheath-tail bat species, four species of leaf-nosed bats, four plain-faced bat species, a planigale, dunnart, possum and kangaroo, birds, lizards and frogs.

The rocks at Rackham's Roost are of two types: the Cambrian rocks that formed the remnant walls (right) and missing roof of the Pliocene cave; and the pinkish, petrified guano deposit on the floor that contains millions of tiny fossilised bones.

Not all of Riversleigh's fossil deposits occur in limestone. The Terrace Site is one of several relatively young fossil-bearing deposits that occur in perched river terraces along Riversleigh's Gregory River. The fossil fauna recovered from the Terrace Site appears to be of Pleistocene age, i.e. between 2 million and 10 000 years old.

A more recent date with destiny: Riversleigh's terrace deposits

A strip of remnant vine forest hugs the banks of the Gregory River, its pandanus and other palms, cluster figs and paperbarks providing the only oasis for miles. Beyond, scorched by the burning Pleistocene sun, spinifex-dotted plains and limestone mesas stretch forever. At the water's edge, an Agile Wallaby cautiously drinks, while beside it freshwater turtles bask on a partly submerged log, and across the river a 2 metre-tall flightless bird, one that looks deceptively like an overweight relative of an Emu, forages for food.

A vibration in the undercut bank of the river startles the wallaby which springs into action and bounds away. The rhinoceros-sized Diprotodon optatum, *the biggest marsupial that ever lived, has lumbered down to the water's edge with her young. Lethargically, the two make their way down towards a sandbank that has recently built up in the bend of the river, and, making the mistake of her life, the heavyweight walks out on to the oozy mud. Almost immediately she sinks to her belly as if in quicksand. She panics and struggles to escape. The youngster, light enough and far enough behind to avoid the deadly trap, watches helplessly as its mother becomes more and more mired.*

Her predicament does not go unnoticed by one of the Pleistocene's most awesome aquatic carnivores. A huge, broad-headed crocodile has left its resting place on a log near a clump of water lilies, sliding silently into the water and stealthily swimming towards the giant marsupial. Within moments the crocodile has seized the adult by the leg, tearing into it and dragging the screaming animal into the river. Torn, bleeding and exhausted, the giant marsupial quickly drowns and the crocodile eats its fill.

The leftovers of the meal join the bones of fish, turtles, crocodiles, water rats, birds, kangaroos, palorchestids and perhaps people on the river bed, eventually forming the fossils found by palaeontologists at the Terrace Site thousands of years later. The deposit, a 3-metre-thick cross-section through a stream bed, is comprised of a coarse conglomerate (containing bone, freshwater mussel shells and charcoal) which grades upwards into finer sands and silts. The bones in the deposit may represent the discarded meals of crocodiles, accidental drownings or seasonal flushing of the surrounding countryside by rain-fed creeks. Some of the bone fragments are very worn which invites an interpretation of long-distance transport and/or re-working of these from older fossil deposits. Others are relatively unworn, suggesting that their owners lived close to the site where their fossilised bones have been found.

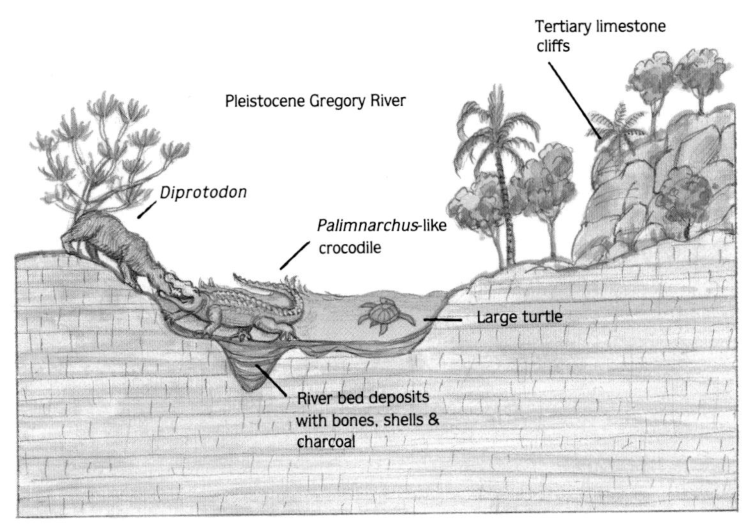

During late Pleistocene time, creatures that came to drink along the edge of the ancestral Gregory River met untimely ends in the jaws of large crocodiles. Their bones accumulated with others in river sands and gravels.

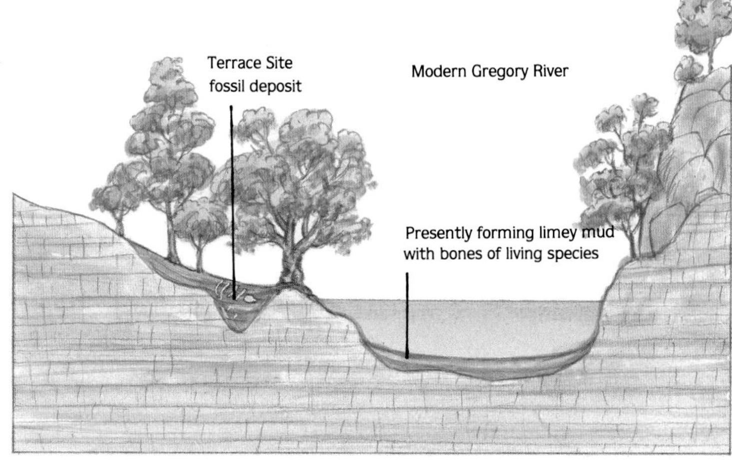

Subsequently, the Gregory River meandered eastward within its valley and cut down well below the level of the stranded, perched Terrace Site with its entombed bones and teeth.

Unexpected paths to immortality: Riversleigh's fossil plants

Fossils form in a vast array of different ways. The mechanisms reviewed here are just a few among many that operated at Riversleigh. But, until this year we had given up on finding plants in any of the Riversleigh deposits. While bones preserve best in a basic (i.e. non-acid) environment, plants commonly require an acid environment. Carbonate deposits, filled as they are with minerals that neutralise acids, are unlikely places to find preserved plants. Hence we were most surprised to find that Dunsinane Site was loaded with plant fossils, the first found at Riversleigh, as well as fossil teeth and bones. Evidently, the chemistry of the site must have been close to neutral (between acidic and basic), thereby giving both plants and bones a chance at immortality and both leaped in. Delightfully, this site didn't read the rule book of preservation. Naturally we hope that future expeditions will reveal more of these wonderfully unconventional sites.

ids n' oids

We classify common objects like vehicles (e.g. planes, boats, trains, cars and bikes) into sub-categories based on relationship (e.g. within 'cars' we recognise as 'natural' groups all Fords, all Holdens, all Mercedes). In the same way, animals are classified in a heirarchy of relationship. Each level has a distinctive technical ending on the word: '-oidea' for superfamilies; '-idae' for families; and '-inae' for subfamilies (which contain genera and these, in turn, species). When informally discussed, a member of these higher categories is referred to as an '-oid' ('-oids', pl.), '-id' ('-ids', pl.) and '-ine' ('-ines' pl.). Thus members of the Diprotodontoidea are diprotodontoids and members of the Palorchestinae are palorchestines.

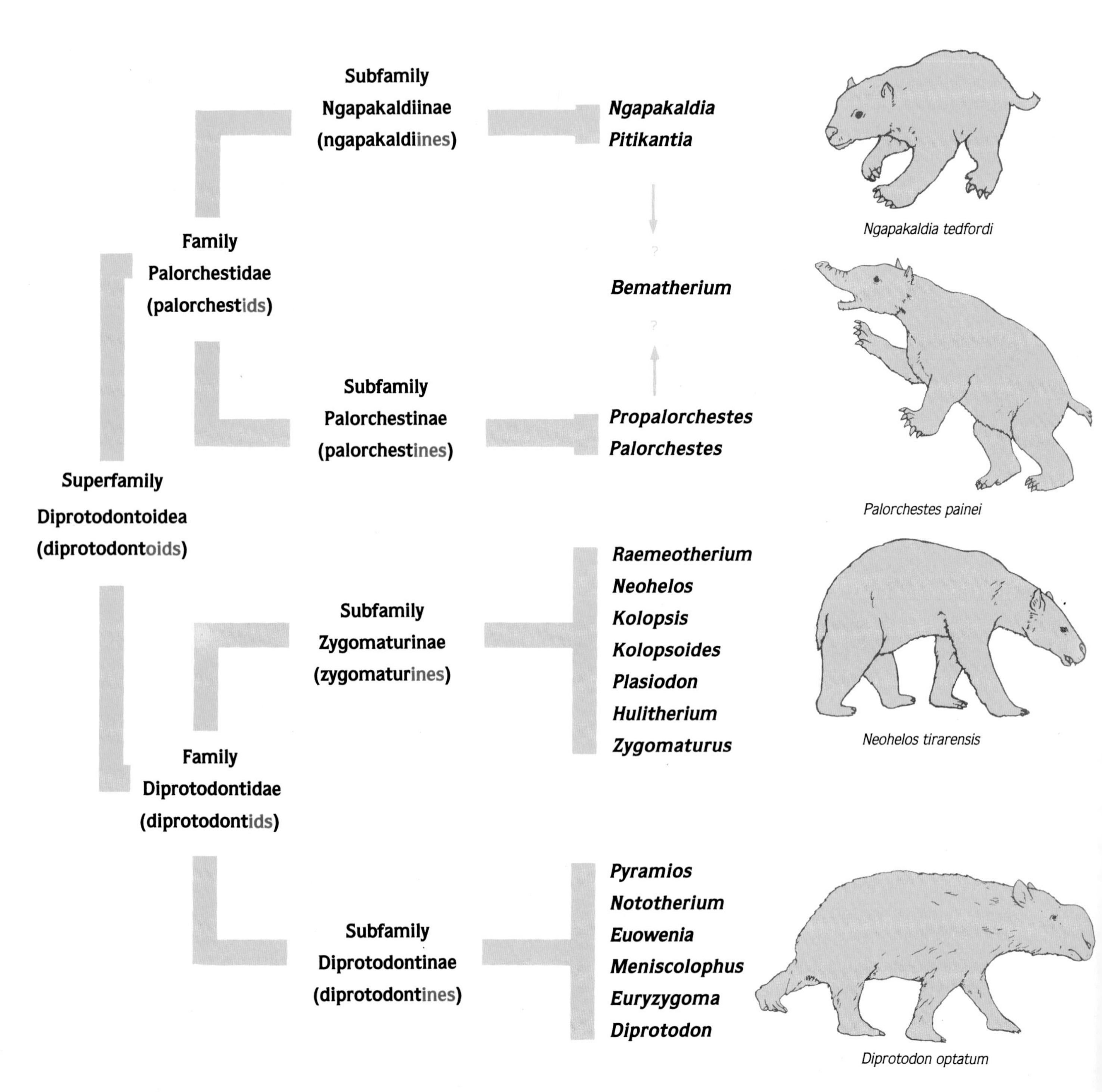

Ngapakaldia tedfordi

Palorchestes painei

Neohelos tirarensis

Diprotodon optatum

SITE D

Approximately 25 million years ago, in the late Oligocene, the lowland areas of Riversleigh were covered by dense, species-rich rainforest. Lakes fed by springs and the regular rains that nourished the region supported large turtles and other reptiles, as well as fish and snails. Slightly strange, slightly familiar kinds of animals filled the forests. Deep-headed, dagger-toothed crocodiles that ran down their prey on the land were lords of the forest floor. Patient pythons patrolled the water's edge where dromornithid 'thunder birds' came to feed. Unfamiliar kangaroos browsed on the shrubs of the forest, never more than an careless moment away from becoming someone else's dinner. Periodic deluges of rain raced bits and pieces of this prehistoric paradise into the lakes where they were buried beside the discarded residues of the lake animals themselves. Over time, compressed by the weight of continually accumulating sediment, the limey mud began to harden. Ground water began a ceaseless molecule-for-molecule exchange so that, within thousands of years, the bones and teeth became fossils. Millions of years later the mineralised bones of these ancestral creatures reappear like genealogical gems on the eroding surface of Riversleigh's grey limestones.

1 Leg bone of a giant dromornithid bird being buried in limey mud of lake floor.
2 Bony fish.
3 Freshwater catfish.
4, 8 Lungfish.
5 Rounded chert nodules of Cambrian age being recemented into floor of the Site D lake.
6 Limey mud, with snails, bones and other detritus accumulating to form the eventual bulk of the Site D limestone.
7 Chelid turtle.
9 Giant python (*Montypythonoides riversleighensis*).
10 Herbivorous palorchestid marsupial.
11 Giant flightless dromornithid bird (*Barawertornis tedfordi*).
12 Potoroine rat-kangaroo (*Gumardee pascuali*).
13 Giant, deep-headed, semi-terrestrial crocodile (*Baru wickeni*).
14 Balungamayine rat-kangaroo (*Wabularoo naughtoni*).
15 Herbivorous palorchestid (*Ngapakaldia* sp.).
16 Lime-rich waters of the Site D lake.
17 Water lilies.
18 Dense, species-rich rainforest.

Throughout this book Balungamayine should read Bulungamayine

Dunphy

CHAPTER 4

Riversleigh's Prehistoric Menagerie

A Rogues' Gallery

Everything we understand about Riversleigh and its prehistoric environments is only as reliable as our understanding of the number and kinds of creatures that frolicked in its ancient habitats. These animals, their needs for food and shelter and their relationships to each other are the keys to understanding how those ancient communities worked and how they evolved into the modern communities which we struggle to conserve today.

Vertebrate palaeontologists normally determine the number of extinct species through morphological comparisons (shape and relative size) of the teeth, jaws and other bones recovered from a fossil deposit. Indeed, the same technique is successfully used to distinguish living species. However, experience also teaches us that if these methods *do* fall short of accurately estimating species diversity, they do so on the side of *underestimating* diversity. In Australia this has been demonstrated several times as a result of molecular systematic studies. For example, where morphologists once recognised only the single 'Little Brown Bat' (*Eptesicus pumilus*) for the whole of Australia, electrophoretic studies of enzymes taken from populations over much of the continent revealed that the 'Little Brown Bat' was actually four quite distinct species (*E. pumilus, E. regulus, E. vulturnus* and *E. sagittula*). In retrospect, morphologists who re-examined the problem did manage to find relatively subtle features of size and shape that distinguished these four. Recent molecular and morphological studies have indicated that there may be as many as *thirteen* distinct species.

As a result, although species-level diversity in Riversleigh's Oligo-Miocene local faunas is remarkably high compared with that of comparable habitats in Australia today, it is likely to be an underestimate of the actual biodiversity of the time.

Determining species diversity of fossil populations

To decide whether two fossil populations represent different species, palaeontologists often examine the degree of difference that separates populations of related living animals whose species distinctions are understood. For example, when we attempt to determine whether the samples of *Hypsiprymnodon* from Riversleigh's fossil deposits represent one or more species, we will first attempt to determine the amount of variation present in *Hypsiprymnodon moschatus*, the living Musky Rat-kangaroo. In the course of this analysis, we will watch for size differences that could be attributed to sexual dimorphism. For example, males of most mammal species are larger than females, sometimes by more than 10%. Similarly, some shape differences are sexually dimorphic. In some bandicoots the males have grossly enlarged canines and posterior premolars. The degree of variation in size and shape in modern populations is then used as a *guide* to estimate species-level distinctions between fossil populations of related animals.

Most commonly, species differences are indicated by unique morphological features present in all individuals of one species but absent in those of another, such as the presence of an entoconid on the lower molars of one species but not on those of the other, or the presence of three rather than five vertical grooves on the sectorial premolar. Sometimes species differences are ones of relative proportion such that, for example, a ratio of posterior premolar length to anterior molar length may be consistently between 1.1 and 1.5 in one living species and between 1.0 and 1.3 in the other. If a fossil sample of individuals closely related to the living taxa reveals ratios spanning 0.5 to 1.5, the palaeontologist ought to become suspicious that more than one species may be represented by the sample. This suspicion would send them to closer examination of the fossil sample to see if there were other morphological differences that might distinguish a larger and a smaller subset of the original sample. If there were, the suspicion would grow that the sample contained two rather than one species.

In practice, palaeontologists first look for constant differences in absolute size or morphology between the individuals of two populations. Whether or not this immediately suggests species distinctions, many features are compared and measured for all specimens. These measurements are then subjected to statistical analysis, commonly using a computer if the data set is large. The aim is to identify statistically *significant* features or combinations of features that may consistently distinguish one population from another.

All of these comparative studies contribute to the ultimate hypothesis about whether or not a particular population is a new species or merely a geographic (horizontal) or temporal (vertical) variant of a previously known species.

What we have done in this chapter is to select from groups represented in the Riversleigh deposits a range of creatures that give a reasonable idea of the region's former biodiversity. However, the catalogue of rogues that follows is *not* a complete list. Such a thing is a long way off because it depends on each of the more than 30 research workers finishing their individual analyses and, impossibly, on comprehensive knowledge of the many thousands of Riversleigh fossil deposits that still await examination. So check again in a decade or two; the list of weirds and not-so-weirds will be a lot longer!

MILLIPEDES, BEETLES, SLATERS

and other many-legged things

One of the most surprising of the Riversleigh discoveries occurred in the laboratory at the University of New South Wales late in 1987. Bill Brown had been acid processing bone-rich limestone from Upper Site and was sorting through the acid insoluble residue for bones and teeth, when he noticed a very peculiar object. Whatever it was, it ranked very high on the scale of things weird. It looked a bit like a

Preserved with its tiny legs intact, a 20 million-year-old slater (isopod) is one of many uncrushed, three-dimensionally preserved arthropods known from Riversleigh.

This fossilised insect from Upper Site, with its head tucked under and delicate wings folded at its side, looks as though it were preserved only days before rather than millions of years ago.

Riversleigh's microscopic 'Venus de Milo' is actually the tail assembly of a millipede.

R.ARNETT

Lilliputian Viking helmet. It was composed of circular ringlets, some of which had 'things' poking out of them, the whole capped with a 'helmet' bearing two tiny horns. It was, as it turned out, the bum of a bug—or, to be more precise, the tail assembly of a larval insect. But all Bill was sure of was that it was something he had never seen before. So he called up from the lab and said, in a matter-of-fact tone as if it happened every day, that he had just found what he thought might be an uncrushed fossil 'bug'. Mike was on the receiving end of this particular announcement and not fooled for a moment. Perfect, uncrushed fossil insects just don't turn up in deposits of 20 million-year-old bones, so obviously it was simply a retarded bug that had gone for a swim in the acid vat the same morning it was found. But Bill was not that easily put off, so at the end of the day he mounted the 'thing' on a plasticine pedestal under Mike's microscope and went home for the day. The next morning, our collective jaws dropped—Bill's bug was exactly what he said it was! After apologising for ever doubting him, we immediately sent him back to the lab to scour all of the previously sorted concentrates for more.

That was the first of many uncrushed, three-dimensionally preserved arthropods to roll out of the Riversleigh acid vats. Slaters, millipedes, ants, fly larvae, cicadas all perfectly formed with tiny legs and wings intact as if preserved mere days before, steadily accumulated.

Fossil deposits that preserve bones and teeth do not normally also preserve soft-bodied fossils. Soft tissues usually decompose long

Abundant freshwater snails in the Oligo-Miocene limestones of Riversleigh are an indication of the wet conditions that prevailed when the fossil deposits formed.

before they can be fossilised. Even insect cuticle, which is a bit harder than the mushy muscle, usually vanishes without trace—and even then only after the beastie is squashed flat by the weight of accumulated sediment. Why these arthropods fossilised so perfectly at Riversleigh is a puzzle. Perhaps it happened in sites where the water was so supersaturated with calcium carbonate that deposition on all sides of the arthropod's cuticle began the moment the beast landed in the water. Then, either the cuticle decomposed from between the encrustations leaving a gap that filled with another acid-resistant compound or the trapped cuticle mineralised in the same way as the bones—molecule for molecule.

In Australia, the record of invertebrates is relatively poor, consisting mainly of flattened smears in shales, mostly too old to be useful in interpreting the history of Australia's living groups. Until the Riversleigh fossils were found, the 120 million-year-old Koonwarra Local Fauna of Victoria provided the most useful glimpse of Australian fossil insects but here, too, the fossils are squashed, paper-thin and very difficult to interpret.

For many Australian arthropod groups, the Riversleigh fossils represent the first pre-modern record. This material is being studied by Peter Jell of the Queensland Museum who has identified to date: millipedes (class Diplopoda); ants (family Formicidae); possible cicada (order Hemiptera); beetles (order Coleoptera), including at least one weevil (family Curculionidae); slaters (order Isopoda); and ostracodes (tiny crustaceans in the class Ostracoda). Most have been found at Upper Site on Godthelp Hill and are of early Miocene age. They represent a mixture of terrestrial and freshwater taxa (some as larvae and adults) that will be important in helping to interpret the palaeoenvironments in which they lived.

The significance of the Upper Site arthropods has been summarised recently by Peter Jell. He noted that in general the Riversleigh arthropods represented the beginning of a database for the history of Australia's insects and myriapods, where none has previously existed. The specimens already recovered from Riversleigh provide the first information about the *Tertiary* history of this continent's insects. This, in itself, is as much a statement about our appalling ignorance as a whole, as it is about the value of the Riversleigh resource to provide critical information.

SNAILS

carpets of molluscan marbles

The world over, fossil vertebrate deposits are commonly associated with fossil snails and Riversleigh is no exception. At least three different kinds of freshwater snails have been identified from the Oligo-Miocene Riversleigh deposits. They are most prevalent in System A sites like Site D, but have also been found in abundance at Tony's Escargot Site, a System B deposit near Neville's Garden. The outlines of snails often appear as logarithmic spirals on the eroded surfaces of the massive grey limestones or, more rarely, as olive-pip-like lumps that stand proud of the rock surface. While they often occur on their own, without bone, they at least serve to herald the presence of Tertiary rocks, rather than the much older snail-free Cambrian limestone of the Riversleigh area. This encourages us to look much more closely at the surrounding rocks for tiny bones or teeth that might otherwise have been missed.

Because of their calcium carbonate composition, snail shells do not usually survive acetic acid processing. Hence it is sometimes very difficult to recover good quality fossil snails from most of Riversleigh's deposits. However, the thousands of snail shells cramming the limestone of Bitesantennary Site emerge relatively unscathed from the acid vats. It seems that the shells, like the fossil bat bones of this site, have been sufficiently well mineralised that they can resist the onslaught of the acid. Perhaps in the case of the Bitesantennary snails, this fortunate coincidence may relate, in some way, to the copious quantities of mineral-rich guano accumulating in the same pools?

LUNGFISH

air gulpers and mud sliders

Among the more intriguing fish fossils found at Riversleigh are those of the lungfish that once lived in the freshwater pools and lakes of the forests. Anne Kemp, from the Queensland Museum, Brisbane, has identified several different types of lungfish from these deposits. One of the extinct kinds appears to be closely related to Australia's only living lungfish, the 1.1 metre-long *Neoceratodus forsteri* found today only in the rivers and lakes of south-eastern Queensland. A second fossil lungfish is a new species, unique to the Riversleigh deposits. A third species appears to be very similar to one previously discovered in the Leaf Locality of the Tirari Desert, South Australia.

The best represented fossil lungfish at Riversleigh is *Neoceratodus gregoryi*, which is also known from Eocene sediments at Redbank Plains in Queensland and from Oligo-Miocene to Pleistocene deposits in central Australia. This species closely resembles the modern species but grew to a much larger size. *Neoceratodus gregoryi*

Fossil lungfish teeth (the jagged-edged tooth plates in the lower centre) on this slab of middle Miocene limestone from Melody's Maze Site are common in the more 'aquatic' deposits at Riversleigh.

Australia's living lungfish, the metre-long *Neoceratodus fosteri* of the rivers of south-eastern Queensland, is small fry compared with some of the giant lungfish that once inhabited our waterways.

made the living lungfish, itself quite an impressive piece of fish, look like a minnow. Anne has estimated, from the size of the largest toothplates (which are 75 millimetres long) and widest rostral bone (85 millimetres), that the largest *N. gregoryi* could have been nearly 4 metres long! This behemoth is thought to have been an omnivore that ate aquatic plants and animals, such as snails and other invertebrates. She suggests that the creature may also have been a scavenger of mammal and turtle carcasses in much the same way as the living lungfish is known to eat dead animals. As it does for the closely related modern lungfish, deep water and adjacent reed banks probably provided food and shelter for this giant lungfish's eggs and young, keeping them from becoming meals for other animals that thrived in the same pools.

Lungfish have the ability to breathe both in water and air (they are gill *and* lung breathers) and belong to an ancient fleshy-finned fish group, one lineage of which is thought to have spawned all four-footed vertebrates more than 350 million years ago. This group of fleshy-finned fish also includes the coelacanth, the 'living fossil' discovered in 1938 off the coast of Madagascar. Before that stunning discovery, coelacanths were known only from fossil remains.

Australia's modern lungfish is one of five species that survives around the world. If fossils identified as this species from Cretaceous deposits in South America are correctly interpreted, it may have been around for a remarkable 70 million years. Lungfish as a group, however, have a fossil record that goes back approximately 400 million years. Lungfish skulls are comprised mainly of cartilage surrounded by a few bony plates for additional protection. The dentition consists mainly of two large crushing or shearing plates of enamel-covered, fan-shaped bone, one in each upper and lower jaw—four in total. The body is covered with large scales beneath a smooth, slimy skin.

Because much of the lungfish skull is made of cartilage rather than bone, most fossil lungfish are preserved as toothplates rather than skulls and jaws. Fortunately, toothplates can be used to reliably identify fossil lungfish species. At Riversleigh, the distinct, fan-shaped toothplates have been found in many sites but are particularly abundant in deeper-water deposits of Systems A and C.

The Gregory River today is host to a wide variety of bony fish, including this catfish held by its captor and imminent consumer Stephan Williams.

One of the most spectacular fossil lungfish accumulations at Riversleigh was found in 1984 at Melody's Maze Site. Covering a broad slab of limestone were dozens of lungfish toothplates, many standing proud of the rock, as well as disarticulated skull elements and even partial skulls—along with turtle skulls and rare mammal jaws. The specimens are particularly well preserved and should enable excellent reconstructions of these intriguing inhabitants to be made.

RIVERSLEIGH'S BONY FISH

eels and catfishes

Although some of the Riversleigh fossil deposits are peppered with fossil teleost (i.e. 'ordinary') fish remains, little is yet known about these creatures. Most are represented by isolated vertebrae which are not at all easy to identify. However, what look like catfish spines have been recovered from the same deposits and an eel jaw and partial fish skull were found at Quentin's Quarry Site. As expected, teleosts are most common in 'aquatic' deposits, particularly those of System C sites such as Ringtail Site but they also turn up, albeit less frequently, in most other sites, including some of the bat-rich deposits like Microsite, Gotham City and

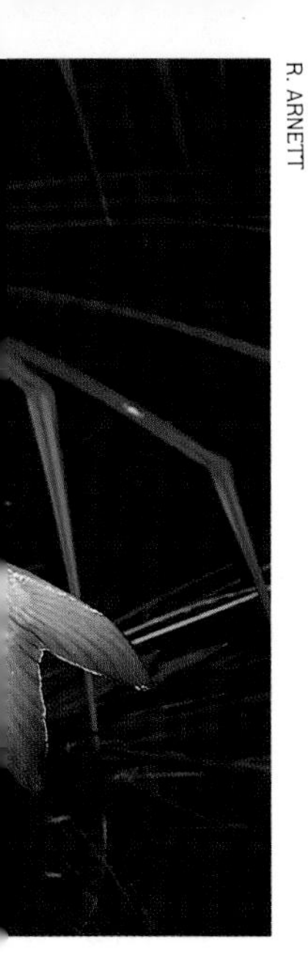

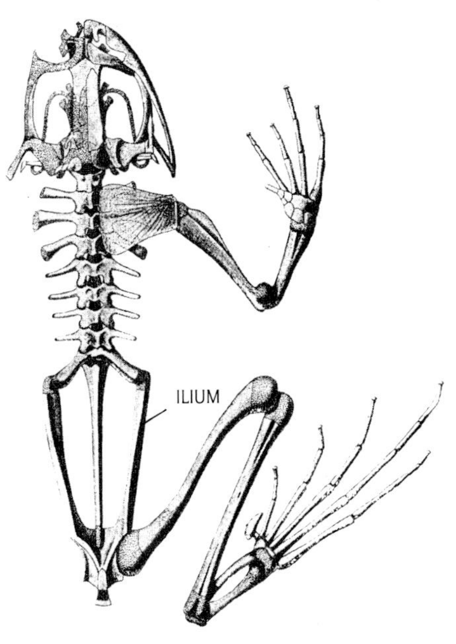

R. ARNETT

Skeleton of a frog with the ilium indicated—a taxonomically distinctive frog bone commonly preserved in the fossil deposits at Riversleigh.

Bitesantennary Sites. In the case of Bitesantennary Site, we are reasonably convinced that the teleosts lived in a circular pool on the floor of a shallow cave opened to the outside by a collapsed roof. The fish presumably entered this pool either as occupants of a stream that flowed through a subterranean system, of which the Bitesantennary Site was a single chamber, or they were introduced to the pool as eggs accidentally transported by wading birds moving between isolated bodies of water. Today inadvertent transportation by itinerant wading birds is the way many small fish are transported between ephemeral bodies of water in the normally dry salt lakes of central Australia.

FROGS

leaping between worlds

Many hundreds of frog fossils have been recovered from Riversleigh. Investigations by frog expert Michael Tyler from Adelaide University, South Australia, indicate that both tree- and ground-dwelling kinds are represented.

Before the commencement of studies on the rich Tertiary deposits at Riversleigh, the fossil record of Australian frogs consisted mostly of specimens recovered from Quaternary cave deposits in South Australia and Western Australia, together with a few isolated records of Tertiary frogs from the Lake Eyre Basin of South Australia.

The 140 millimetre-long Green Tree Frog, *Litoria infrafrenata*, is Australia's largest living frog. Recently, Riversleigh has revealed an even larger tree frog among the more than 20 new kinds of fossil frogs identified.

HAL COGGER

Following the 1983 discovery of the vast, vertebrate-rich limestone sequences at Riversleigh, fossil frogs began to leap out of the Oligo-Miocene deposits. These include nearly complete skeletons from deposits such as Upper Site (one partial skeleton includes a large portion of an intact skull), as well as a stunning number of ilia (the longest bone of the frog pelvis—two per frog). To date, nearly a thousand ilia have been recovered representing at least 500 individual frogs.

To put this into perspective, Michael Tyler has identified Riversleigh as the richest frog fossil site in the world, with a fauna of possibly as many as 20 species. At least eight species of hylid tree frogs in the genus *Litoria* and six leptodactylid ground frogs have been identified.

There are approximately 50 living, Australian species of *Litoria* found across Australia. These vary widely in body form, size, colour, habitat and biology. Perhaps the best known are the green tree frogs, which include what is probably the world's largest living tree frog, the 140 millimetre-long *Litoria infrafrenata* of northern Australia and New Guinea. However, Riversleigh has recently revealed an even larger tree frog than this living giant. This most exciting beast was substantially bigger than *Litoria infrafrenata* and represents a new species. It has been recovered from a number of deposits, including Gag Site.

The Riversleigh fossil leptodactylids include species of *Limnodynastes*, *Crinia*, *Kyarranus* and *Lechriodus*. Living species of *Limnodynastes* (a name which means 'Lord of the marshes') are found right across Australia. Most species of *Crinia* live in or near permanently damp sites in south-eastern and south-western Australia. The three living species of *Kyarranus* are confined to the border area of New South Wales and Queensland, mainly in high altitude, cold, wet forests. The only living Australian species of *Lechriodus* (there are three New Guinean species) lives in rainforest in coastal areas from northern New South Wales to southern Queensland. These modern habitats for the relatives of Riversleigh's fossil frogs may give some idea about the former habitats of the extinct species.

A high proportion of the ilia of frog species found in the Riversleigh deposits belong to small creatures. Small frogs are particularly vulnerable to dehydration and consequently are predominantly found in areas that remain moist throughout the year. In Australian frog communities, a high frequency of small frogs occurs only in areas of high and seasonally

Barrington Tops in northern New South Wales is home to a variety of frogs, some of which are distantly similar to types found in the older deposits at Riversleigh.

reliable rainfall, such as the extreme south-west of Western Australia and the northern periphery of Arnhem Land in the Northern Territory.

The quality and quantity of fossil frog material recovered to date is enabling investigation of changes in the composition of the frog fauna across the time span of the Riversleigh deposits. The dependence of frogs upon moisture at ground level means that they are excellent indicators of environmental conditions and shifts in the predominance of any one taxon across time can be assessed against the specific ecological requirements of the taxon involved.

For example, a study by Mike Tyler, Sue Hand and Veronica Ward of the frequency of the Riversleigh fossil frog *Lechriodus intergerivus* in the various deposits has revealed that more individuals occur in older deposits than younger ones and less individuals in deep-water deposits than in shallow-water deposits. Living species of *Lechriodus* are restricted to temperate and subtropical rainforests of New Guinea and Australia. They have speculated that the decline in numbers of *L. intergerivus* in Riversleigh's younger Oligo-Miocene deposits may reflect a decline in some aspect of the moist, rainforest habitat that resulted from broad changes in forest and animal communities occurring across northern Australia during the middle and later Miocene. Water depth, too, may have been a limiting factor for the distribution of *L. intergerivus* in the Riversleigh rainforests. Living species of *Lechriodus* choose small bodies of water in which to deposit their eggs.

Recently there have been reports from around the world of serious declines in, and even local extinctions of, frog populations. These include several examples from eastern and south-eastern Australia. Mike Tyler has been invited to contribute to an assessment of the extent of the phenomena and to help determine whether frogs are useful indicators of current global weather trends (i.e. Greenhouse changes). He believes that the Tertiary changes in frog dominance at Riversleigh may prove highly relevant as an historical perspective within which the significance of modern climate-related changes in diversity can be better understood.

CROCODILES

giving everything else a run for its life

Today, Australia has only two species of crocodile: the Saltwater Crocodile (*Crocodylus porosus*) and the Freshwater Crocodile (*C. johnstoni*), both found in northern Australia. However, according to Paul Willis of the University of New South Wales, this low diversity is a peculiar feature of little more than the last 10 000 years. At any given time throughout most of its Mesozoic and Cainozoic history, Australia has played host to at least three crocodile species in its lusher habitats.

Fossil crocodiles have been recovered from Riversleigh's Oligo-Miocene and Pleistocene deposits. Related crocodiles are also turning up in the Tertiary deposits of South Australia (Oligo-Miocene forms from Lake Palankarinna in the Lake Eyre Basin and Lake Pinpa in the Frome Embayment), the Northern Territory (late Miocene taxa from the Alcoota Local Fauna and middle Miocene taxa from the Bullock Creek Local Fauna) and Queensland (early Tertiary taxa from Murgon and Rundle). While most of these fossil sites have revealed two or three taxa in total, Riversleigh has produced nine species representing five genera. These include species of *Crocodylus, Pallimnarchus, Baru* and two new genera.

H. COGGER

Crocodiles have a long and distinguished history on this continent, one that extends back at least 110 million years.

One of the most rewarding finds has been the first fossil of the Freshwater Crocodile. A

R. ARNETT

A long, narrow lower jaw (above) recovered from the late Pleistocene Terrace Site represents the earliest record for the fish-eating Freshwater Crocodile (below).

nearly complete left dentary of this species turned up in the Pleistocene gravels of Riversleigh's Terrace Site in 1986. Since then a second specimen from Pleistocene sediments at Floraville Station in north-western Queensland has also been identified. Although these fossils are from individuals only slightly larger than the average modern Freshwater Crocodile (which rarely exceeds three metres in length), their morphology suggests that the Freshwater Crocodile evolved within Australian waters from an ancestral population of Saltwater Crocodiles.

From the Terrace Site, a second crocodile species is also known. This is believed to be a species of *Pallimnarchus*. Fossil material referable to this genus had already been found in Pliocene and Pleistocene deposits throughout what were the inland waterways of eastern Queensland. It is also now known from similar-aged sites in Western Australia, South Australia and New South Wales.

The largest crocodile yet recorded from Riversleigh is a new species of *Baru* about to be named after Tony Wicken, Pro-Vice-Chancellor of the University of New South Wales. It was a five-metre juggernaut (approximately equal in size to the infamous 'Sweetheart', a Northern Territory Saltwater Crocodile that delighted in eating boats until it was accidentally killed in 1979), known only from Riversleigh's Site D and its stratigraphic equivalents. It was a freshwater carnivore that may have grabbed its dinner guests along the margins of the water and killed them by lacerations of its huge, recurved, razor-edged teeth. This killing strategy would have differed from that employed by the living Saltwater Crocodile which either spins its prey in order to dismember them or drowns them by holding them below water. One of the most spectacular specimens of this species of *Baru* recovered to date is an almost complete skull recovered in 1976 from Site D. *Baru darrowi*, a close relative and perhaps descendant of this Riversleigh croc, occurs in the middle Miocene Bullock Creek Local Fauna of the Northern Territory.

Smaller crocodiles, one to two metres long, may have been more significant as predators in Riversleigh's rainforests. These were very deep-headed and apparently terrestrial crocodiles that could have run down their prey, terrier-like, on land. At least two genera of these terrestrial crocodiles are represented in the Riversleigh deposits. In a single limestone block from Ringtail Site two small but beautifully preserved overlapping skulls represent two of these crocodiles. Each appears to represent a distinct genus. Although highly specialised in quite

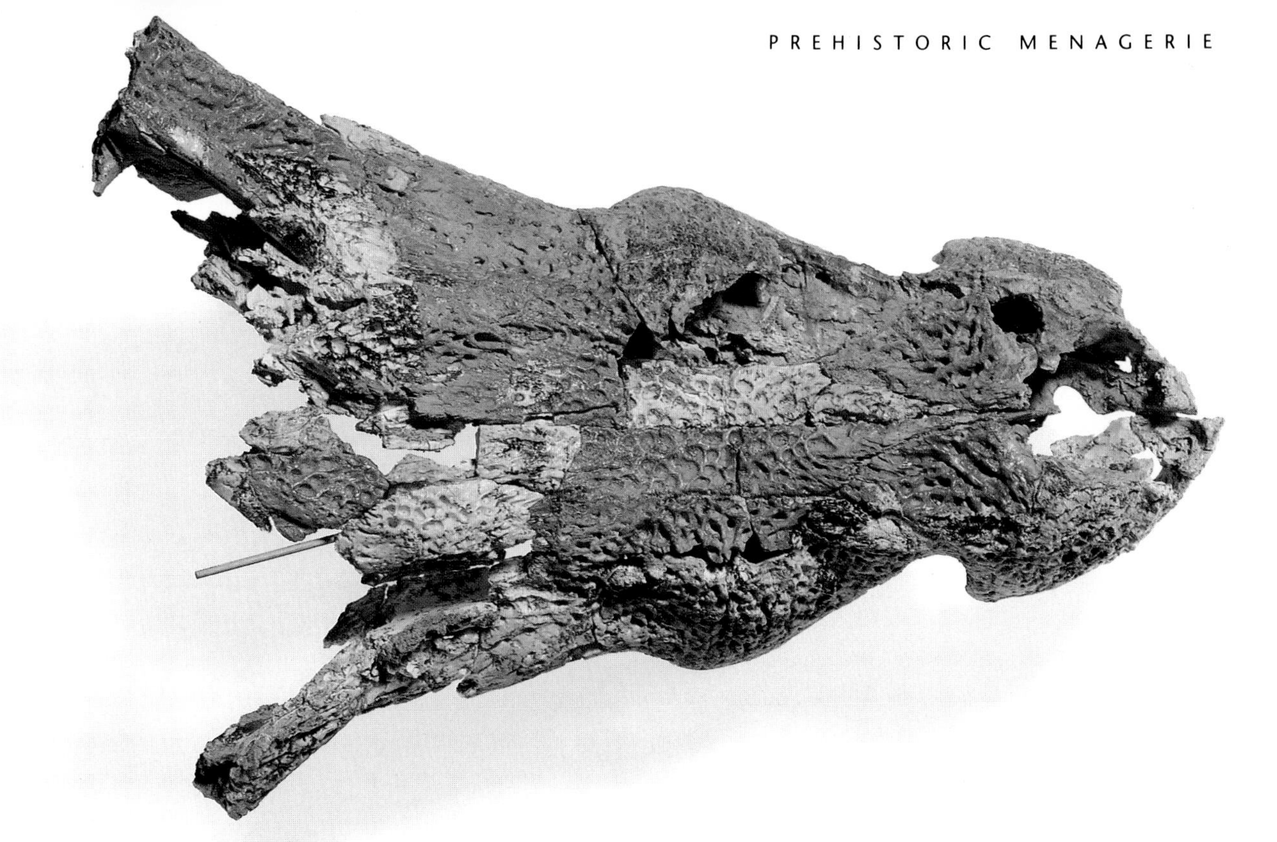

R. ARNETT

Upper (above) and palatal views of the front half of a skull of Riversleigh's largest crocodile, *Baru wickeni*, from Site D.

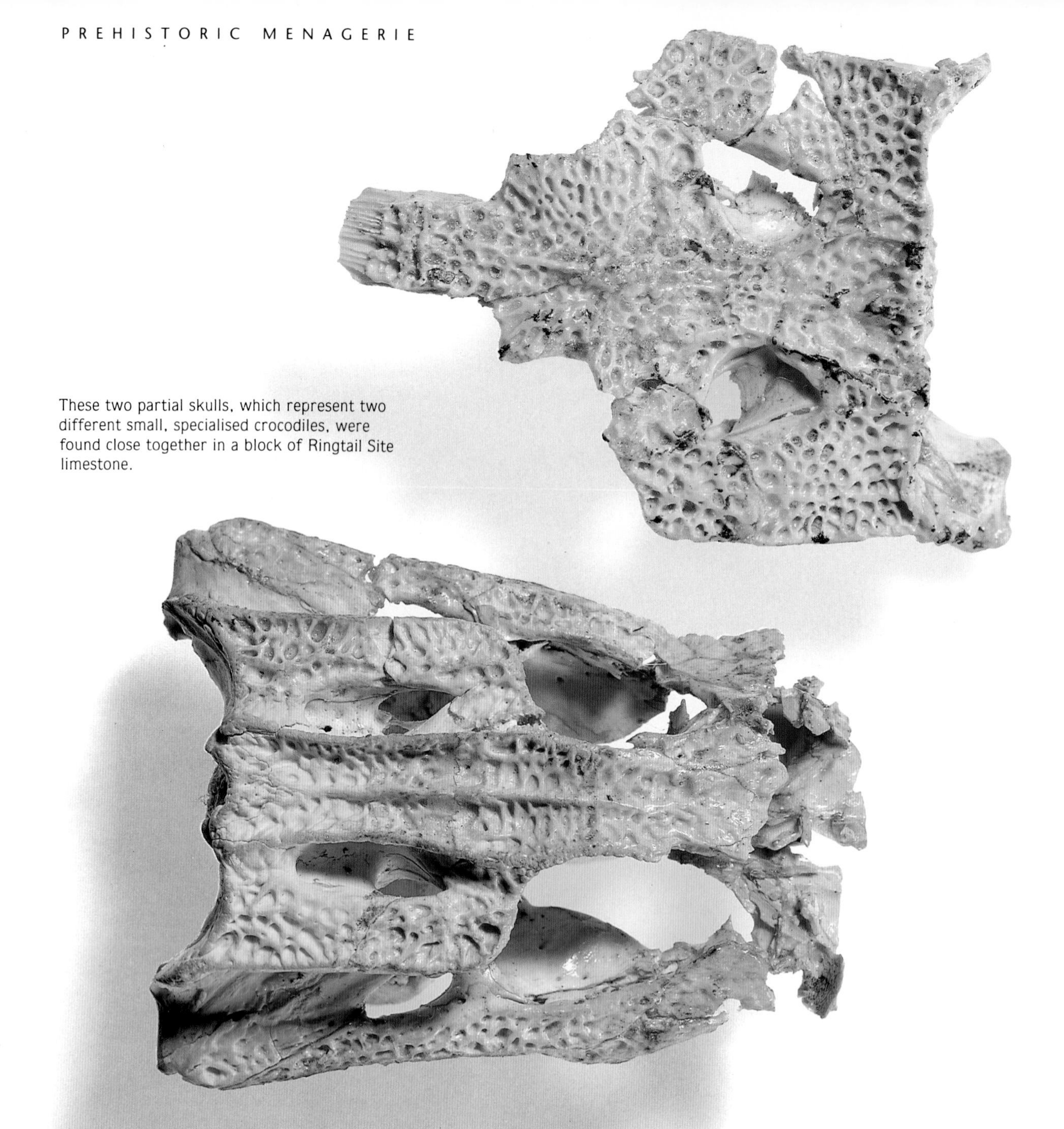

These two partial skulls, which represent two different small, specialised crocodiles, were found close together in a block of Ringtail Site limestone.

R. ARNETT

different ways, these animals appear to be reasonably closely related to each other.

In fact, apart from the clear indications that Riversleigh's crocs can be placed into at least three different groups, all appear at a higher level to be more closely related to each other than they are to crocodiles from other areas of the world, such as alligators or gavials. An intriguing, albeit tentative, conclusion of Paul's study is that Australia had an endemic suite of crocodilians throughout most of its Cainozoic history. These unique but now totally extinct forms were evidently replaced by species that either invaded from elsewhere (the Saltwater Crocodile is believed to have entered Australian waters from the north) or were descended from these invaders (e.g. the Freshwater Crocodile).

The extinct Riversleigh crocodiles appear to have occupied a much wider range of ecological niches than any do today. Overall, they included semi-aquatic ambush predators, specialised fish-eaters and terrestrial carnivores. As terrestrial carnivores, some of these Riversleigh crocodiles appear to have adopted ecological roles similar to those of the monitor lizards (varanids) and smaller mammalian carnivores. With the many large pythons, crocodiles may have been the dominant carnivores of many of Australia's Cainozoic terrestrial ecosystems.

A block of Camel Sputum limestone revealing the white cross-sections of vertebrae of a giant python-like madtsoiid snake.

SNAKES

legless layabouts coming to grips with dinner

To judge from the quantity of remains in the fossil deposits, Riversleigh's Oligo-Miocene rainforests must have been festooned with snakes. Snake material from Riversleigh has more than doubled the number of extinct Australian taxa described and extended the record back for most groups.

John Scanlon of the University of New South Wales has identified four families from Riversleigh's many Oligo-Miocene faunas, three of which are major groups in the living fauna. These are: 1. the venomous elapids, 2. the pythonids, 3. the burrowing, blind typhlopids, 4. the madtsoiids, extinct snakes only known from South America, Africa, Madagascar and Australia.

Missing from Riversleigh's Tertiary record so far are colubrids (which have no pre-Pleistocene record from anywhere in Australia), acrochordids (which are known from the early Pliocene of Australia) and boids (which occur in New Guinea but not mainland Australia). But considering that it was not until 1988 that the first of Riversleigh's elapids was recognised, the three missing groups may yet put in an appearance.

Most of Riversleigh's snakes are represented by vertebrae, rather than skull material. This is perhaps not surprising considering that a python, for example, has up to 600 vertebrae, but only one head! However, we have found situations in which chains of python vertebrae snake their way through adjacent blocks of limestone ending with the glimpse of a skull and jaws. These specimens are highly prized. However, even isolated snake vertebrae are extremely useful because they are very distinctive. Details of structure enable

GLADA

Volunteer Karen White with one of Riversleigh's larger living snakes, a 3-metre Olive Python. These are common along the banks of the Gregory River.

D. MAITLAND

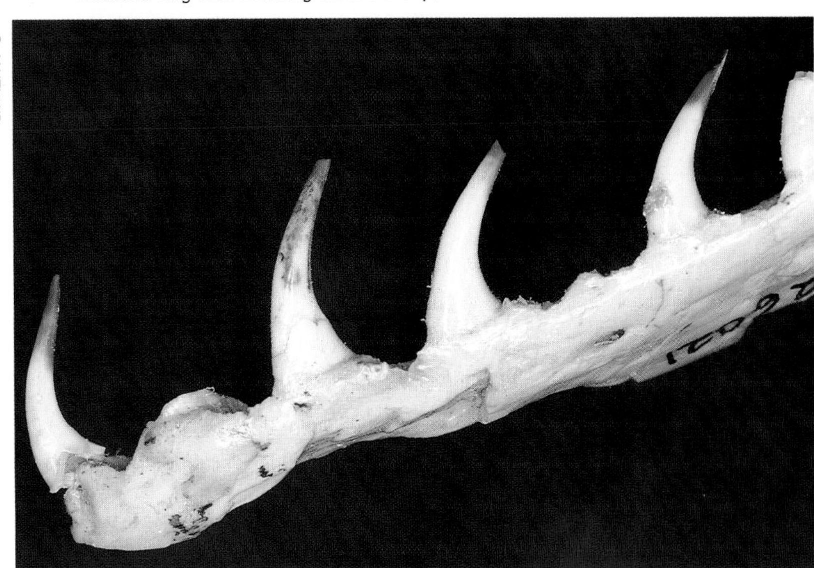

The lower jaw of a very large madtsoiid snake from the middle Miocene Gag Site showing needle-sharp, recurved teeth.

determination of the family, genus and even species to which the vertebra belongs.

Pythonids. Most of Riversleigh's snakes were either pythons or madtsoiids. Giant pythons up to 7 metres long, with girths only slightly smaller than dinner plates, would have been reasonably common. According to John Scanlon, a Riversleigh giant collected at Site D named *Montypythonoides riversleighensis* by Meredith Smith and Mike Plane in 1985 is closely related to living pythonids, although no species of this genus appears to have survived the Tertiary. Another large python, *Morelia antiquus*, from the middle Miocene Bullock Creek Local Fauna, was named at the same time. *Morelia* is the modern genus that contains, among other living species, the Diamond Python (*M. spilotes*).

Madtsoiids. This group of presumably non-poisonous constrictors originated in the Cretaceous and occupied the southern continents into the Tertiary. They persisted in Australia until the late Pleistocene, long after they had died out elsewhere. Australian madtsoiid fossils are the most complete and informative in the world. One Riversleigh madtsoiid evidently grew to a length of more than six metres, putting it in the size range of the Anaconda of South America. The Riversleigh madtsoiid appears to have been an ancient relative of the enormous and well preserved 6 metre-long *Wonambi naracoortensis*, named by Meredith Smith on the basis of Pleistocene material from fossil deposits in Victoria Fossil Cave and Henscke's Quarry, near Naracoorte, South Australia. Prior to 1985, this was the only

Reconstruction of one of the last surviving madtsoiid snakes, the 6 metre-long late Pleistocene *Wonambi naracoortensis*, found in caves of southern Australia.

RECONSTRUCTION BY P. SCHOUTEN

fossil snake named on the basis of Australian material.

Elapids. Elapids, poisonous non-constrictors, occur on almost every continent. They are believed to have colonised Australia from south-eastern Asia. The age of the Riversleigh elapid material may present a challenge to current understanding about the genetic history of this group in Australia. The 100 or so elapids that occur in the Australasian region (i.e. Australia, New Guinea and islands such as Fiji and the Solomons) appear to be closely related. A 'molecular clock' date, based on the interpreted time of divergence of albumin molecules in the various groups, suggests that they all shared a common ancestor about 12 to 15 million years ago. Many of the elapid fossils that John Scanlon recognises from Riversleigh are, however, more than 20 million years old. Consequently, they could be taken to present a challenge to the molecular clock date.

However, it is just possible that the Riversleigh elapid material could represent an ancestral lineage that existed 5 million years *before* it began to diversify to produce the various lineages now found in the Australian region. The molecular clock date might not then be falsified because it only applies to the time of initial diversification. A single ancestral elapid lineage demonstrated to be 50 million years old need not falsify the common divergence date. Only discovery that more than one of the modern lineages had representatives prior to 15 million years ago would do this. And until John finishes his analysis of the Riversleigh elapids, we will not know if this is the case.

Typhlopids. Typhlopids are small, non-venomous, burrowing, blind snakes that have greatly reduced eyes consisting of dark spots beneath the scales on the head. They feed principally on ants and termites and live under ground, rocks and logs. Twenty-two living species are found in Australia and all are placed in the genus *Ramphotyphlops*. Typhlopid snakes are rarely found as fossils anywhere in the world but in some Riversleigh deposits, including Upper Site, vertebrae and the minute ribs are plentiful. There is no other Australian fossil record for this group.

Acrochordids and colubrids. Australian acrochordids (aquatic file snakes) and colubrids (water snakes and tree snakes) are not known from Australian deposits older than Pliocene and Pleistocene respectively. File snakes, which are characterised by their extraordinarily loose skins with rough, ridged scales and nasal openings on top of their snouts, are common in the modern Gregory River at Riversleigh, as they are in all the river systems of northern Australia open to the sea.

Acrochordids almost certainly entered Australian freshwaters from the oceans to the north, at least by early Pliocene time based on vertebrae from the Bluff Downs LF of north Queensland. They are aggressive colonisers of any accessible fish-rich freshwater bodies and are distributed widely around the northern margins of the Indian Ocean. The oldest file snakes are from Siwalik sediments of middle Miocene age in southern Asia. However, these are relatively derived members of the group which suggests that more 'primitive' file snakes

H. COGGER

The earliest record for the venomous elapid snakes of Australasia now comes from the Oligo-Miocene deposits of Riversleigh.

H. COGGER

The 'primitive', blind burrowing snakes had no fossil record in Australia before they were discovered in the Oligo-Miocene sediments of Riversleigh.

The fish-eating, non-venomous File Snakes, although common today in the Gregory River, have been found in none of the fossil deposits of Riversleigh.

were in the area of the Indian Ocean well before the middle Miocene. If file snakes were present in the oceans north of Australia during Oligo-Miocene times, their absence from Riversleigh's Oligo-Miocene units suggests that Riversleigh's depositional basins were not at that time open to the sea. The ancestral northward flowing Gregory River probably did not establish itself until late Miocene or early Pliocene time.

As investigations of Riversleigh's Pliocene to Pleistocene local faunas proceed it seems highly likely that we will eventually recover remains of late Tertiary and Quaternary colubrids and file snakes. But so far, no show.

TURTLES

reptilian tanks by the tonne

Almost all of Riversleigh's fossil sites contain the remains of turtles. Dominantly aquatic fossil sites at Riversleigh—such as System C's Ringtail and Quentin's Quarry Sites—have been particularly productive hunting grounds for these reptilian tanks. Some sites have yielded skull and shell remains from a variety of turtles in association with fish (especially lungfish), invertebrates (e.g. gastropods) and, commonly, mammals. From a few of Riversleigh's Systems A, B and C sites, complete skulls and shells have been recovered (e.g. from Melody's Maze and Bob's Boulders Sites). With some of these shells, partial skeletons have been found.

Perhaps one of the most amazing fossil turtle finds at Riversleigh has been Crusty Meat Pie Site where literally thousands of individual turtles have been compacted together. Death assemblages of this kind are, as far as we are aware, unique within Australia and invite 'unorthodox' palaeoecological explanations.

Turtle remains have also been found in the Pleistocene gravels at Riversleigh's Terrace Site. A new giant chelid, *Elseya lavaracki*, excavated from this site is one of few 'megafaunal' turtles known from Australia.

Turtles are today a distinctive component of

The shell of a fossil chelid turtle from the Miocene Bob's Boulder Site at Riversleigh.

GLADA

Jim Lavarack braces the delicate shell of a huge extinct freshwater turtle found in the late Pleistocene Terrace Site before it is secured in a hard plaster jacket for protection.

R. ARNETT

Australian freshwater ecosystems. They are hunters and scavengers and play a crucial role in recycling biological materials in the vast river systems that today characterise most areas of the continent, from the arid interior to the rainforest-covered highlands of north-eastern Queensland. The fossil materials from Riversleigh indicate that the region had a diverse and well established freshwater turtle community in the Oligo-Miocene. Riversleigh's fossil turtle researcher Arthur White, a Research

Freshwater turtles, common scavengers today in the Gregory River, have filled this ecological role for millions of years at Riversleigh.

Some of the fossil accumulations of turtles at Riversleigh are puzzling, such as this compressed bundle of shell pieces found at Crusty Meat Pie Site.

Extinct, probably terrestrial, horned turtles such as this species from Lord Howe Island have also been found at Riversleigh.

J. FIELDS

Associate of the University of New South Wales, is confronting a growing mountain of new information about the structure and evolution of Australia's freshwater turtle communities.

Riversleigh's turtle remains appear to represent the following groups: freshwater chelids, also known as side-necked turtles; meiolaniids, extinct giant horned turtles; and possibly soft-shelled trionychids.

Chelids. There are five genera of living chelids (*Chelodina, Elseya, Emydura, Pseudemydura* and *Rheodytes*) and many modern species. Current hypotheses suggest that these turtles radiated with the development of Australia's river systems. However, until now, apart from a few reasonably preserved middle Tertiary specimens from Tasmania, there was little in the way of a fossil record that could test this hypothesis. Little could be said with confidence about the history of any of the chelid groups, particularly the enigmatic *Pseudemydura*. Now hundreds of chelid specimens have been recovered from Riversleigh, including representatives of at least *Chelodina, Emydura, Elseya* and *Pseudemydura*. These remains include the oldest fossil of the modern Australian genus *Chelodina*, a group believed to be closely related to South American chelid turtles and the first (and oldest!) fossil remains of *Pseudemydura* and *Elseya*.

Before the Riversleigh discovery, *Pseudemydura* was only known from a single endangered species, the Swamp Turtle (*P. umbrina*), surviving in two swamps north of Perth, Western Australia. It may have the distinction of being the rarest reptile in Australia. When Gene Gaffney, from the American Museum of Natural History, was rummaging through turtle bits accumulated from Ringtail Site, he was stunned to find the posterior part of a skull of an undoubted species of *Pseudemydura*. In size and most aspects of its shape, it is similar to the living *P. umbrina*. The early Miocene *Pseudemydura* from Riversleigh is particularly interesting because this genus appears to represent a primitive stock descended from a common ancestor shared with the remaining groups of Australian chelids.

Meiolaniids. Giant horned turtles (family Meiolaniidae) were previously well known from the Pleistocene of Australia and the Pleistocene and Holocene of Lord Howe Island and New Caledonia. From these Pleistocene sites, massive specimens with shells 1.5 metres long and estimated body weights of half a tonne had been found. These turtles had horn-like projections growing from the top of the skull and spiked or clubbed tails. Meiolaniids are also represented in the fossil record of South America and Madagascar and are considered to be an undoubted Gondwanan group. On the basis of shell fragments, their presence has also been reported from central Australian Oligo-Miocene deposits.

Now, partial skulls, tail rings (these support the casing and club of the tail) and limbs of a new, distinctively primitive horned turtle are turning up regularly in Riversleigh's Oligo-Miocene terrestrial and freshwater deposits. In some Riversleigh sites (e.g. Pancake Site on Godthelp's Hill), it is now evident that meiolaniids are very common, with cross-sections of very large, thick turtle shells and limbs being visible in many of the weathered limestone tors.

But no carettochelyids! The most aquatic of Australia's freshwater turtles, the Pig-Nose Turtle, which survives today in far northern Australia and New Guinea, has not yet been recognised among Riversleigh's turtles. The fossil history of carettochelyids is unclear but in Australia the record appears to extend back to at least the Quaternary, although in other areas of the world it extends back to the Cretaceous.

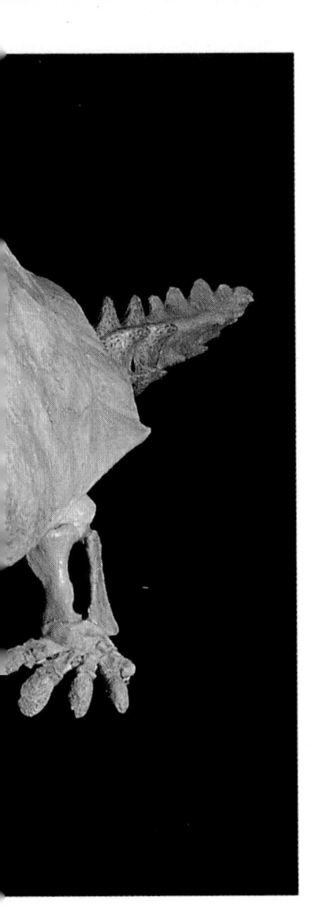

H. COGGER

Until discovery of a species of *Pseudemydura* at Riversleigh, the living Swamp Turtle of Western Australia was thought to be the only representative of this distinctive group.

LIZARDS

modern groups in ancient settings

All the major groups of lizards alive today are represented by Riversleigh fossils: skinks, dragons, goannas, geckoes and legless lizards. Of these, the skinks and dragons are the most common and best known.

Prior to the 1980s, virtually nothing was known about the history of skinks and geckoes in Australia, despite the fact that these two groups represent major components of almost all Australian terrestrial ecosystems. What little was known at the time, particularly about the Oligo-Miocene lizards of central Australia that had been collected by Reuben Stirton and his colleagues, was summarised in a review paper by Richard Estes. With discovery of the diverse Riversleigh vertebrate assemblages, opportunities to clarify the history of these groups really arose for the first time.

Mark Hutchinson from the South Australian Museum in Adelaide found that diverse assemblages of skinks, geckoes and legless lizards existed in Riversleigh's Oligocene-Pliocene habitats, the span of time crucial to understanding the history of the modern fauna.

Modern Australian lizards are highly endemic and very diverse. The most distinctive Australian skinks are the relatively gigantic bluetongue

R. ARNETT

This living Marbled Velvet Gecko of Riversleigh reflects the long history of the group as evidenced in the limestones of Riversleigh.

lizards of the genus *Tiliqua*. The first Riversleigh skink to be named (by Glenn Shea and Mark Hutchinson) was *Tiliqua pusilla*, a relatively small extinct species perhaps 15 cm in length that appears to be related to the living pygmy bluetongue, *T. adelaidensis*. Other skink remains found at Riversleigh are very similar to other living species suggesting that Australia's lizards have changed less than its mammals over the same period of time.

The first Tertiary legless lizard from Australia has been found at Riversleigh's Neville's Garden Site. Lizards of this kind are normally classified in a family of their own (the Pygopodidae) but recent studies suggest that it would be more appropriate to regard them as specialised, limbless geckoes.

Prior to the Riversleigh discoveries, the only fossil dragon from Australia was a Pliocene agamid lizard similar to some species of *Amphibolurus*, the bearded dragons. Now, a large collection of agamid remains from Riversleigh's Oligo-Miocene deposits is being studied by Jeanette Covacevich at the Queensland Museum in Brisbane and her colleagues.

Most of the specimens they have examined to date have been assigned to the extant genus of water dragons, *Physignathus*. Some resemble the modern Australian Eastern Water Dragon (*P. lesueurii*). The remainder, the bulk of the material, represents other distinct species of the same genus (*Physignathus* sp. nov. or *Physignathus* sp. *cf. P. lesueurii*).

The discovery at Riversleigh of species of *Physignathus* confirms an earlier surmise that agamids were present on this continent at least as early as the late Oligocene. And it corroborates the current hypothesis that Riversleigh was a lot wetter than it is today. In Australia, no modern species of *Physignathus* occur far from the moist east coast.

Covacevich and her colleagues have described another, more distinctive Riversleigh agamid from Wayne's Wok Site as *Sulcatidens quadratus*. This creature has highly distinctive teeth and, as a result, its relationships to other agamids are difficult to work out.

The high numbers of individuals but apparently low species diversity of Riversleigh's Oligo-Miocene agamids suggest that the modern pattern of low agamid diversity in wet areas versus high diversity in the drier parts of the continent appears to have been established at least as early as the Oligo-Miocene.

Covacevich and her colleagues are also curious about an apparent ecological difference between Riversleigh's fossil agamid assemblages

Water dragons were once common in the fossil deposits of Riversleigh. Some were distantly related to the living Eastern Water Dragon.

R. MORRISON

Alongside the fossilised leg bones of this giant flightless dromornithid, affectionately known as the 'Big Bird' of Site D, are small pebbles that ground up food in its gizzard.

and those of modern Australia. In Riversleigh, a highly diverse crocodilian fauna co-existed with a large population of water dragons. Today, only two species of crocodiles occur in Australia and only in one location do water dragons and crocodiles co-exist (in the extreme north of the species range of the Eastern Water Dragon).

BIRDS

Riversleigh's fossil feathered fliers

Study of Australia's fossil birds is a venerable discipline. It was a research topic that attracted the attentions of a few of Australia's first palaeontologists in the early part of the nineteenth century, including Sir Richard Owen from the British Museum and Charles W. De Vis of the Queensland Museum. Early collections made in the late 1800s at Lake Callabonna, South Australia, revealed enough of the peculiar dromornithid birds to enable, for the first time, reconstruction of most of a skeleton and partial skull of these giant, flightless birds.

Despite an early burst of enthusiasm, particularly involving the fossil birds of the Plio-Pleistocene of central and eastern Australia, there was relatively little effort put into this research area until the Oligo-Miocene sediments of central Australia were explored by Reuben Stirton, Alden Miller and their colleagues in the 1950s and 1960s. This led to Miller's studies of the systematics of several groups of aquatic birds such as the Australian flamingoes, anhingas and pelicans. This, in turn, led in the 1970s and 1980s to studies by Pat Rich, Gerry Van Tets and Bob Baird of the dromornithid and casuariid ratites from all of the then known Australian Tertiary and Quaternary fossil deposits (including Riversleigh), and later, to the study of non-ratite groups, such as owls, owlet-nightjars, plains wanderers, megapodes, pigeons, native-hens, pelicans and falcons.

The Tertiary fossil deposits discovered on Riversleigh have provided a wealth of new fossil bird material including not only dromornithids, casuariids and wading birds, but also many song-birds and representatives of other orders previously rare in Australia's Tertiary record.

Walter Boles of the Australian Museum in Sydney has been studying these Riversleigh fossils. According to Walter, the age of the Riversleigh deposits and their diversity of bird remains make them some of Australia's most important sites for fossil birds. Australia's Tertiary bird sites, particularly those of central Australia, are dominated by waterbirds. Riversleigh is important because, while it also has waterbirds, it has a large number of terrestrial forms, thus providing the earliest

R. WILLIAMS

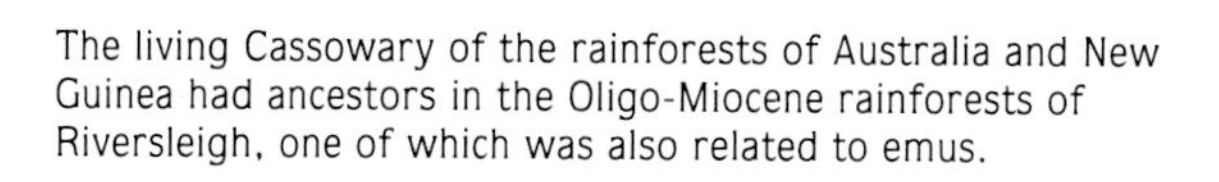

The living Cassowary of the rainforests of Australia and New Guinea had ancestors in the Oligo-Miocene rainforests of Riversleigh, one of which was also related to emus.

The most distinctive fossil bird bones at Riversleigh are those of large to gigantic flightless birds, such as ancestral emus and dromornithids.

glimpses of bird communities that were not tied to open water. These more balanced fossil bird assemblages enable us to develop a better understanding of the structure and evolution of Australia's bird communities through middle and late Tertiary time.

Because birds lead different lifestyles and have different requirements to other animals, they can supply unique information about palaeoecology, palaeoenvironments and palaeogeography. Walter's study should provide key pieces to the puzzle of reconstructing the environments and animal communities that existed in Riversleigh's Oligo-Miocene rainforests.

Dromornithids. By far the most conspicuous bird fossils found in the Riversleigh rocks are the enormous bones of dromornithids, also known as mihirungs or thunder birds. At least two different genera are represented at Riversleigh, including one species each of *Barawertornis* and *Bullockornis*. Dromornithids, known only from Australia, are now all extinct but survived until at least as recently as 26 000 years ago and were evidently known to Aborigines. Their ranks include perhaps the heaviest bird that ever lived, the three metre-tall, 400-kilogram *Dromornis stirtoni* from the late Miocene Alcoota Local Fauna of central Northern Territory. Although very large and decidedly flightless, dromornithids are evidently not closely related to other ratites such as emus, cassowaries, or even ostriches. Indeed, although controversial they are perhaps most closely related to galliform birds, the group that contains the domestic chook.

Dromornithids must have been a conspicuous part of Australia's Tertiary landscape because they form a significant component of many fossil sites around Australia. The most comprehensive review of the whole group was published by Pat Rich in 1979. Yet, despite their abundance and considerable attention, very little is known about their origin, relationships, functional anatomy or natural history.

Although dromornithids had lost the ability to fly, their bones retained the characteristic internal structure of bird bones that remain hollow but strong enough to carry the bird's weight by means of a system of struts that produce a honeycomb-like appearance in cross section. Some Riversleigh sites are particularly rich in dromornithid bones. Sticky Beak Site, for example, is named for the many dromornithid bones it contains. Similarly, Bone Reef and Burnt Offering Sites are riddled with the bones of massive birds. These are often found associated with the bones of large aquatic animals, such as turtles, crocodiles and lungfish, and it is possible that, like modern cassowaries, they were good swimmers and spent at least some of their time in water.

The most famous Riversleigh bird fossil is the individual known affectionately as 'Big Bird'. The remains of the dromornithid Big Bird were found in a large limestone boulder at Site D in 1986. Preserved in the rock along with Big Bird's bones are a mass of small, rounded pebbles or 'gizzard stones' swallowed to help the enormous bird grind up its food, in the same way a domestic chook uses sand-sized pebbles to help grind up its food prior to digestion.

Casuariids. Cassowaries and emus today comprise a distinctive part of the Australo-Papuan bird fauna. Emus are well represented in the Australian fossil record with several fossil species having been described. Cassowaries are less well represented: a single bone has been recovered from Wellington Caves in New South Wales and some fossilised leg and toe bones have been found in New Guinea. At Riversleigh, the family is represented by an Oligo-Miocene species that shares characteristics of both emus and cassowaries—as Walter Boles quips, it is a kind of 'cassomu' or 'emuwary'. Smaller than the smallest living member (the Dwarf Cassowary of New Guinea), *Dromaius gidju* was first described by Chris Patterson and Pat Rich from the Kutjamarpu LF of central Australia. Walter's researches suggest that it may have been a less efficient runner than the living Emu but probably a better runner than any of the three living cassowaries.

GLADA

Part of the hindleg of one of Riversleigh's fossil dromornithids (right) is compared with the same bone in an Ostrich to give an idea of the enormous size of these gigantic Oligo-Miocene birds.

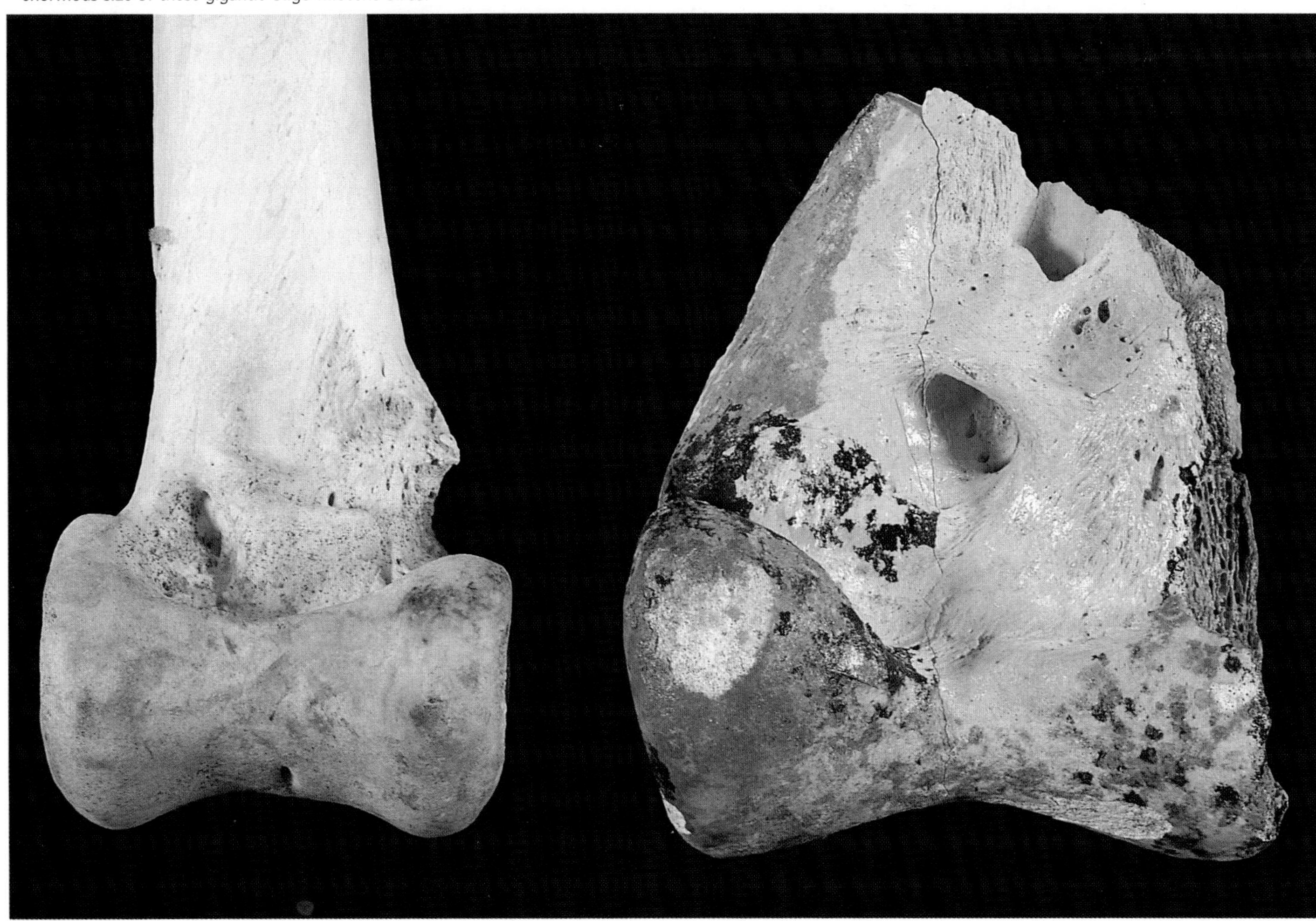

A.FARR

P. GERMAN

Modern Galah, one of the most common Australian cockatoos. While most live in the drier areas of Australia, some, like the Riversleigh cockatoo, are found in rainforest.

Cockatoos. Until an almost complete upper mandible (or beak) emerged from RSO Site at Riversleigh, the oldest records of cockatoos were of Pleistocene age (less than 2 million years old). Known as the 'Land of Parrots', Australia is home to more than 50 parrot species. A number of parrot subgroups are characteristic of the Australian region; one of these is the cockatoos (family Cacatuidae). The Riversleigh fossil cockatoo which is about the size of the modern Galah is more than 20 million years old. Although its precise relationships are in doubt, it is indistinguishable from the smaller species of *Cacatua*, the genus that contains the Galah. In Australia today, only one cockatoo species lives primarily in rainforest, the others mainly being birds of the open forest or woodlands, but outside Australia there are several smaller cockatoos that are normal inhabitants of rainforest.

Passerines. The Riversleigh bird fossils provide an important start to developing understanding about the evolutionary history and palaeobiogeography of a number of Australian groups. One example is the passerines or song-birds (Passeriformes), the group that comprises half of all living species. Contrary to long-standing belief, it has recently been suggested that these birds had a southern origin. They are not known from the extensive Tertiary deposits of the Northern Hemisphere until the late Oligocene. In contrast, they form a significant component of the bird remains from Riversleigh, with some from late Oligocene sites being equal in age to the oldest passerines thus far reported in the world. Collectively, these deposits promise to be of great importance in the understanding of passerine origins.

A very old Australo-Papuan passerine group is the genus of logrunners, *Orthonyx*. Because of

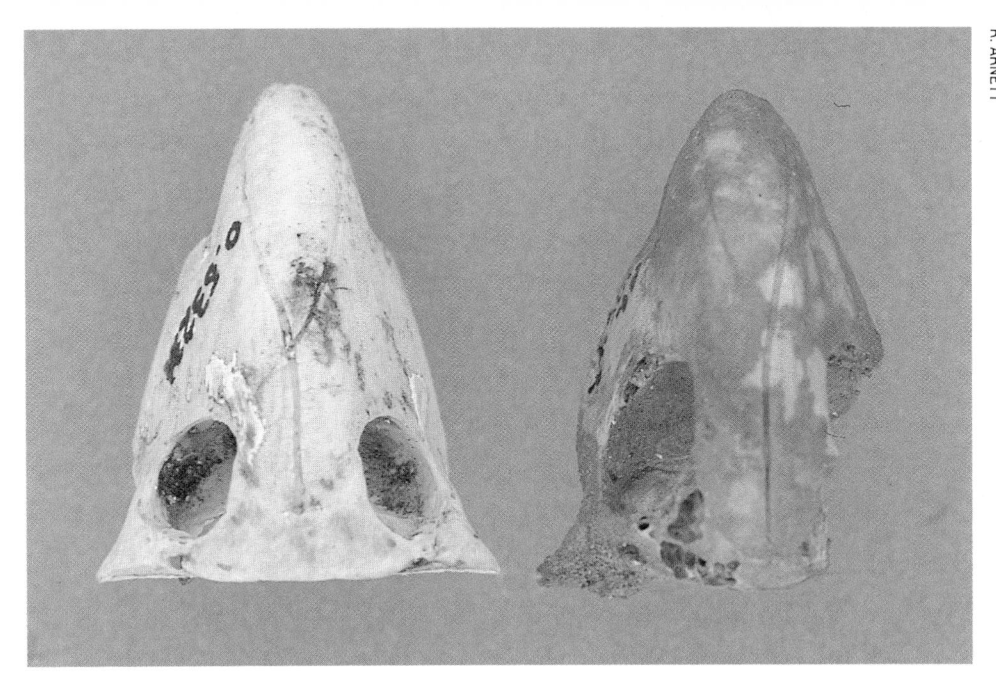

R. ARNETT

Australia's first Tertiary cockatoo, the brown beak (right) from RSO Site, and a modern cockatoo beak.

Songbirds, such as this Eastern Yellow Robin, have such delicate bones that they are rarely fossilised. Riversleigh's carbonate sediments, however, have preserved many different kinds.

F. KRISTO

a very peculiar method of feeding, these birds have very distinctive skeletons. Logrunners feed on the ground by scratching for food with their feet but do so by leaning back on the long spines of their tails, scratching their foot backwards and then kicking out to the side. The large powerful feet and legs require well-developed muscles from pelvis to thigh bone. The shape of its leg bones reveal the fossil's pedigree. The Riversleigh species is smaller than either of the two living species, which are found in rainforest in parts of New South Wales, Queensland and New Guinea, and also differs from a Pleistocene species from South Australia.

A comparably significant discovery has been that of the first fossil lyrebird (*Menura*), spectacular passerines that today live in the rainforests of eastern Australia.

Another group of passerines represented at Riversleigh are swifts and possibly swiftlets (equal or similar to the living genus *Collocalia*). Swiftlets in the genus *Collocalia* are a group of echo-locating birds that build nests on the walls of caves in many parts of south-eastern Asia and north-eastern Australia.

Predatory birds. Another Riversleigh bird with a very distinctive mode of feeding was a large predatory bird that evidently had the capacity to rotate its hind leg to reach deep into tree hollows to gaff and retrieve its prey. In this case it seems that, although the fossil bird shared its morphology (distinctive articular surfaces on the same leg bone) and presumably its behaviour with three other predatory birds—*Polyboroides radiatus* and *P. typus* of southern

Today, logrunners are restricted to rainforest habitats in New South Wales, Queensland and New Guinea.

Africa and Madagascar and *Geranospiza caerulesens* of South America—it was probably not closely related to and twice the size of any of these living groups.

Other birds. Besides these representations, Riversleigh has also produced rails, storks (which include some forms that seem to have their closest relatives outside Australia) and many other bird bones yet to be identified.

Feathered links to other lands. The changing relationships of the Australo-Papuan bird fauna, in contrast to those of the rest of the world, can be documented through the fossil record. Among the most interesting examples are the megapodes and frogmouths (Podargidae), two families that are considered highly characteristic of the Australo-Papuan avifauna. Yet there are no pre-Pleistocene records for either from Australia or any of its surrounding islands. In contrast, both have been recorded from Eocene-Oligocene deposits in France. This suggests that the present distributions of these groups may be vestiges, with the modern restricted occurrences being little more than historical artefacts. Should either megapode or frogmouth remains be found at Riversleigh, they would help to refine understanding about this palaeobiogeographic enigma.

E. & N. TAYLOR

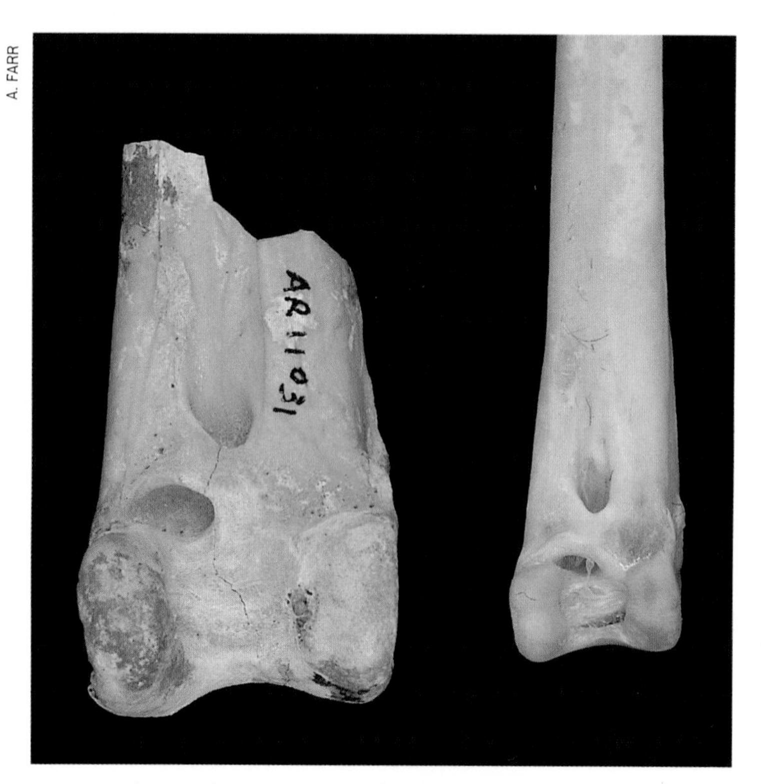

A. FARR

Large raptors are uncommon in ecosystems and not often found as fossils. One discovered at Riversleigh (left) appears to be very similar to non-Australian harrier hawks (right) that can rotate their hind legs to gaff prey in tree hollows.

Logrunners have very distinctive bones. Those found at Riversleigh (white, below) are similar to those of living species (brown, above).

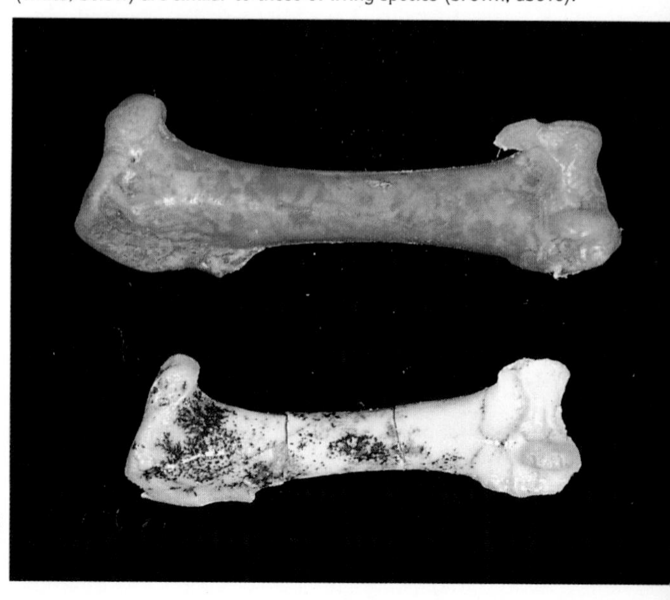

K. LOWE

MONOTREMES
egg-layers with half a history

The monotremes of Australia and New Guinea are the only mammals known to lay eggs. This fact, in combination with their relatively 'primitive' anatomy, has led them to be regarded as 'living fossils', a sort of halfway house between reptiles and other living mammals (marsupials and placentals).

There are two living groups: the platypuses (ornithorhynchids) and the echidnas (tachyglossids). The only Cretaceous mammal known from Australia is also a monotreme, *Steropodon galmani*, a platypus-like animal that might be more appropriately placed in a family of its own.

Prehistoric platypuses. Of all Australia's mammals, the egg-laying, 'duck-billed' Platypus (*Ornithorhynchus anatinus*) has to be the most remarkable. This one creature combines the mystery of a survivor from the world of the dinosaurs—a representative of the most primitive mammals known—with a kaleidoscope of anatomical bits so incongruous that the first zoologists to see one presumed it was a hoax.

Until 1971, an understanding of the history of Australia's Platypus was non-existent. Apart from the fact that it was a monotreme and only known from Australia, all else was mystery. Then three major discoveries occurred: two molar teeth of *Obdurodon insignis*, a 25 million-year-old platypus from the Tirari Desert; a jaw fragment with three teeth of *Steropodon galmani*, a 110-million year-old monotreme from opal deposits at Lightning Ridge, New South Wales; and, from 15-20-million year-old Riversleigh deposits, a complete skull as well as a nearly complete dentition of a second species of *Obdurodon*.

In 1971, while one puzzled group of palaeontologists stared in confusion at a strange tooth found in the late Oligocene deposits of Lake Palankarinna, another group goggled at a similar tooth found at the same time in similar-aged deposits in the Lake Frome Embayment. These were the first teeth of a fossil platypus to be found and were named *Obdurodon insignis*. The living Platypus, in contrast, has only rudimentary vestiges of teeth which it replaces with horny pads when it is still a juvenile. Naturally, these central Australian discoveries caused a great deal of delight as well as anticipation but despite many tonnes of processed matrix later, central Australia has failed to produce more than a few isolated teeth, a fragment of a lower jaw and a portion of a pelvis.

RECONSTRUCTION BY P. SCHOUTEN

Steropodon galmani, from 110 million-year-old opal deposits at Lightning Ridge, New South Wales, is the world's earliest known monotreme.

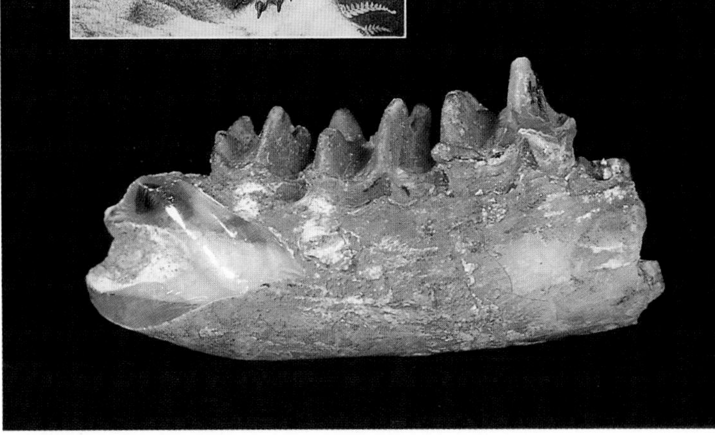
J. FIELDS

The opalised jaw fragment of *Steropodon galmani* contains three well formed teeth.

Discovery of the opalised jaw of *Steropodon galmani* in the dinosaur-rich early Cretaceous deposits of Lightning Ridge was one of the most exciting moments in the history of Australian palaeontology. It was the first and so far only known Mesozoic mammal from Australia. It may also have been the largest Cretaceous mammal anywhere in the world, overlapping in size some of the smallest dinosaurs, although less than twice the size of the living Platypus. It had well developed teeth whose shape indicates that even by this early date monotremes had achieved many of the distinctive dental features that distinguish them from other groups of mammals.

RECONSTRUCTION BY J. MUIRHEAD

When we first discovered Ringtail Site at Riversleigh, one of the System C localities in Ray's Amphitheatre, apart from the ringtail possum jaw and some bats, we saw little in the way of mammals although the site was riddled with the bones of turtles and lungfish—clearly an aquatic palaeoenvironment. But in the lab we were delighted to find isolated molars of the first fossil platypus from Riversleigh—a new species of *Obdurodon*. Consequently, the site was high on our 'hit' list for the 1985 expedition but nothing prepared us for the discovery that occurred that year: a whole, perfect skull, the first and only ornithorhynchid skull known apart from that of the modern Platypus. It was spotted as a braincast, complete with 'fossilised' blood vessels, when pieces of the skull were broken away in the course of fracturing blocks of limestone. After recovery and restoration of the broken pieces and dissolution of the surrounding limestone in dilute acetic acid, the skull that emerged took our breath away.

Left. The Riversleigh platypus, *Obdurodon dicksoni*, was slightly larger than its nearest relative in central Australia. It was found in Ringtail Site, an aquatic rainforest deposit that also contained abundant lungfish and turtles.

The Riversleigh *Obdurodon* material has provided a great deal of new information about the structure, relationships and palaeobiogeography of platypuses in general. At this point, besides the complete skull (which is the logo of the Riversleigh Society Inc.), there is a complete dentition (except for the last lower

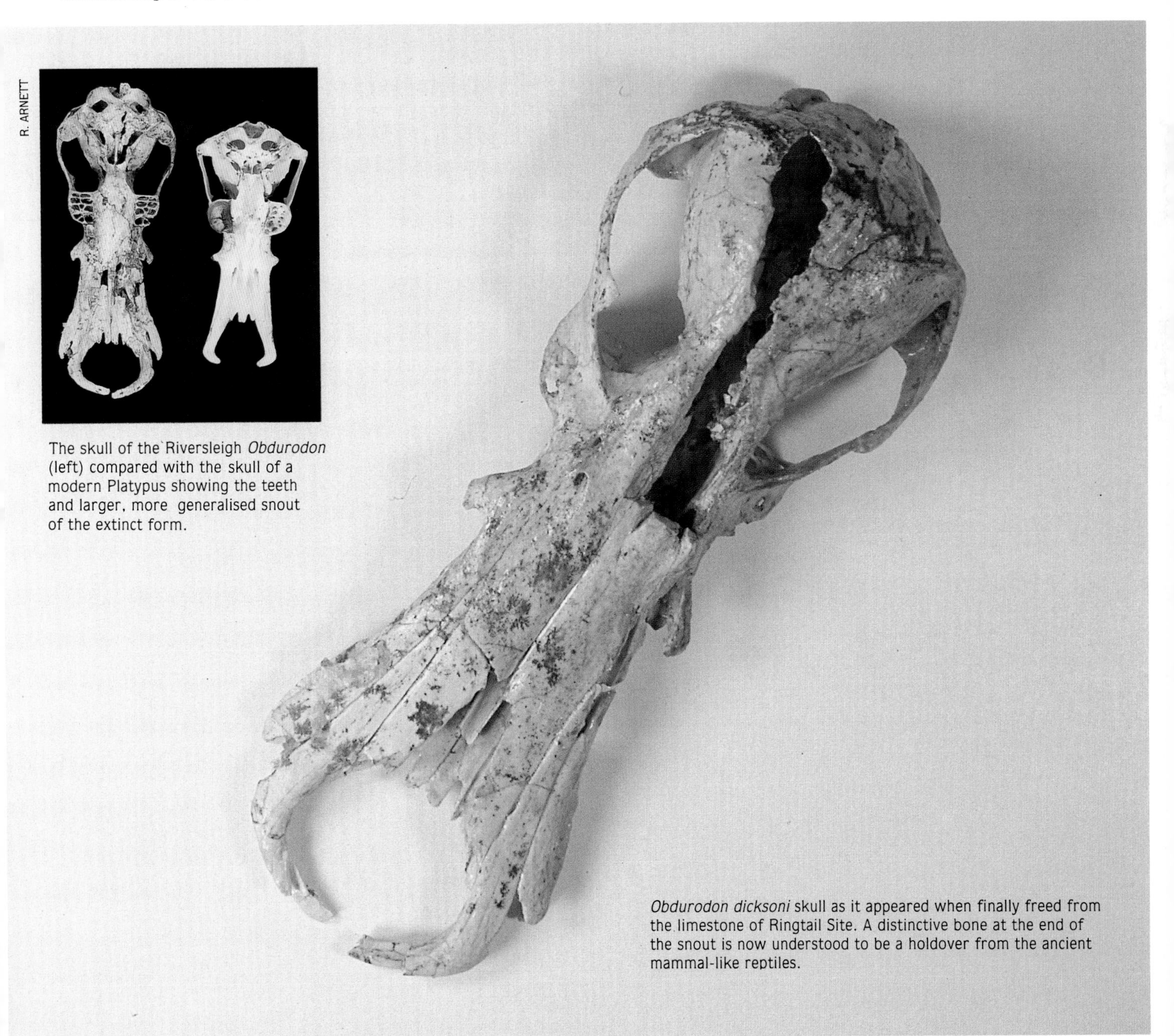

R. ARNETT

The skull of the Riversleigh *Obdurodon* (left) compared with the skull of a modern Platypus showing the teeth and larger, more generalised snout of the extinct form.

Obdurodon dicksoni skull as it appeared when finally freed from the limestone of Ringtail Site. A distinctive bone at the end of the snout is now understood to be a holdover from the ancient mammal-like reptiles.

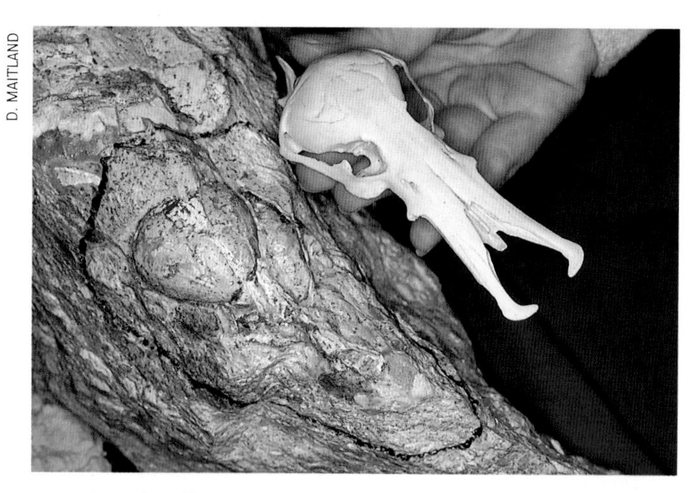

D. MAITLAND

A modern Platypus skull is here held alongside the Riversleigh *Obdurodon* which, when first found, was visible only as a brain cast and a cross-section through a tooth.

molar) and a partial lower jaw. It should not be long before we recover limb and girdle elements for this most primitive-known group of Cainozoic monotremes.

So far, specimens of the Riversleigh *Obdurodon* have come from the so called 'aquatic' sites: Ringtail Site, Bob's Boulders Site, Quentin's Quarry Site and most recently Neville's Garden Site. Here these platypuses were eyeball to eyeball with abundant turtles, lungfish, crocodiles and fish, as well as the occasional drowned marsupial and bat. Neville's Garden is a bit unusual among these sites because of its abundance of terrestrial mammals. This might be explained if there were extensive shallow margins around a pool with a deeper centre where lungfish and other aquatic vertebrates thrived.

Discovery in 1991 and 1992 of *Obdurodon*-like monotreme teeth in 61-63 million year old sediments in Patagonia, southern Argentina, has shocked everyone who thought platypuses were uniquely Australian. Clearly they were once part of a Gondwanan fauna that must also have been present on Antarctica, but survives today only in Australia.

Because the living Platypus, the only surviving descendant of *Obdurodon*, has become highly specialised with loss of its functional teeth, overall reduction in body size and

The modern Platypus lives only on the waterways of eastern Australia where it uses special electric sensors in its bill to detect invertebrate prey.

F. KRISTO

T. FLANNERY

One of the mysteries of the Riversleigh record is the complete absence of echidnas, such as this Long-beaked Echidna from Papua New Guinea.

A kaleidoscope of seemingly incongruous anatomical bits, the modern Platypus was originally thought to be a hoax.

'simplification' of most of its cranial anatomy, we are concerned that the lineage is in decline. If palaeontology teaches us anything about the fate of evolving mammal lineages, it is that when a group loses its generalised body form and edges too far out on its evolutionary limb in terms of specialisation, that limb is in increasing danger of falling off. While modern ecologists may regard the Platypus to be in no danger, the fossil record of rapid decline over just the last 15% of its known history suggests a less optimistic view. We would caution that the Platypus's resilience and future are not things to be taken for granted solely on the basis of its seemingly stable modern populations. This is an animal just surviving in the twilight of a long and remarkable history.

To add to these concerns, today's Platypuses persist only in the permanent river systems of eastern Australia, sometimes in remnant rainforests but more often in the rivers that drain the open forests of the Great Divide. With humans abusing the chemistry of the waterways with fertilisers and other noxious substances and deforestation and pollution of the rivers' watersheds (which increase the violent floods that exterminate the Platypus's burrow systems), its candle may be in imminent danger of going out.

The apparent absence of echidnas from all of the known Riversleigh deposits is an enormous puzzle to us. Although it is possible one will turn up in the future, we have looked through a lot of material. What makes it even more puzzling is the fact that the living Short-beaked Echidna (*Tachyglossus aculeatus*) occurs in all habitats in Australia from deserts to lowland rainforests to alpine snowlands.

DASYURIDS

flesh-eaters on the way up

The dominant mammalian carnivores in Australia today are dasyurids. Living kinds range in size from the 4-gram Long-tailed Planigale to the 8-kilogram Tasmanian Devil. There are 87 known species, of which 63 are living. These range throughout Australia's and New Guinea's diverse habitats with most of the more 'primitive' kinds (e.g. species of *Murexia*) living in the rainforests and the more derived kinds (e.g. species of *Dasyuroides*) in the deserts.

In our most recent summary of the fossil record of this group, we pointed out that most of the extinct dasyurids appear to represent weird groups without close relationships to any of the living genera. For example, the most distinctive was *Ankotarinja tirarensis* from the Oligo-Miocene Ditjimanka Local Fauna of Lake Palankarinna in South Australia which we placed in its own subfamily. Not until the early Pliocene, barely 4.5 million years ago, did modern genera (*Planigale* and possibly *Antechinus*) make their first appearance. As a result, without evidence to the contrary, one might have been tempted to suspect that living dasyurids shared a relatively recent common ancestor, perhaps no longer than 10-15 million years ago.

The Riversleigh dasyurids presently being studied by Steve Wroe of the University of New South Wales present a complex mix of new information. Like those from the Oligo-Miocene deposits of central Australia, most belong to lineages unrepresented by living taxa, although a couple appear to be similar to unnamed taxa in the Kutjamarpu Local Fauna from the Tirari Desert. However, one Oligo-Miocene Riversleigh dasyurid from Panorama Site tells a quite different story. When it turned up as a lower jaw in an acid vat in Sydney, it seemed to Mike to be a decidedly *Antechinus*-like taxon, a genus common today throughout

Small dasyurids, such as this Brown Antechinus, are uncommon in the Riversleigh deposits but one found at Panorama Site appears to be near the common stem for two modern groups.

G. HOYE

Australia but without a fossil record prior to the Plio-Pleistocene. Because Steven Van Dyck of the Queensland Museum was researching the biodiversity of living *Antechinus* and *Antechinus*-like dasyurids, he was invited to study this new Riversleigh beast. It wasn't long before he reported the exciting news that the Australian fossil record had, at long last, produced an archaic dasyurid that *did* have something to say about the origins of the modern dasyurid groups!

This animal, which has yet to be named, appears to lie near, if not on, the common stem of all modern *Antechinus* (known only from Australia) and all modern *Murexia* (known only from New Guinea). It had already been concluded that living species of these two genera were closely related from Steven's studies of teeth and skulls and Peter Baverstock's studies of the relationships of blood proteins. The structurally intermediate Riversleigh dasyurid not only supported their earlier hypotheses but it may provide an indication of the maximum age for the two modern lineages—about 20 million years old.

Many of Riversleigh's dasyurids have yet to be studied in detail but general aspects of this record are already clear. First, contrary to our naive expectation, all fall within the size range of living species. Based on the record of mammalian carnivores on other continents (e.g. canids, felids etc.), we had expected to find at least a few horrendously large dasyurids! But it would appear that their size and morphological diversity rose from a low in the Oligo-Miocene to a high today. This was all the more surprising because we had come to expect the diversity of most of Riversleigh's Oligo-Miocene mammals to be higher than that of the same groups today. Searching for a reason, dasyurid diversity may have been slow to blossom because in Riversleigh's Oligo-Miocene rainforest communities there were so many *other* kinds of carnivores, including marsupial lions, thylacines, snakes, crocodiles and even carnivorous kangaroos.

Also, contrary to initial expectations, none of New Guinea's extant genera are represented in any of Riversleigh's assemblages. Because we have tended to view New Guinea's highland rainforests as a 'refuge area' for Australia's Tertiary groups, we half expected to find species

Tiny planigales, such as this Long-tailed Planigale, first appear in the Riversleigh record in the Pliocene Rackham's Roost deposit.

of *Murexia, Myoictis, Neophascogale* and *Phascolosorex*, genera that today are confined to the forests of New Guinea. Our failure to do so may be saying something important about when the forests of New Guinea became isolated from those of northern Australia.

Further, *none* of the Riversleigh dasyurids was as common in those older habitats as many small dasyurids are today. Tentatively, Jeanette Muirhead attributes this limited representation to the otherwise extraordinary abundance and diversity of similar-sized peroryctid bandicoots in the same faunal assemblages. It is possible that species of the two groups competed for food on the rainforest floor, the numbers of one limiting the numbers of the other. But why did Riversleigh's older bandicoots have the edge over dasyurids in terms of biodiversity when the opposite is true now? Curiously, although modern bandicoot biodiversity is lower today than that of dasyurids, bandicoot *biomass* in many of Australia's open forests and the closed forests of New Guinea significantly exceeds that of dasyurids in the same habitats. The reasons for these changing patterns may become clear when Jeanette's research is complete.

Finally, it is clear that by Pliocene time, a major turnover had occurred in the kinds of dasyurids in the Riversleigh area. The weird, archaic groups were gone, or at least unrepresented in the Rackham's Roost deposit, and in their place were species of the modern genera *Planigale* and *Sminthopsis*. Perhaps descendants of the archaic groups hung on in rainforests that persisted for a longer time in areas of eastern Australia?

W.A. MUSEUM

A mummy, nearly 5000 years old, of the now extinct Thylacine *Thylacinus cynocephalus* from Thylacine Hole cave on the Western Australian Nullarbor. When found, the tongue and left eyeball were still intact.

THYLACINIDS

flesh-eaters on the way down and out

Dasyurids, numbats (of which, as far as we are aware, none occur at Riversleigh) and thylacinids comprise the superfamily Dasyuroidea and the order Dasyurimorphia. All are at least dominantly carnivorous or insectivorous. Of dasyurimorphians, the recently extinct Thylacine ('Tasmanian Tiger' or 'Tasmanian Wolf', *Thylacinus cynocephalus*) was the largest, although the late Miocene *T. potens* may have been more heavily built.

The dog-sized and superficially dog-like Thylacine was the last of its family. It survived on the Australian mainland until perhaps 80 years ago in the Kimberley region and at least 3000 years ago in the southern half of the mainland. The species also occurred in New Guinea until at least 9920 years ago. In Tasmania, the last certainly known individual died in captivity in 1936. The reason for its extinction on the mainland and New Guinea may have been the arrival of domestic dogs. In Tasmania, the government and sheep farmers systematically and wilfully pushed the species over the precipice of existence.

Before Riversleigh's fossil record began to unfold, the Tertiary fossil record of thylacinids was poor, although perhaps fairly indicative of diversity in this group during the later part of the Tertiary. There was only one distinctive species, the late Miocene *Thylacinus potens* from

D. KIRSHNER NPWS

Riversleigh's record of Thylacines is strikingly diverse, ranging from tiny, specialised forms to large, more primitive kinds. One of the larger kinds is compared here (above) with the lower jaw of a Tasmanian Thylacine (below).

R. ARNETT

the Alcoota Local Fauna of the Northern Territory. This was not particularly surprising because mammalian carnivores are normally at least ten times as rare as the species upon which they feed. Only in situations where many hundreds of individuals drawn at random from the community were preserved would we expect to obtain a reasonable understanding of carnivore diversity—come in Riversleigh.

At last count, Jeanette Muirhead had recognised five different thylacines from Riversleigh's Oligo-Miocene faunas, thereby nearly trebling previous known diversity for the continent. They range in size from one as small

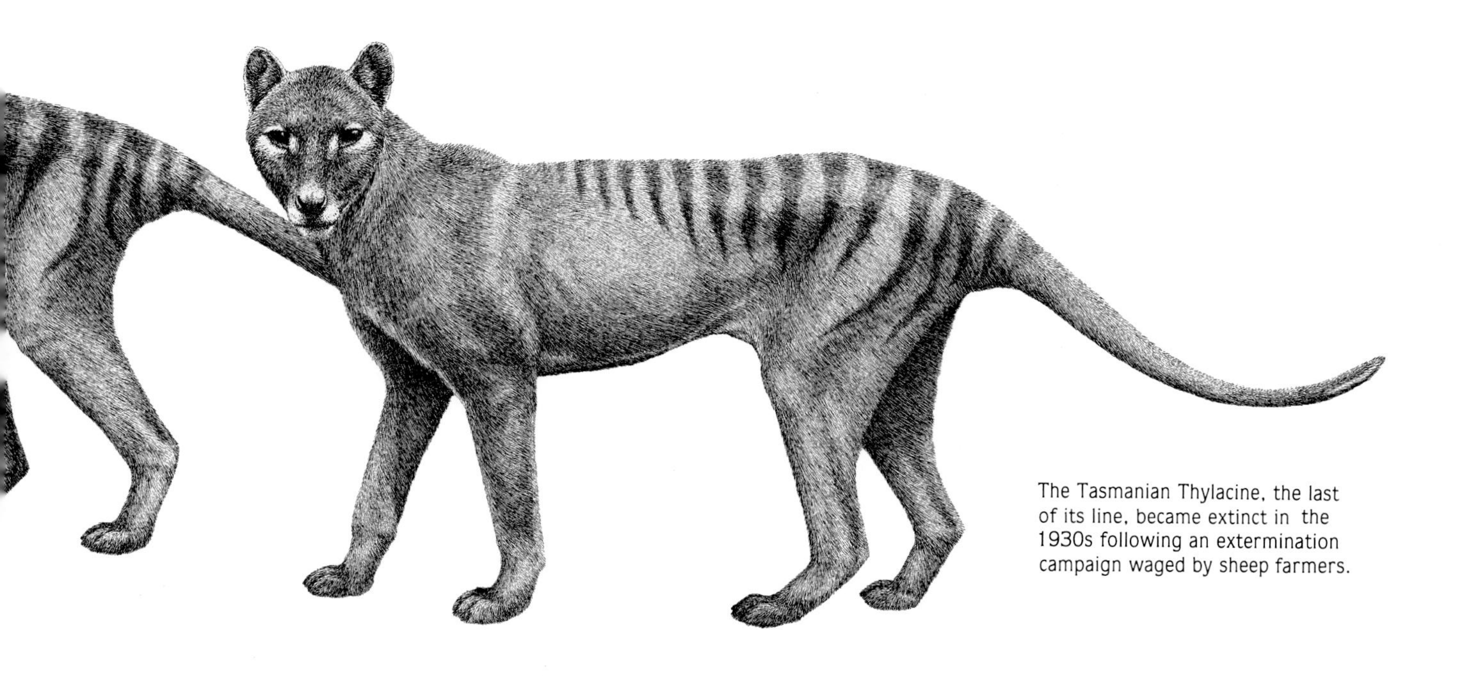

The Tasmanian Thylacine, the last of its line, became extinct in the 1930s following an extermination campaign waged by sheep farmers.

as a large domestic cat to one perhaps three-quarters the size of a German shepherd dog. The smallest is only slightly smaller than the largest of Riversleigh's dasyurids but the largest is larger than any dasyurid known, living or extinct.

Of these new beasts, *Nimbacinus dicksoni* is very primitive, in that it lacks distinctive features characteristic of non-Riversleigh thylacines (e.g. it retains a distinct metaconid on its lower molar teeth) . A jaw fragment indistinguishable from *N. dicksoni* was also found at Bullock Creek in the Northern Territory. Some of Riversleigh's thylacines are so primitive that they resemble primitive dasyurids almost as much as they do other thylacinids, a fact that corroborates current wisdom that thylacinids are most closely related to dasyurids. Others, including the tiniest, are almost as specialised as *Thylacinus cynocephalus*, indicating that the family is a lot older than the age of the Riversleigh deposits.

Interestingly, if we had discovered and understood the fossil deposits of Riversleigh in 1883 instead of 1983, we would have foreseen the extinction of the Tasmanian Thylacine. All the signs were clear. The first time we encounter thylacines, in local faunas of Riversleigh's Systems A and B at approximately 20 million years of age, they are diverse (at least two genera and five species of different size and shape) and widespread (at least mainland Australia). By the time we get to the System C local faunas at approximately 15 million years of age, diversity had dropped to two species both of which had carried through from System B time. By Alcoota time, approximately 8 million years ago, taxonomic diversity had dropped steeply to just one species. This situation persisted through the Pliocene and Pleistocene but by the late Holocene the single remaining species had also become extinct on the mainland and New Guinea. The steep declines in diversity and geographic range were clear, the warnings engraved in stone. Had we been able to comprehend those hieroglyphic alarm bells before 1900, the extinction of the last fragile species, in its last besieged stronghold, just might have been averted. But it was not, so this magnificent creature, the last of its ancient line, was vaporised into memory.

'THINGODONTA'

off the scale of the unexpected

No tally of Riversleigh's more interesting creatures would be complete without noting the discovery of 'Thingodonta' (*Yalkaparidon*). At this point, although two species are recognised—*Y. coheni* from System B and *Y. jonesi* from System C local faunas— with one of them (*Y. coheni*) being known from an almost complete skull, we still understand precious little about these beasts.

We can be certain that they represent a very ancient group of Australian mammals, one that had probably differentiated before Australia split from Antarctica about 45 million years ago. But, if so, why hasn't some trace of the group turned up before in the Oligo-Miocene assemblages from central Australia? Perhaps Australia was differentiated into zoogeographic regions by ecological barriers that *partially* segregated the central from the northern biotas. This would help to explain the apparent endemism of some of Riversleigh's genera like *Yalkaparidon, Yingabalanara, Ekaltadeta* and *Wabularoo*, while others such as *Wakaleo, Wakiewakie, Namilamadeta* and *Ngapakaldia* were shared between the two regions. The northern tropical rainforests of the Oligo-Miocene may have acted as a refuge for archaic groups like Thingodonta which, for some reason, were unable to persist this long in the central areas of the continent.

The ecological role of Thingodonta is similarly puzzling. Because no living mammal has a total dentition similar to that of species of Thingodonta, there is no easy way to be confident about what they ate. We have speculated that the combination of powerful, sharp, protruding lower incisors and the crescent-shaped molars which lack any 'incusive' surface for horizontally shearing food, suggests that they ate something that required initial puncturing because of a tough skin followed by very little 'chewing' because their molars were only capable of vertical shearing. Suitable foods that first came to mind included eggs and caterpillars.

A better suggestion might be worms. This *could* explain the absence of at least long-beaked

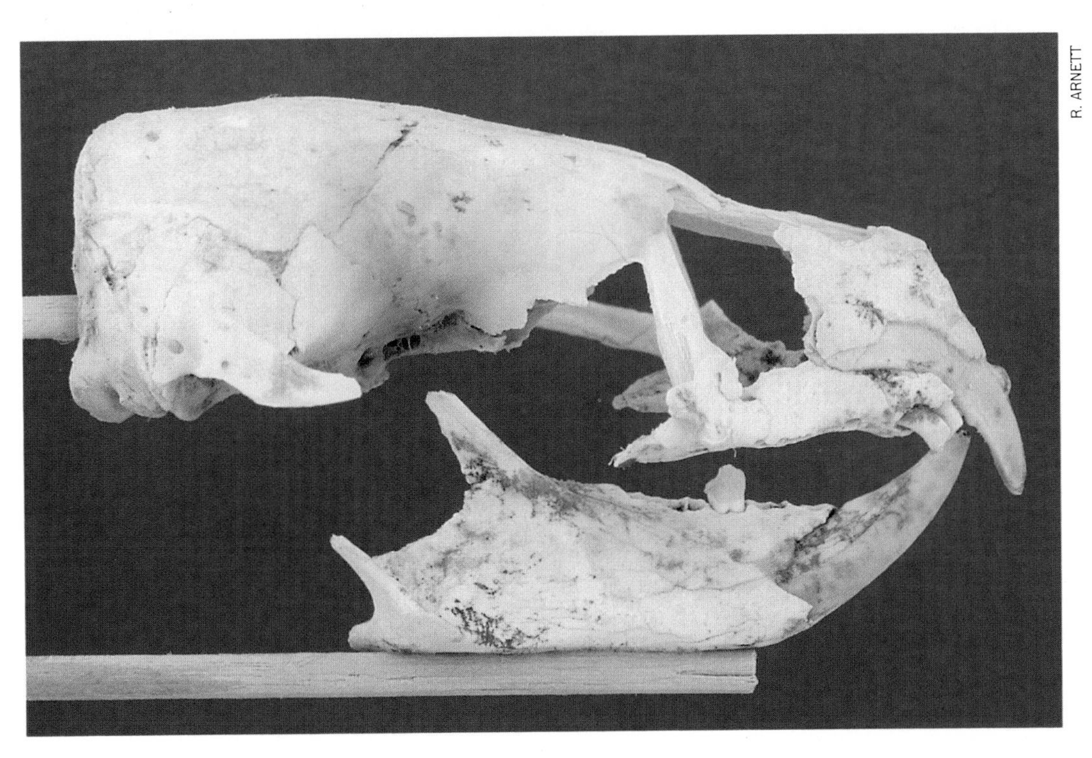

R. ARNETT

The skull and lower jaw of one of the weirdest Riversleigh mammals, the aptly named 'Thingodonta'.

echidnas from any of Riversleigh's Oligo-Miocene deposits because in New Guinea these are specialised worm-feeders. Perhaps long-beaked echidnas became a feature of Australian ecosystems only after Thingodontans became extinct. When (not if!) we discover what the front limbs of Thingodontans looked like, we

RECONSTRUCTION BY P. SCHOUTEN

Because we are uncertain about how the teeth of 'Thingodonta' were used, we can only guess about the foods it ate.

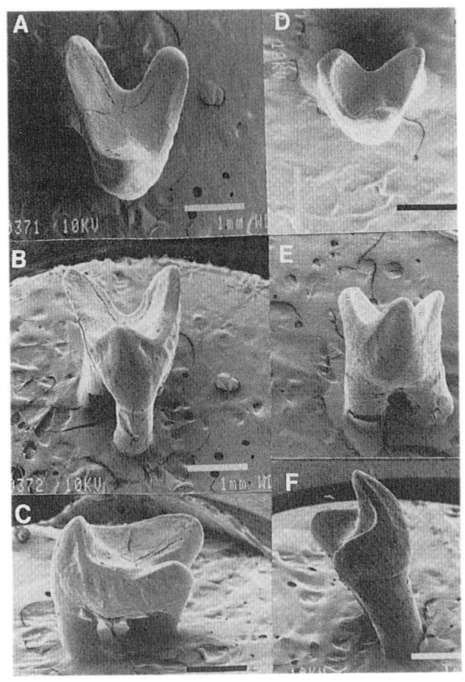

The teeth of 'Thingodonta' were the first parts of the creature known and the reason for its name.

might discover that they were specialised for digging which would at least not contradict the notion that worms were on their menu.

The closest match of Thingodontan molar teeth to those of *any* other mammals are with those of the Madagascan Streaked Tenrec, a placental mammal that lives in rainforests—and eats worms. Although molar shape here must be convergent (because Thingodontans are marsupials), this may be an important clue. But if they did focus on worms, what did they use their remarkably specialised incisors for? However we add it up, there are loose ends and the mystery persists.

The extent to which Thingodontans differ from all other known mammals is reflected by the fact that when they were first described by us, they were placed in a unique high-level taxon of their own: the order Yalkaparidontia. There are only four other orders of marsupials known from Australia: Dasyurimorphia (dasyurids, numbats and thylacines); Peramelemorphia (bandicoots); Notoryctemorphia (marsupial moles); and Diprotodontia (previously known marsupials with two hypertrophied, procumbent lower incisors). Riversleigh's yalkaparidontians are the only new order of marsupials discovered this century.

NOTORYCTIDS

old moles in green holes?

Until 1985, the only marsupial mole known from Australia was the living *Notoryctes typhlops*, from the deserts of central Australia. It is so specialised that it was placed in its own family, the Notoryctidae and, recently, in its own order, the Notoryctemorphia. No notoryctid fossils of any kind were known.

Soon after discovery of Riversleigh's System B local faunas in 1985, strange upper molars with very reduced paracones began to appear in the acid vats. Soon other bits of the beast turned up, including lower jaws, maxillae and even limbs. All quickly led to the realisation, as unlikely as it seemed, that a 'primitive' marsupial mole used to plough through the Oligo-Miocene rainforests of north-western Queensland.

Its teeth indicated that it was not as specialised as the living species and, more interestingly, demonstrated the way in which the teeth of the modern Marsupial Mole had evolved—through suppression of the paracone. Without the Riversleigh fossil, no one knew how the specialised molar shape of the living animal had developed.

But what was this 'primitive' mole doing in an ancient rainforest? At first we had a lot of difficulty with this question. The front limbs indicated that it burrowed, like its living, more

M. GILLAM

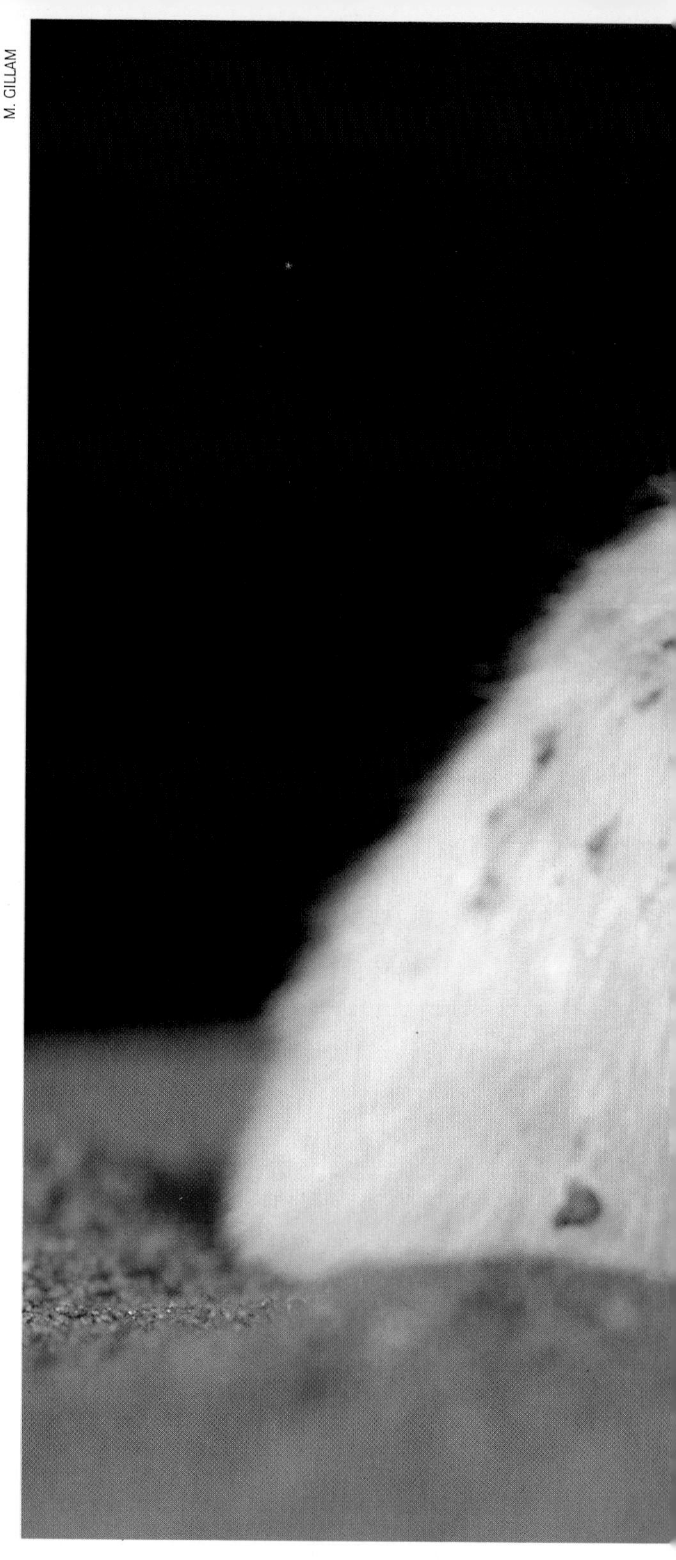

The living Marsupial Mole is confined to the deserts of central Australia. Riversleigh has produced a primitive mole that may have become pre-adapted for desert life by burrowing through the leaf litter of ancient rainforests.

specialised relative. But whereas the modern mole has no trouble in borrowing through poorly consolidated desert sand, the floors of Australian rainforests seem an unlikely situation for a burrowing mammal. These normally have a very thin layer of humus overlying a solid substrate such as hard clay (which is why land 'reclaimed' from beneath rainforests quickly becomes useless for crops).

Miranda Gott, who studied this Riversleigh mole for her Honours project at the University of New South Wales, suggested the possibility that perhaps it used its flattened forelimbs for swimming rather than burrowing. But we should not abandon the burrowing hypothesis too quickly. Firstly, *some* placental moles, such as a number of African chrysochlorids, are burrowing inhabitants of swamps and rainforests, although others are restricted to deserts like the Marsupial Mole. Secondly, in New Guinea at least some of the rainforest floors are *not* hard because they develop a thick layer of moss and plant roots in which many kinds of animals live. Perhaps this kind of floor was present in areas of the Oligo-Miocene rainforests at Riversleigh? Whatever, it would appear probable that however the mole managed to use its flattened forelimbs within those ancient rainforests, it became 'pre-adapted' for a burrowing mode of life in the dry habitats that eventually came to replace the forests of central Australia.

BANDICOOTS
a plethora of perameloids

Today bandicoots make up only a small percentage of the mammals in Australian habitats. Part of the reason for what we now recognise to be low diversity, relative to that in the Riversleigh local faunas, is certainly post-European extinctions in central and eastern Australia. Many ecologists suspect this loss mostly reflects European interference with the land management programs routinely used by Aborigines, such as 'fire-stick farming'. These practices of Aborigines may have been necessary to maintain sufficient habitat diversity.

However, even before Europeans arrived it was rare to have more than three or, at most, four types of bandicoots in any single habitat in Australia. Curiously, the habitats with the highest bandicoot diversity just before Europeans arrived were the semi-arid grasslands of the continent. For example, on the western Nullarbor Plain, the Pig-footed Bandicoot (*Chaeropus ecaudatus*: Peramelidae), the Greater Bilby (*Macrotis lagotis*: Thylacomyidae), the Short-nosed Bandicoot (*Isoodon obesulus*: Peramelidae) and the Western Barred Bandicoot (*Perameles bougainville*: Peramelidae) might be found together. The fact that only one species per genus, representing two families, co-existed

RAY WILLIAMS

Peramelid or 'ordinary' bandicoots, such as this modern Long-nosed Bandicoot *Perameles nasuta*, began to dominate Australian ecosystems only after the decline of the continental rainforests.

K. ATKINSON

in this way is itself interesting and suggests that in order for this number of species to co-exist, they had to be significantly different from one another.

In contrast, in New Guinea, the highest bandicoot diversity occurs today in the mid-montane rainforests, where up to five species in three genera (*Microperoryctes*, *Peroryctes* and *Echymipera*, all peroryctids) can occur together, all of which are basically much more similar to each other than the sympatric Australian species. In these rainforests, most of the basically similar species are differentiated by size.

T. FLANNERY

The long, tapered snouts of New Guinean peroryctids (shown here) are typical of these 'forest bandicoots'. The group is well represented in Riversleigh but only one species survives in Australia today.

Bandicoots are exceeded only by bats as Riversleigh's most common fossils. Here the skull and lower jaw of a new type emerges from a block of Upper Site limestone.

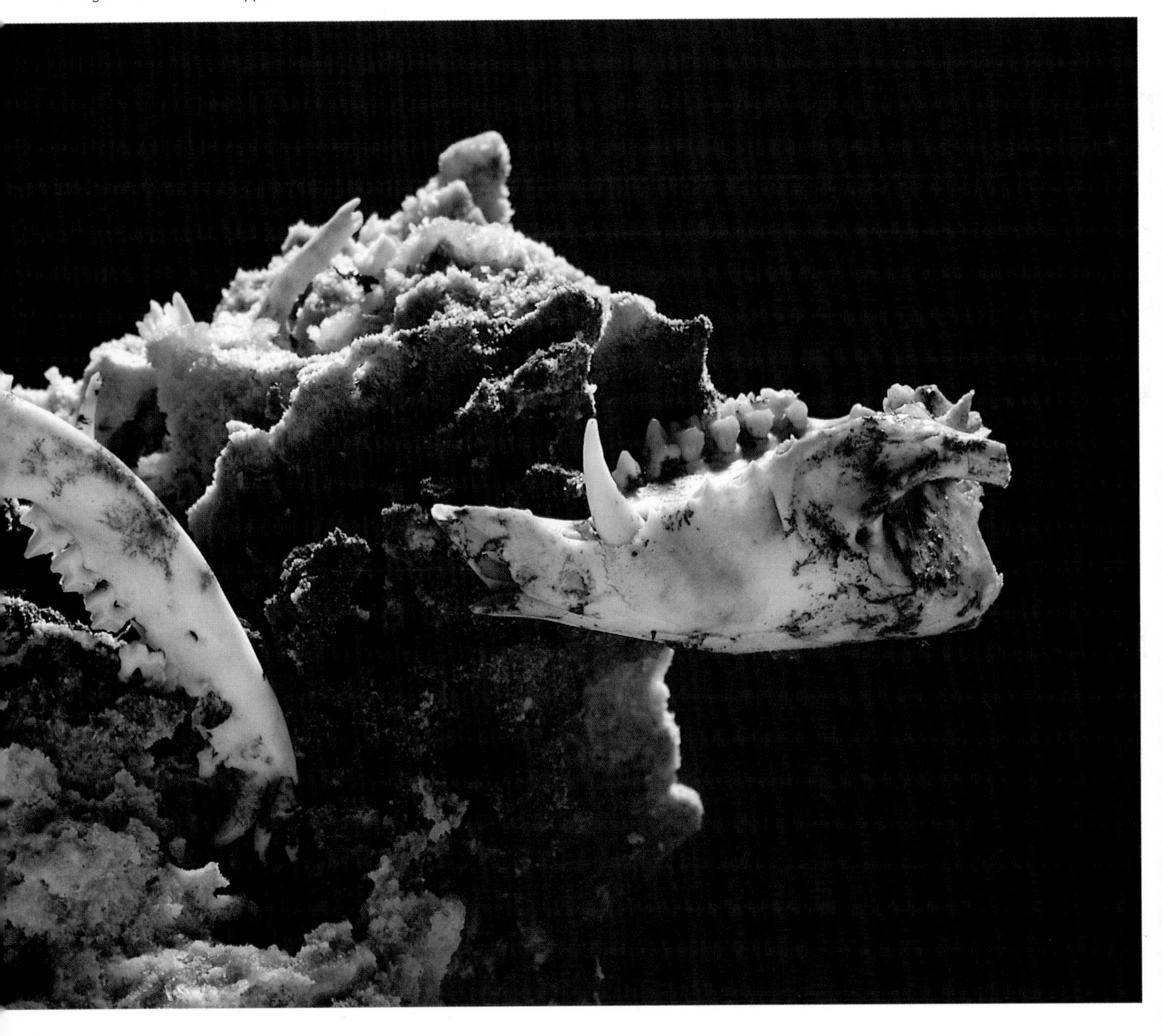

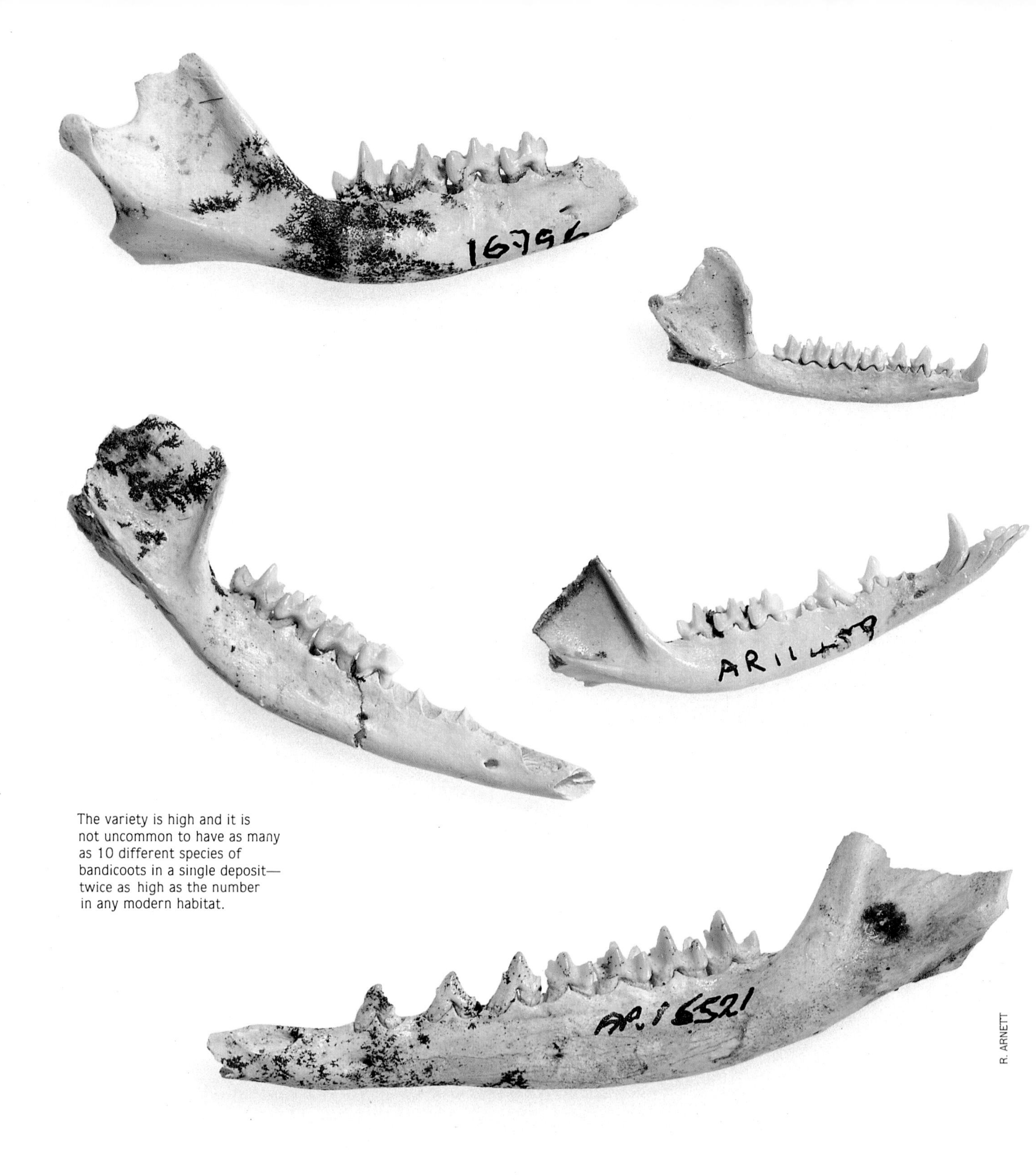

The variety is high and it is not uncommon to have as many as 10 different species of bandicoots in a single deposit—twice as high as the number in any modern habitat.

R. ARNETT

Which brings us to Riversleigh's prehistoric bandicoot hordes! Essentially, they reveal patterns more like those of New Guinea's montane rainforest communities than any assemblage from modern Australia. Camel Sputum Site has produced approximately 10 different species, twice the highest-known diversity for any Australian/New Guinean habitat. How could this be?

We suspect four reasons. Firstly, as noted elsewhere, Riversleigh's Oligo-Miocene rainforests appear to have been generally far more species-rich than any that persist in this area of the world today so in this sense the high number of sympatric bandicoot species should not come as too great a surprise.

Secondly, Jeanette Muirhead's studies of Riversleigh's bandicoots and dasyurids suggest that the relatively low diversity of Riversleigh dasyurids may have enabled or been the result of higher bandicoot diversity. In the main, modern bandicoots are omnivorous/carnivorous to herbivorous while similar-sized dasyurids are carnivorous to omnivorous which provides considerable room for competition and, probably, competitive exclusion. For whatever reason, bandicoots appear to have done better out of whatever competition occurred between the two groups for toe-holds in the ancient rainforests.

Thirdly, in those rainforests there were no murid rodents. Whereas bandicoots form a constant 'background' of specimens in Riversleigh's Oligo-Miocene deposits, in modern Australian ecosystems that role is filled by rodents. The omnivorous to herbivorous murids now make up approximately 25% of the mammal species known from both Australia and New Guinea. Although their initial entry into Australia may well have been via non-rainforest corridors in the latest Miocene, it is probable that sometime between Pliocene and late Pleistocene time, at least three murid lineages managed to invade Australian and New Guinean rainforests. It seems likely that these relatively recent invasions would have been at the expense of some of the smaller lineages of bandicoots.

Finally, some of Riversleigh's Oligo-Miocene bandicoots appear to have finely partitioned up what were undoubtedly highly diverse and probably constant food resources by means of what are called feeding guilds. For example, in the diverse Camel Sputum bandicoot assemblage, there were peroryctid-types, VD-types (a family yet to be named by Jeanette Muirhead) and peramelid-type bandicoots. Within each of these groups, while dental morphology is basically similar, species are spaced apart in size, each step in the progression being separated from the last by a significant

Curiously, one of the habitats in Australia that sustained the highest variety of bandicoots was the pre-European Nullarbor where as many as four species coexisted.

size difference. This elbow room for specialisation must have been sufficient to enable the many otherwise-similar species to co-exist. Of course competition may also have been avoided by vertical partitioning of resources, e.g. perhaps some species were arboreal, and/or time partitioning, with some active by night and others by day.

By Rackham's Roost time, diversity of Riversleigh's bandicoots appears to have fallen significantly. Most of the older types that abounded in the Oligo-Miocene seem to have vanished from this region, there being only a single species of bandicoot, a possible peramelid, so far recovered. In contrast, the Rackham's Roost deposit has produced more than 12 species of rodents.

MARSUPIAL LIONS
magnificent meat merchants of the Miocene

If a Kenyan youngster announced to his parents that he wanted to camp overnight in the field behind the house, some anxious looks would be exchanged. Visions of huge lethal claws, slobbering jaws and broken, glistening bones left to dry in the sun would flash through their minds—for there be lions in that place!

The Pleistocene Marsupial Lion was a leopard-sized, flesh-eating relative of the wombats.

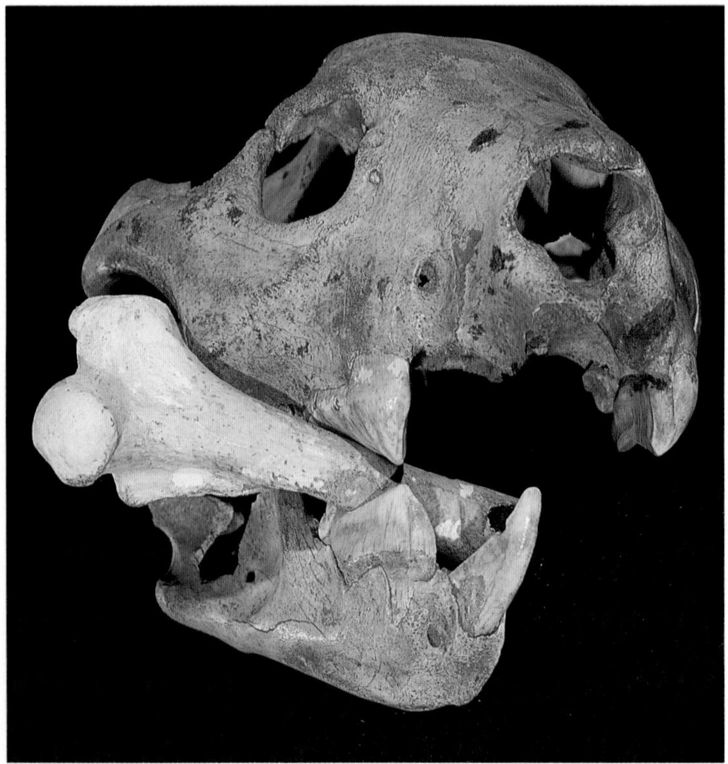

F. COFFA

No such anxiety meets a modern Australian child's request for a nocturnal adventure. But such complacency would not do for an Australian child of the late Pleistocene or an adventurous kangaroo of the Miocene—for then there were also lions, albeit beasts of a very different kind. These were marsupial lions, members of the highly distinctive thylacoleonid family.

Until recently, the only one that was well known was the Pleistocene Marsupial Lion, *Thylacoleo carnifex*, a leopard-sized marsupial very distantly related to, of all things, wombats! Despite a long-standing debate about what they ate, all recent studies of structure and function concur that it was meat, as well as possibly bones. In fact, a study of carnivorous mammals of the world concluded that *T. carnifex* was the most specialised mammalian carnivore that had ever evolved, mainly because of its incredibly huge, sectorial meat-cutting premolar.

The older fossil record of marsupial lions is spotty with the cat-sized *Priscileo pitikantensis* being known from late Oligocene time, the dog-sized species of *Wakaleo* being known from early Miocene (*W. oldfieldi*), middle Miocene (*W. vanderleuri*) and late Miocene (*W. alcootaensis*) time, and other species of *Thylacoleo* being known from late Miocene (small dog-sized *T. hilli*), to Pliocene (leopard-sized *T. crassidentatus*) time. One of the most magnificent recent finds was a complete skull of *W. vanderleuri* from the middle Miocene deposits of Bullock Creek in the Northern Territory.

Because marsupial lions were carnivorous, it comes as no surprise to find that their remains are not commonly encountered in most fossil deposits, particularly from the older part of the Tertiary record. The first sign we had that they had prowled the forest floors of Riversleigh was a maxilla literally plucked from the back of the mouth of a crocodile! It belonged to a species of *Wakaleo* most similar in size to *W. oldfieldi* from the Kutjamarpu LF of central Australia. Since then, many additional Riversleigh specimens referable to *Wakaleo* have turned up but we have not, as yet, determined either its phylogenetic relationships to other species or precisely how many species the material represents—a challenging project looking for a lover of lions.

From Low Lion Site at Riversleigh, the palate of a very primitive type of marsupial lion turned up, one that links two of the younger groups.

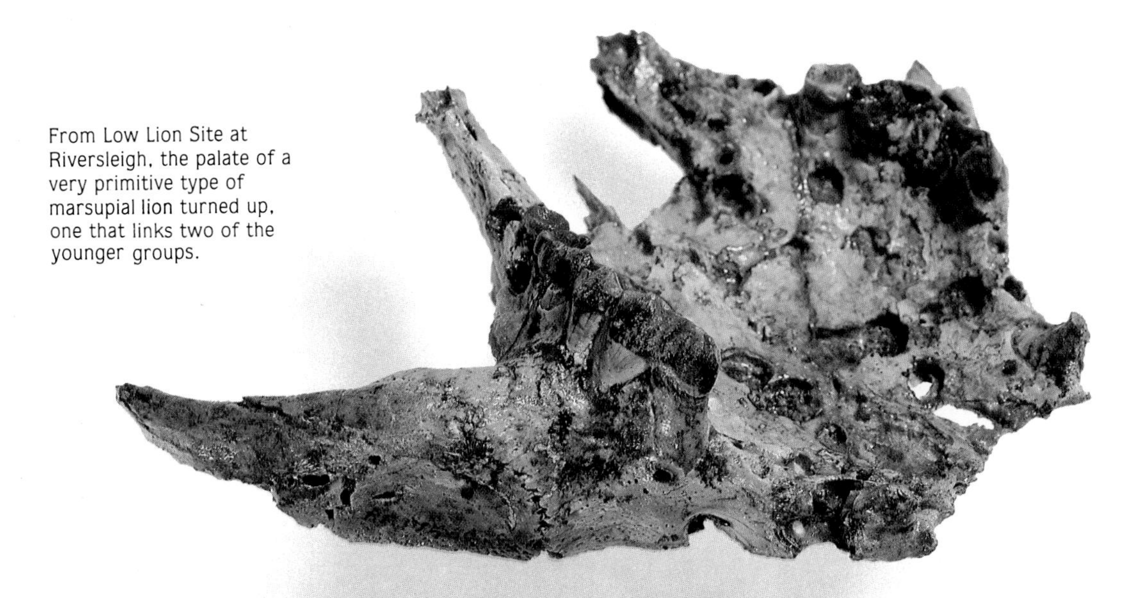

R. ARNETT

During the 1989 Expedition, Alan Rackham was carrying gear near the base of the steep slope below Bitesantennary Site when he put his foot on a loose slab of rock. As it started to move, his eye caught sight of a curious mass of bone projecting from its surface. On closer inspection, he instantly realised what it was—the upside down snout of a small marsupial lion, one from a stratigraphic level below any we had sampled before. The teeth and palate were still safely imprisoned in the petrified mud that had been its burial shroud for 20 million years but a mere thousand years of exposure on the hillside and the bounding feet of hill-climbing kangaroos had taken their toll on the upper part of the skull. When the slab was prepared later that year, the blue teeth revealed much more than your average lion! It appeared to be a 'missing link' between the two great lineages of marsupial lions, the *Wakaleo* and the *Thylacoleo* branches of the family tree. A little step for Alan Rackham, a giant step for Australian palaeontology!

But even more surprising was discovery at Upper Site of the perfect skull of an incredibly tiny and very primitive marsupial lion, one probably no larger than a house cat. While detailed study has yet to be carried out, it appears to be most closely related to the oldest marsupial lion known, the late Oligocene *Priscileo pitikantensis*. The latter was discovered in 1961, as a fragment of a maxilla and a few post-cranial bones, by Reuben Stirton and his colleagues in the Tirari Desert of South Australia. It was formally named more than a quarter of a century later by Bonnie Rauscher who studied the animal for her MSc project. Coincidentally, the much better preserved Riversleigh skull was discovered while Bonnie's paper was in press, too late to help with the

An unexpected surprise from Upper Site was the skull of a very primitive marsupial lion, one barely the size of a Brushtail Possum.

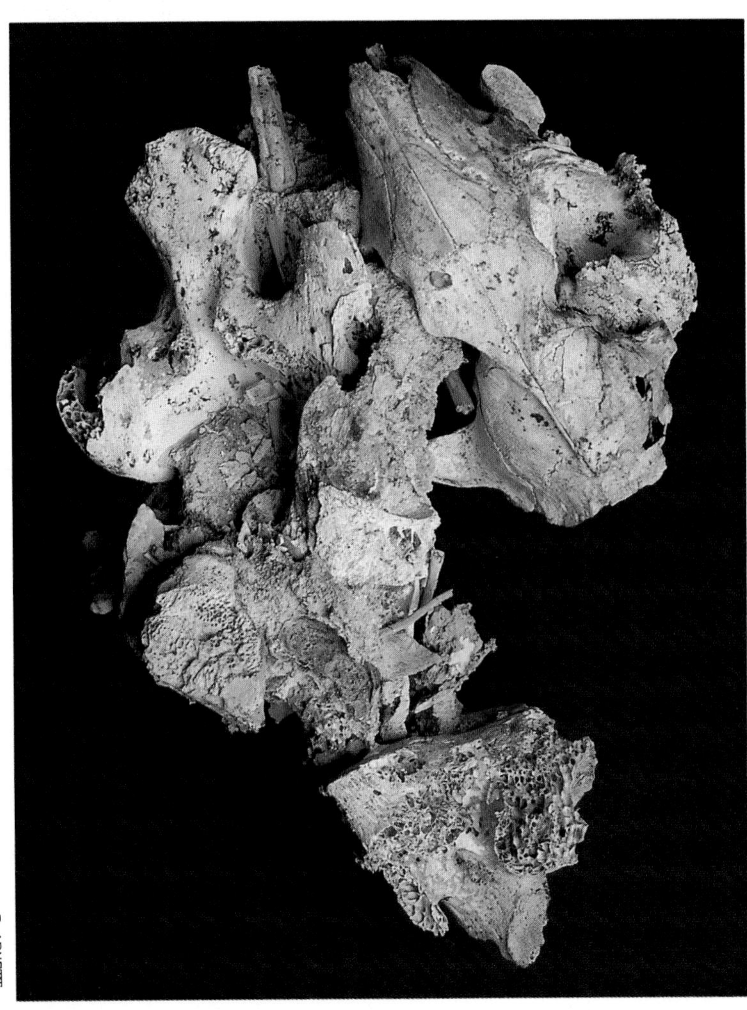

R. ARNETT

interpretation of *P. pitikantensis* but in plenty of time to form the basis of a new study that will test and significantly modify current understanding about the relationships and evolutionary history of the whole family.

The Riversleigh skull also demonstrates that the 'marsupial lion niche' was much broader than we had previously been led to think, based only on the Miocene to Pleistocene record. Clearly, as with Riversleigh's bandicoots, the evolution of feeding guilds enabled more than one kind of creature to co-exist in the same habitat. Finally the skull demonstrates, like the Platypus, that an essentially primitive lineage can survive well beyond the time when its earlier representatives gave rise to different, more 'advanced' lineages. In this case, while the Riversleigh *Prisciteo*-like marsupial lion could easily have given rise to the species of *Wakaleo* (as well as *Thylacoleo*), at Riversleigh the two lion lineages overlap in space and time.

After their spectacular appearance in the Riversleigh deposits, the *Prisciteo*-type lions vanish forever, perhaps ultimately losing in the grand struggle to survive to their larger, more widespread and certainly more diverse descendants, the species of *Wakaleo*.

R. WILLIAMS

KOALAS
mindless munchers of gum leaves—forever?

One of the dubious claims to fame of the living Koala (*Phascolarctos cinereus*) is that its brain has been said not to fill up the small space in the skull where it does its thing. But if this lack of 'fulfilment' is in any way correlated with its fate through time, mindlessness may translate into evolutionary endurance, for koalas as a group (family Phascolarctidae) have persisted for at least the last 25 million years in more or less the same state of diversity and morphology as they do at present.

Most of central Australia's Oligo-Miocene deposits sport one or two species. Their teeth and what little is known of their skulls, although indicating three different genera (*Perikoala, Madakoala, Litokoala*), appear to have been reasonably similar to the living animal. Clearly they were all arboreal, tree-leaf-eaters, like their living counterpart.

In Riversleigh the koala fossil record, recently studied by Honours student Karen Black, exhibits a bit of the expected and a lot of unexpected. Certainly there are species of *Litokoala* which, because of its better preservation than the central Australian forms, help to clarify the relationships of this group to other koalas including the living species. There is also one new genus that is more primitive than any others so far found. Another

The modern Koala is the sole living representative of a very old lineage. Riversleigh's record of Koalas includes a form that links primitive central Australian forms with the modern Koala.

(with several species) appears to represent what koalas would look like if they tried to develop teeth like goats! Their high-crowned and complex molars indicate that this lineage must have specialised on something very different from eucalypt leaves.

Few fossil koalas from elsewhere in Australia are known from more than a few teeth or a piece of a jaw. Some of Riversleigh's koalas, in contrast, are known from partial skulls. One of these, representing one of the weird forms, is the only fossil koala that can be compared in detail with all previously described forms including the modern Koala.

All of the Riversleigh koalas, and for that matter most of the central Australian koalas, are rare compared with representatives of other marsupial families. What could be the reason for

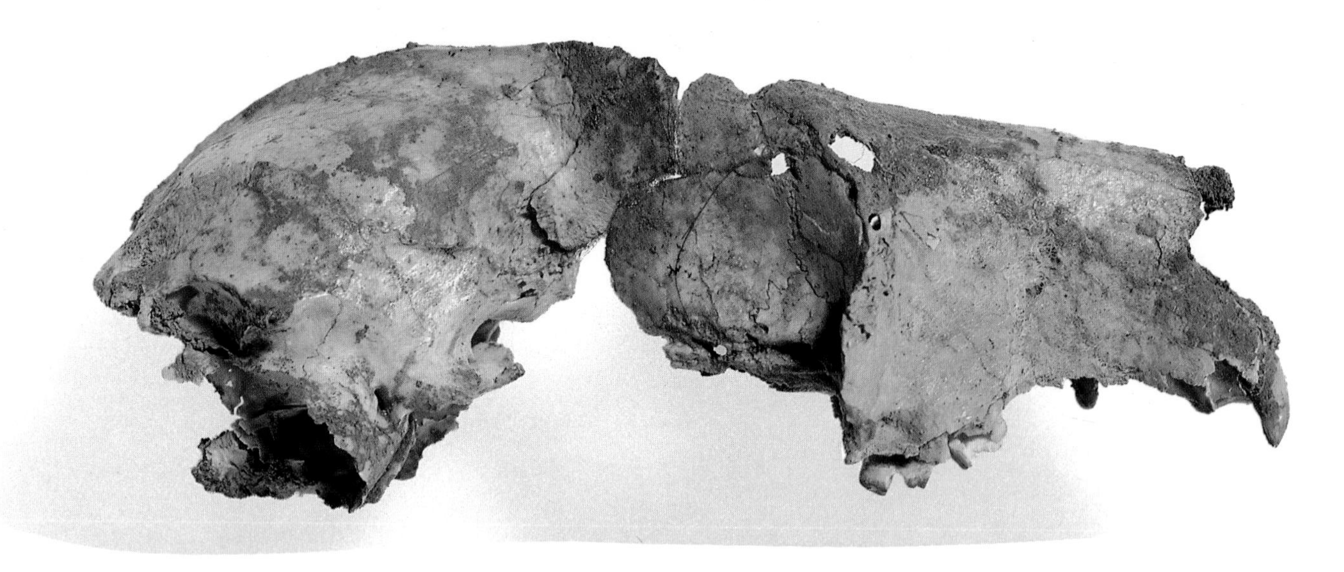

R. ARNETT

A skull of a primitive Koala from Riversleigh.

this relative rarity when today Koalas in their preferred habitats are often quite common?

Pressed to come up with a plausible scenario that could explain the presence and abundance of the living Koala in Australia's open forests, we have previously suggested the following. Because of their relative rarity, we suspect that the early koalas specialised on a relatively uncommon resource in Australia's Oligo-Miocene rainforests—which may well have been eucalypts. Today there are still a few eucalypts in the rainforests but of course they now dominate the open forests and woodlands of the rest of the continent. This may explain why modern koala densities seem very high compared with those indicated for the Oligo-Miocene closed forests of Riversleigh and perhaps central Australia. Because modern rainforest eucalypts appear to specialise on poor soils, when the Australian climates began to deteriorate in the late Tertiary, these once rare gums with their equally rare koalas, may have been pre-adapted to persist and spread in the rapidly expanding areas of nutrient impoverished soils.

It is likely that at least one of Riversleigh's Koalas had already specialised on gum trees that occurred within the rainforests.

The phascolarctid fossil record adds information and, ideally, perspective to questions about the conservation status of the living Koala. While historical changes in Koala distribution and density are important to consider when the status of the living Koala is assessed, our perspective of long-term changes suggest that in terms of density, where modern Koalas occur now, they are more abundant than they ever were in the Tertiary. Further, because most Tertiary local faunas have between one and two species at most, the existence of one relatively abundant species today does not in itself suggest that the group is in decline; in fact the fossil record suggests the opposite conclusion. Finally, considering dental morphology, the living Koala is little more specialised than its ancestors, which suggests that it has not edged particularly far out along its evolutionary limb. Providing humans do not continue to chew away at the Koala's preferred habitats, phascolarctids should have no problem out-lasting humans.

WOMBAT-LIKE BEASTS APLENTY

anticipating the grasslands to come

There are four kinds of living wombats, three hairy-nosed species (*Lasiorhinus latifrons*, *L. krefftii* and *L. barnardi*, the latter sometimes being regarded as a variety of *L. krefftii*) and the Common Wombat (*Vombatus ursinus*). During the Pleistocene there were many more, ranging from tiny kinds the size of small dogs to the cow-sized *Phascolonus gigas* with massive chisel-like upper incisors. The Pliocene, too, witnessed many kinds of wombats differing mostly in size and in the shape of their incisors. Among all of these late Tertiary and Quaternary wombats, only one stands out as really different: the late Pleistocene *Warendja wakefieldi*, which appears to be a long-surviving member of a relatively primitive wombat lineage.

Wombats in general are one of the two major marsupial responses to the spread of grasslands in Australia, the other being the macropodine ('ordinary') kangaroos. Whereas these kangaroos responded to the development of this resource through alteration of dental eruption sequences and 'molar progression' (so that new, unworn teeth were constantly moving into the area of the mouth where the grass was processed), the wombats' response was evolution of high-crowned and eventually

J. HOPE

Modern wombats (left) are dwarfed by the giant Pleistocene wombat *Phascolonus gigas* (right). Riversleigh's wombat-like animals were diverse but more primitive.

F. KRISTO

Common Wombats live in forests but feed on grasses. The absence of true wombats from Riversleigh's early and most of its middle Miocene sites contributes to the conclusion that grasses were rare.

rootless teeth that continually erupted throughout their lives. In this way, despite the abrasive nature of a grass diet, old living wombats never become 'gummers'.

The oldest fossil wombat, *Rhizophascolonus crowcrofti* from the middle Miocene Kutjamarpu LF of central Australia, although clearly a wombat, is known from one isolated premolar that formed the basis for the initial description and two subsequently collected molars. It appears to demonstrate an intermediate condition between the rootless Pliocene and Quaternary wombats and the hypothetical wombat ancestor that had low tooth crowns and distinct roots. This Kutjamarpu taxon has tall, rather wombat-like crowns but *closed* roots on all three teeth so far known (the Latin prefix *Rhizo* means 'rooted'). This suggests it was beginning to specialise on some food that was abrasive enough to merit the need for more enduring teeth.

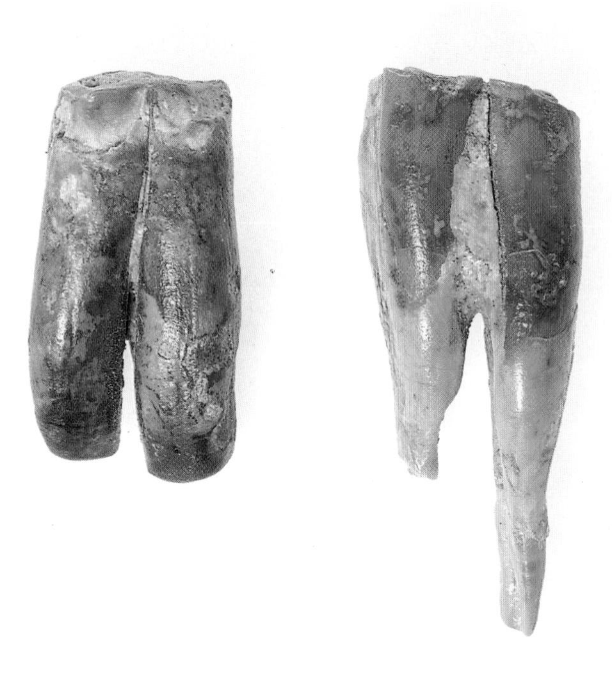

The teeth of primitive wombats (such as these from the Leaf Locality in central Australia) have roots that indicate that, unlike modern wombats, their teeth did not grow throughout life.

Riversleigh's wombat-like animals, although not common, are very diverse.

In view of this, it has not surprised us to learn from the studies of Honour's student Alan Crickman that a species of wombat very similar to *Rhizophascolonus crowcrofti* has turned up in the youngest of the middle Miocene units on the Gag Plateau. What *did* surprise us was to find what looks like a primitive species of *Warendja* in Encore site also on the Gag Plateau. Here and only here it occurs in abundance with a unique, very large species of the carnivorous kangaroo genus *Ekaltadeta* and a very large hipposiderid bat. Taken together, we are wondering if Encore Site might not be younger in age (possibly early late Miocene) than other System C sites. Although it is not clear that the Riversleigh *Warendja*-like animal ate grass, it had ever-growing teeth adapted to feeding on some form of abrasive vegetation.

However, the whole story is a bit more complicated than this. Riversleigh's wombats include many kinds that are far more primitive than either *Rhizophascolonus* or *Warendja*. None of these are as yet named although Alan Crickman is nearing the completion of his studies. One, from the early Miocene Boid Site East LF, was a Brushtail Possum-sized animal with a wombat-like skull and dental formula but rooted, relatively short-crowned teeth. Another, from Upper Site, was about the size of modern wombats but again had rooted teeth with crowns that were far more complex than those of any previously known wombat. Because it was the first species found at Upper Site, we expected to see much more of the creature when the material was prepared—but no such luck; all that has been recovered so far is seven isolated teeth.

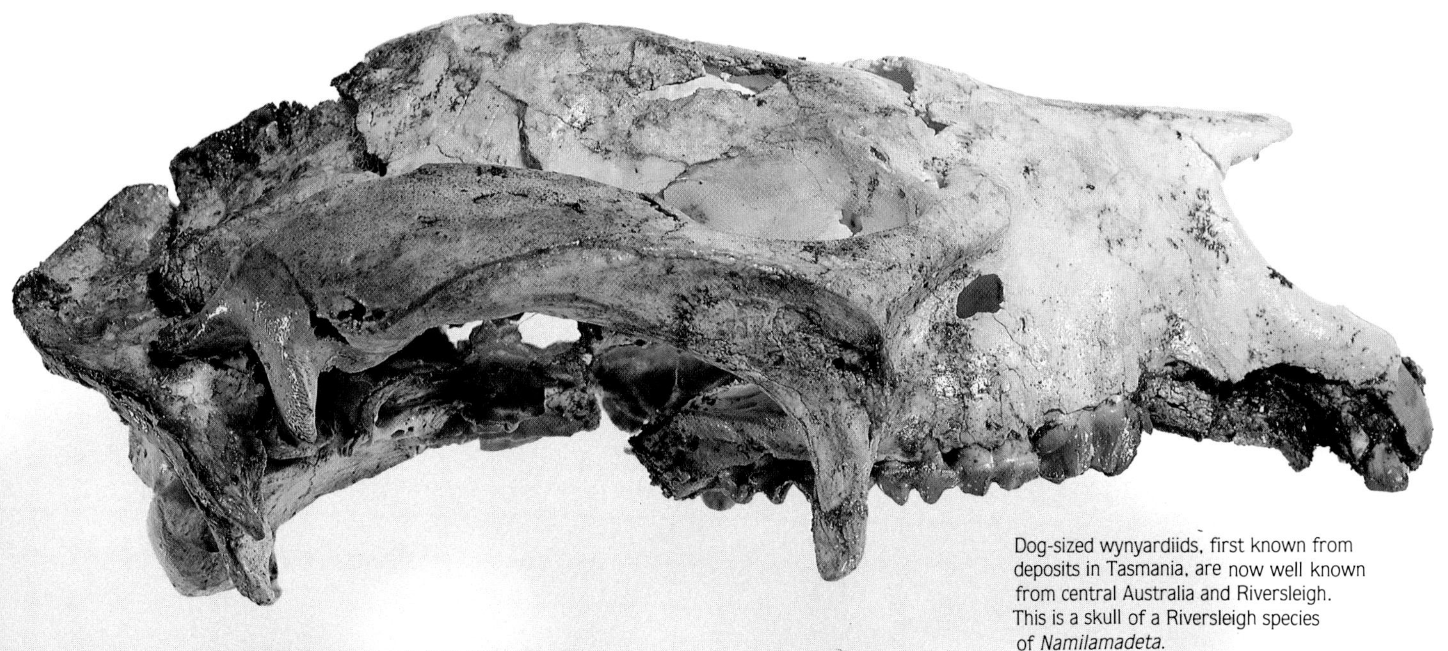
Dog-sized wynyardiids, first known from deposits in Tasmania, are now well known from central Australia and Riversleigh. This is a skull of a Riversleigh species of *Namilamadeta.*

WYNYARDIIDS
their last stand?

The first-known Tertiary mammal from Australia was an enigmatic creature known as *Wynyardia bassiana*. It was found on the beach at Wynyard, Tasmania, in a slab of rock that tumbled down from the sea cliff above. Its age was eventually determined to be earliest Miocene but its nature took far longer to determine because, unfortunately, it had arrived at the base of the cliff minus its teeth.

Discovery by Richard Tedford and his colleagues of wynyardiid-like marsupials in the late Oligocene Namba Formation of South Australia presented a much more substantial look at this group of creatures. Tom Rich and Mike Archer then described yet a third wynyardiid as *Namilamadeta snideri*, based on a partial skull they had found in the Namba Formation. Most recently, Neville Pledge named another wynyardiid *Muramura williamsi* on the basis of two skeletons he had found at Lake Palankarinna in South Australia. All of these wynyardiids are late Oligocene to early Miocene in age.

When Riversleigh began to reveal the first of its rich faunal assemblages between 1983 and 1984, there was not a trace of a wynyardiid. But in 1985 when Henk and some of the volunteers discovered Godthelp Hill, wynyardiids were suddenly everywhere! As we now realise, the Godthelp Hill sequence (part of System B) is older than the sediments we had worked in 1983-84 at Ray's Amphitheatre (part of System C) and this abrupt disappearance of wynyardiids within the Riversleigh region between the two systems was one of our first clues that something important had happened at the end of System B time.

Once these wynyardiids began to turn up, we recalled an enigmatic, toothless jaw that we had collected in 1976 from the still older Site D (part of System A). Although it was only a section through the rear part of a dog-sized jaw, on re-examination, it was clearly a wynyardiid. As a result, we now work with the simple hypothesis that wynyardiids persisted in the region during the time of accumulation of Systems A and B, as they presumably did in

A Riversleigh wynyardiid from the Upper Site Local Fauna.

central Australia and Tasmania, but became regionally extinct at Riversleigh before the System C deposits began to accumulate. They may have vanished from central Australia at the

R. ARNETT

A partial skull of a Riversleigh wynyardiid showing the distinctive teeth of this group.

same time, but without a better biostratigraphic record, we cannot be certain when the family completely vanished.

As for what kind of wynyardiids Riversleigh had, so far they all seem to be referable to *Namilamadeta*, the genus previously only known from the Tarkarooloo LF of South Australia. But in contrast with the central Australian wynyardiids, the ones from Riversleigh appear to represent at least two and possibly five species, as well as many developmental stages from possible pouch young to old adults.

One of the best specimens of a Riversleigh wynyardiid was discovered quite by accident. Having realised in 1985 and 1986 how rich Upper Site was, we resolved to chase out more of the deposit in 1987—only to discover that the deposit appeared to thin out into the side of the hill. But preparator Syp Praesouthsouk was unwilling to give up so easily. He suggested that we use light explosives to remove a mass of the relatively sterile overlying limestone and chase the underlying level towards the heart of the hill, which we began to do. But when the light explosive had done its work and we pried the fractured face with crowbars, the first shaft of modern daylight illuminated a perfect cross-section of a dog-sized skull, with the back of the last molar just showing. From this glimpse, the distinctive pattern of a wynyardiid was instantly evident. This was a skull of *Namilamadeta* sp., the first known! The matching blocks of limestone were retrieved and subsequently prepared in the laboratory in Sydney.

Since then, this skull and all of the other Riversleigh wynyardiid material has been studied by Neville Pledge in the South Australian Museum, along with all other wynyardiid material known from Australia. When his studies are complete, he will have determined exactly how many different kinds there were, what their distribution was in space and time, what the broader relationships of this mysterious group are to all other known groups of marsupials and, hopefully, determined something about their lifestyle, habitats and position in the ecosystems of ancient Australia.

Study of the shape of the bones and the musculature of wombats has enabled Peter to reconstruct the muscles of the head.

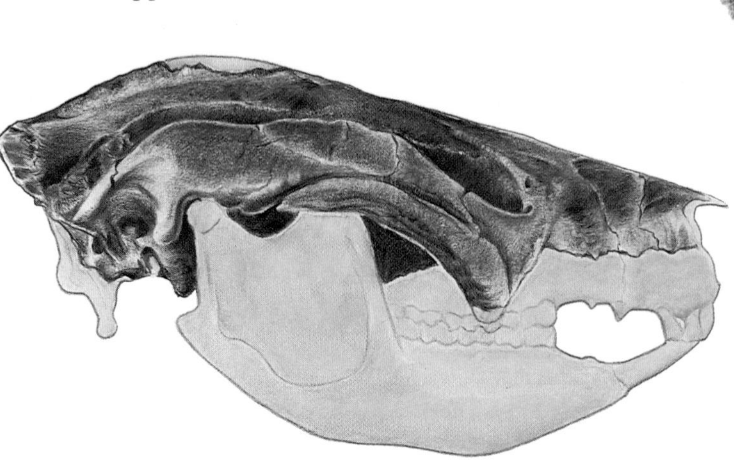

The first known wynyardiid skull fragment from Tasmania has been reconstructed here by palaeontologist/artist Peter Murray.

Consideration of the superficial muscles of wombats enabled Peter to complete the restoration of the head of this now extinct group of marsupials.

DIPROTODONTOIDS

doomed vegetarians of the forests and plains

Ever since at least the late Oligocene, two groups of large herbivorous marsupials appear to have competed for access to ground plants: the kangaroos and the diprotodontoids. The kangaroos, as discussed later, started the race for dominance as smaller, more diverse and less specialised feeders. Their race was rapidly run but in all sorts of different directions and this fact was probably the key to their success, for by late Miocene time one of those directions was into the heart of the newly developing grasslands and Australia's future. The diprotodontoids, on the other hand, were more conservative, perhaps because the ones we see in the late Oligocene had already run previous races and settled on a winning combination—for survival in a land of forests and soft vegetation. By the time deserts swept over half of the continent in the late Pleistocene, the death knoll was sounding for the last of Australia's diprotodontoids—perhaps one of the exhausted giants mired in the baking muds of Lake Callabonna. The kangaroos had won.

The many diprotodontoids known ranged in size from small sheep to large hippopotamuses. There were two families: the more conventional, somewhat wombat-shaped diprotodontids; and the much less conventional, sometimes trunked palorchestids. In fact, until a few 'missing links' began to roll out of the Riversleigh deposits, these two groups were so distinct that it was a bit difficult to find reasons for keeping them together in the superfamily Diprotodontoidea.

In turn, there are two subfamilies of the diprotodontids: the zygomaturines, whose members have multi-cusped last upper premolars; and the diprotodontines, whose members have much simpler premolars. The zygomaturines are the most common and most diverse of the diprotodontoids found at Riversleigh. *Neohelos tirarensis* may have been as common and widespread in the Miocene habitats of Australia as cattle are today. Some zygomaturines (*Nimbadon*, represented by skulls and articulated skeletons found in 1993) were no larger than labrador dogs, others were as large as small cows. All were driven by amazingly small brains and probably a constant need to fill their bellies with soft leaves. They may have travelled in herds for protection from the flesh-eaters of the forest, such as the marsupial lions, thylacines and giant pythons.

As time went on, the smaller zygomaturines vanished or gradually changed into larger forms,

Reconstruction of *Diprotodon optatum*, the largest marsupial that ever lived. The teeth and bones of this animal have been found at Terrace Site.

an evolutionary pattern commonly encountered in the fossil record of many groups of vertebrates. By Pliocene time, zygomaturines were almost exclusively represented by very large species of *Zygomaturus*, one of which (*Z. trilobus*) survived long enough into the Pleistocene to provide dinners for Australia's first humans. Curiously, at least one lineage of

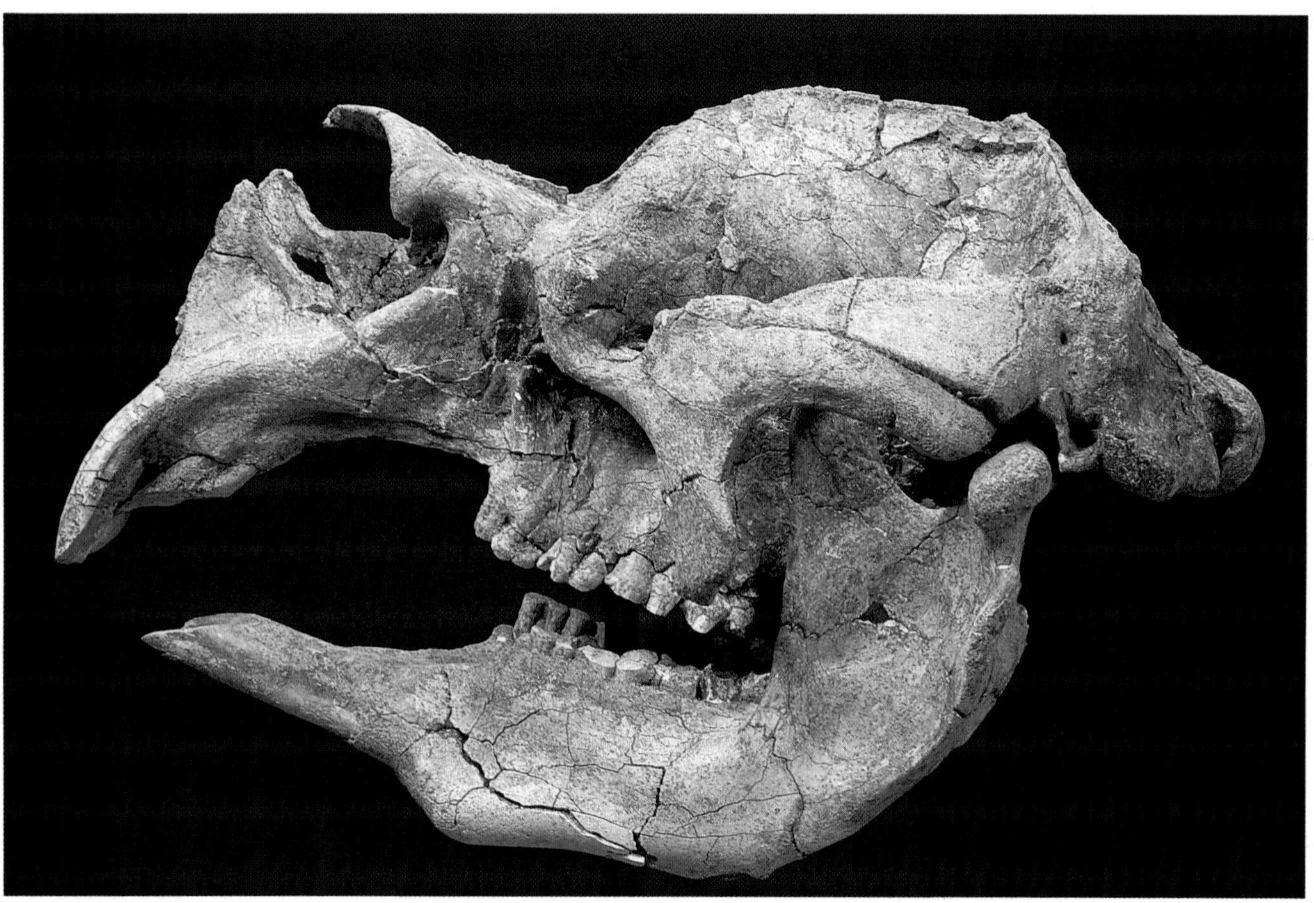

The skull and lower jaw of a rhinoceros-sized *Diprotodon optatum* from Pleistocene deposits in New South Wales.

H. HUGHES

Alex Ritchie (left) supervises the excavation of a skeleton of *Diprotodon optatum* which showed, on recovery, marks suggestive of it having been butchered by prehistoric hunters.

giant panda-like zygomaturines (*Hulitherium*) appears to have survived until the beginning of the Holocene in the mountains of New Guinea.

The diprotodontines underwent a similar but less diverse history although by the time their race was run in the late Pleistocene, their last contestant was the largest marsupial that ever lived: *Diprotodon optatum*. This browsing giant was nearly 3 metres long and 2.6 metres high at the shoulder and must have weighed at least as much as a rhinoceros. Could it have been semi-aquatic? Gut contents from individuals collected in the fossilised muds of Lake Callabonna in South Australia suggest that it was a browser, unlike the grazing kangaroos that would have watched the last of these giants die.

Bones and teeth of *Diprotodon optatum* have been found in Riversleigh's Terrace Site, as well as from most peripheral areas of the continent, sometimes with the unmistakable marks of stone tools inscribed on their surfaces.

Prior to some of the discoveries at Riversleigh, we had no idea about when or how the first diprotodontines evolved. The previously oldest known member of this group was the very large *Pyramios alcootense* from the late Miocene Alcoota LF. Now, from several System A and B local faunas (e.g. Riversleigh's Alsite), we have recovered a much smaller, very 'primitive' diprotodontine that betrays the

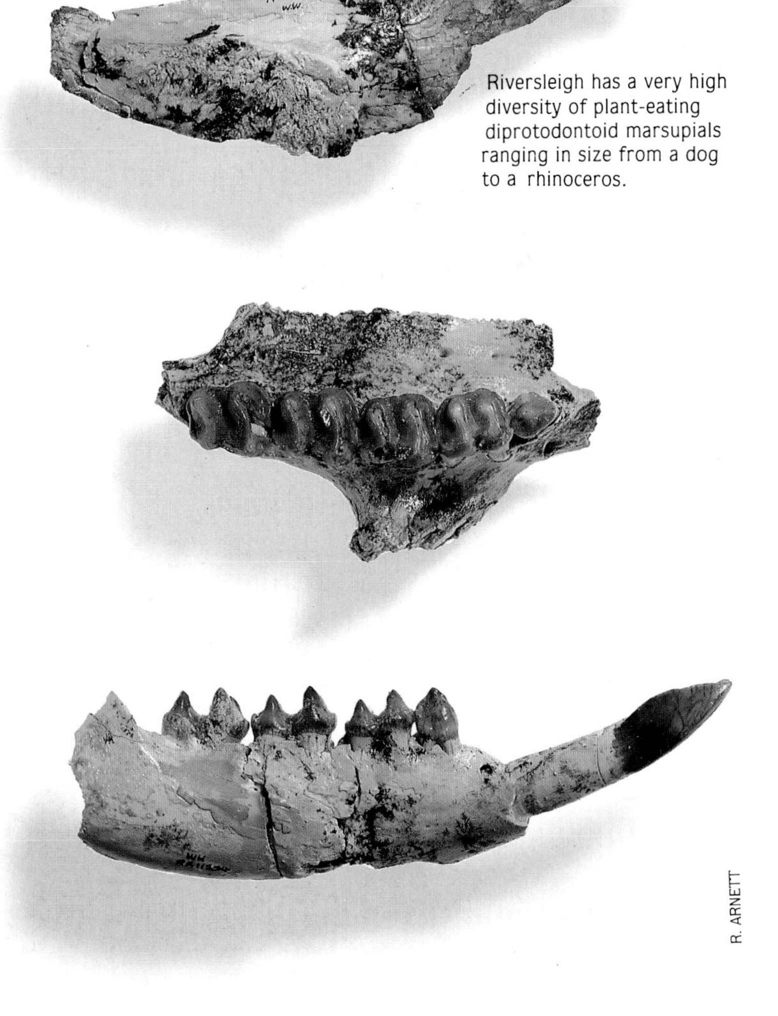

Riversleigh has a very high diversity of plant-eating diprotodontoid marsupials ranging in size from a dog to a rhinoceros.

R. ARNETT

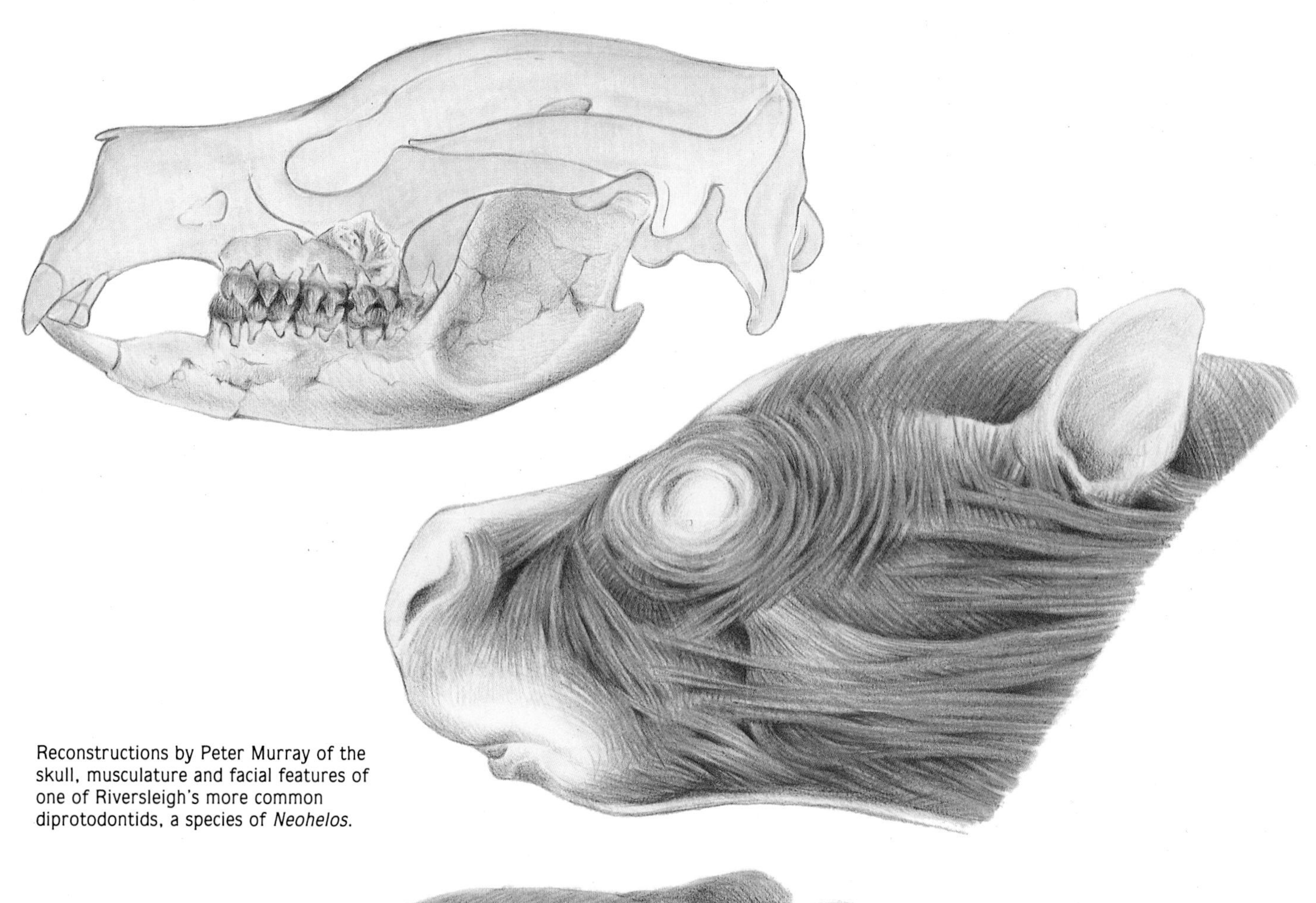

Reconstructions by Peter Murray of the skull, musculature and facial features of one of Riversleigh's more common diprotodontids, a species of *Neohelos*.

The Vice-Chancellor of the University of New South Wales, John Niland, discovered whole skulls of new kinds of diprotodontids here at AL90 Site in 1993.

S.WILLIAMS

Among the more unexpected diprotodontoids are some that show features suggestive of 'links' between otherwise distinctive groups.

R. ARNETT

origins of the group. It has a very simple diprotodontine-like premolar but zygomaturine-like molars, i.e. it's a made-to-order missing link that is also older than any of the more derived forms (e.g. *P. alcootense*), to which it probably gave rise.

The palorchestids are the other half of the diprotodontoid radiation—their weird half brothers. The Pleistocene and last-surviving species, *Palorchestes azael*, was about the size of a horse. In body shape it resembled no other creature known. It had huge koala-like claws, enormously powerful forelimbs, a long ribbon-like tongue and a large elephantine trunk. Until the skeleton is adequately studied, speculation about what it did for a living will continue. Although we are certain that it was a herbivore, we are not at all certain about the kinds of plants it ate. In contrast to its diprotodontid cousins, its high-crowned, complex teeth probably *could* have withstood the abrasive effects of a grass diet but this simple interpretation feels hollow when its peculiar limbs, claws and trunk are considered. Consequently, we have elsewhere suggested that it may have pulled up shrubs with the powerful forelimbs and eaten the fibrous tubers that developed from the roots of some; or perhaps it used its claws to strip the bark from trees to gain access to the softer inner layers of the trunk?

The oldest named palorchestids are the species of *Ngapakaldia* and *Pitikantia* previously only known from the Oligo-Miocene deposits of central Australia. But now a species of *Ngapakaldia* has turned up in the older Riversleigh deposits enabling biocorrelation between the two regions. Many other previously unsuspected small palorchestids have turned up

Although best known from the Oligo-Miocene of central Australia, Riversleigh's older units have produced representatives of the distinctive palorchestid genus *Ngapakaldia*.

The plant-eating species of *Palorchestes* appear, from their skull structure, to have sported small, mobile trunks. This particular and best known species comes from the late Miocene Alcoota Local Fauna of the Northern Territory.

RECONSTRUCTION BY PETER MURRAY

F. COFFA

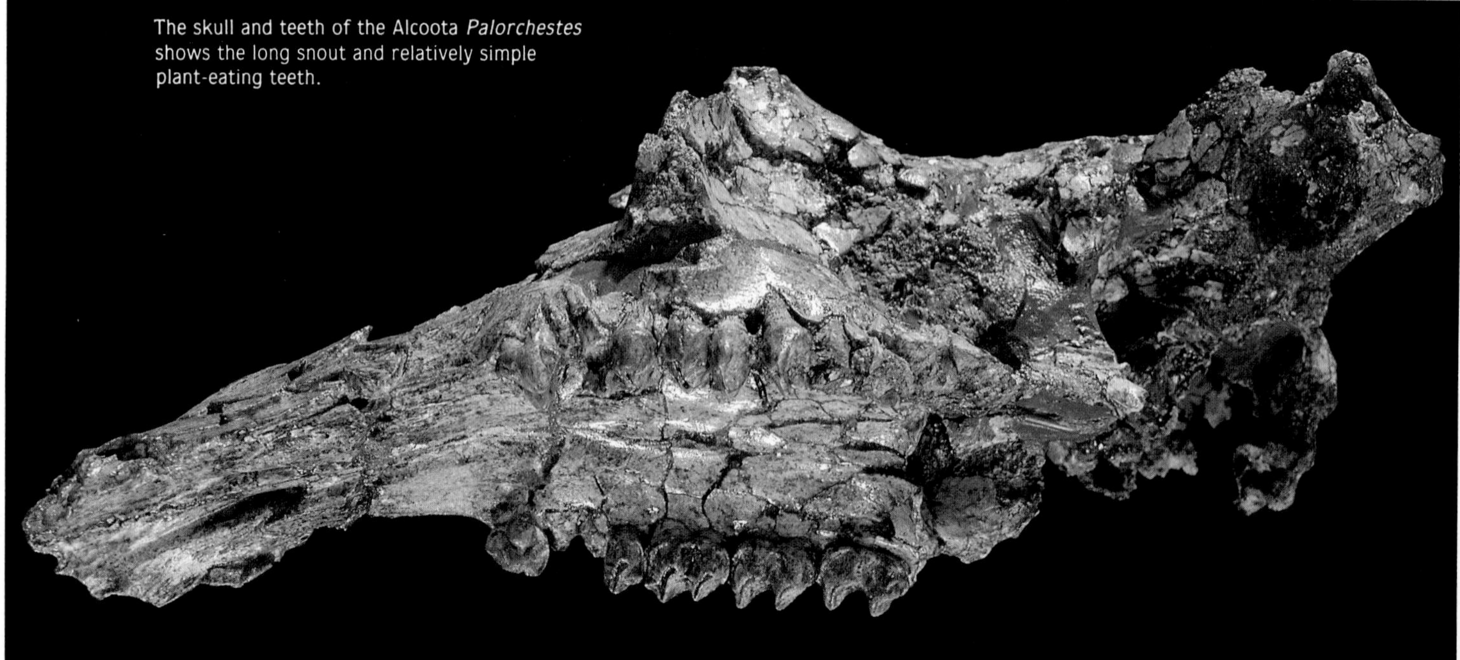

The skull and teeth of the Alcoota *Palorchestes* shows the long snout and relatively simple plant-eating teeth.

in the local faunas of Systems B and C. As well, a few months ago, we found the first Riversleigh tooth of *Propalorchestes*, a taxon previously only known from Bullock Creek LF of the Northern Territory. Study of these Riversleigh animals is a major task, in part because of their considerable diversity and in part because all samples are small. This in itself is interesting because it suggests that whatever they were doing in Riversleigh's Oligo-Miocene rainforests, there wasn't much of it to be done!

Other Riversleigh diprotodontoids betray mixtures of what were previously thought to be diprotodontid and palorchestid features. We interpret this blend of features to indicate that the common ancestor of palorchestids and diprotodontids had premolars like those of palorchestids, molars and incisors like those of primitive zygomaturines and ear regions (the middle ear of the skull) rather like those of *Ngapakaldia* species. That common ancestor, however, must be considerably older than any of central Australia's or Riversleigh's mammal-producing units.

KANGAROOS

a rainforest full of experiments

If Riversleigh's fossil record tells us any single thing about Australia's kangaroos it is that today's standards are decidedly abnormal. The Red Kangaroo, which has become a kind of symbol for the continent's fauna, is yesterday's child, as is the Eastern Grey Kangaroo, the Euro, all of the other great kangaroos and possibly the entire mass of ordinary wallabies from Red-necks to Nail-tails. As far as we know, none of these modern lineages have a generic pedigree that extends any further back than 4.5 million years (at which time we see the ancestors of the Agile Wallaby, Swamp Wallaby and Euro alongside a host of strange cousins) and most are unknown before the Pleistocene.

How could this be? Quite simply, until Australia's grasslands invaded and pushed back the crumbling walls of most of Australia's ancient rainforests, there were few if any opportunities for grazers. Since that

Although thought of as typically Australian, the Red Kangaroo of arid inland Australia is decidedly atypical in the light of the whole history of kangaroos.

F KRISTO

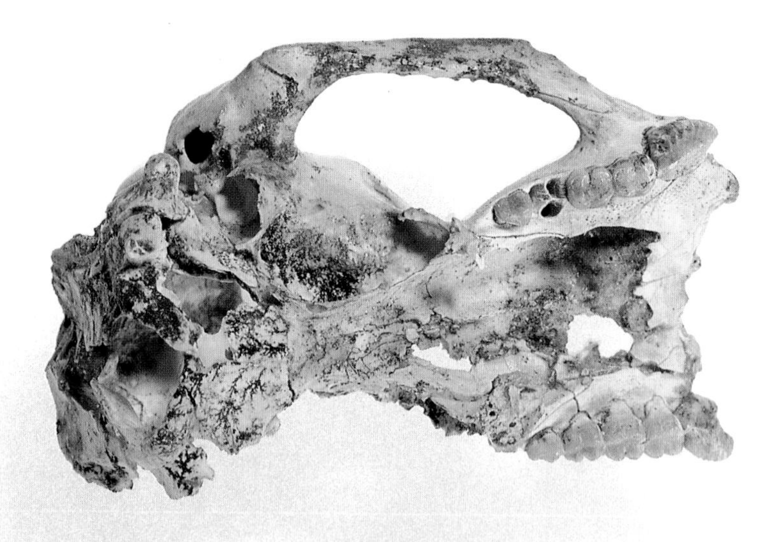

R. ARNETT

The most primitive living kangaroo is the Musky Rat-kangaroo (*Hypsiprymnodon moschatus*). Riversleigh produced a partial skull of *H. bartholomaii* that appears in all respects to be ancestral to the modern form.

fundamental change does not appear to have occurred until the late Miocene, earlier kangaroos could not have dreamed of grass salads.

Today there are two families of kangaroos: the normally omnivorous rat-kangaroos in the family Potoroidae; and the 'ordinary' herbivorous kangaroos in the family Macropodidae. However, there are suggestions from molecular biology and palaeontology that one living species, the Banded Hare-wallaby, may represent yet a third equivalently distinct group.

In the oldest of Riversleigh's forests, the alert, bright-eyed rat-kangaroos were in their heyday. They were everywhere, in every possible guise, from omnivores to leaf-eaters. Even ferocious 'cannibals' masqueraded behind beguilingly fuzzy faces. This was clearly a time of massive experimentation, perhaps 12 million years before the macropodid grazers began to elbow them aside. In the variety so far recovered, we have recognised at least four distinct groups of rat-kangaroos:

1. the hypsiprymnodontines, of which one descendant survives today as the Musky Rat-kangaroo in the Atherton Tableland's tropical rainforests.
2. the potoroines, descendants of which, such as bettongs and potoroos, make up the remainder of the living potoroids.

The living Musky Rat-kangaroo, alone among modern roos, gallops rather than hops, has twins and retains the fifth toe on the hind foot. In this sense, it is a 'living fossil'.

J. ROBINSON

B. BROWN

The dentition of extinct (shown here) and living species of musky rat-kangaroos includes a distinctively shaped, 'buzz-saw' premolar.

The skull of *Bettongia moyesi* when it was first found at Riversleigh.

3. the balungamayines, a massive radiation of now totally extinct forms.
4. the now totally extinct propleopines, robust monsters that coveted the flesh of their neighbours.

Of these, the propleopines were certainly the most unlike any living kangaroos. The best-known Riversleigh propleopine is *Ekaltadeta ima*, a feisty bit of unpleasantness that probably stood about 1.5 metres tall. It is, so far, the only species of the genus known, which is itself very surprising. We would have expected that an animal as distinct as this could have fitted into the ecosystems of other areas of Oligo-Miocene

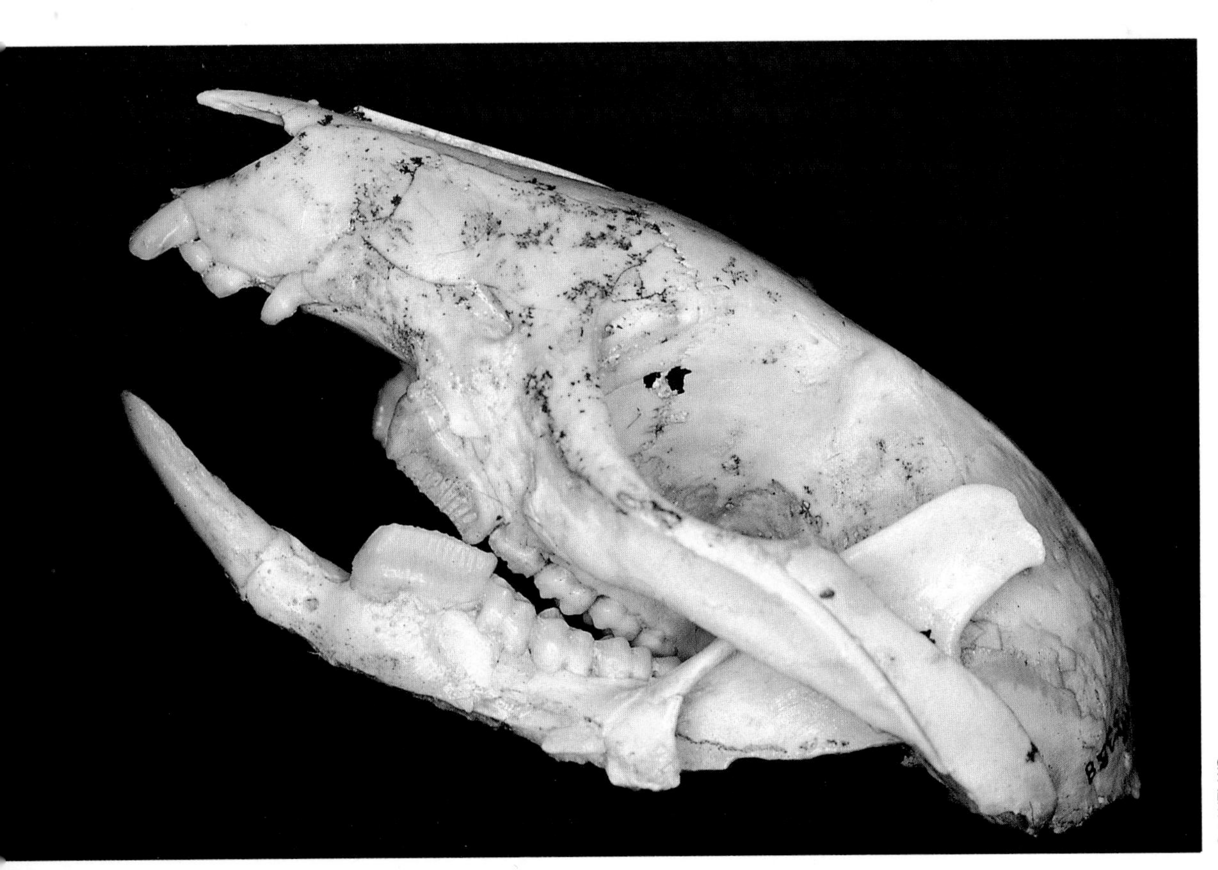
D. MAITLAND

The skull of the same Riversleigh bettong once the acetic acid had stripped the limestone from its entombed nut. The lower jaws were found to be still in place.

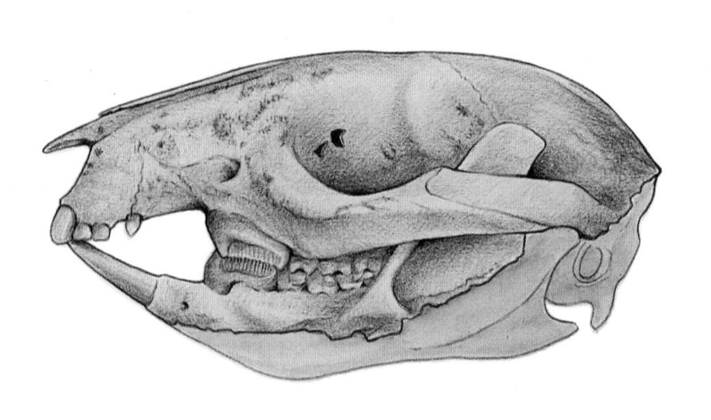

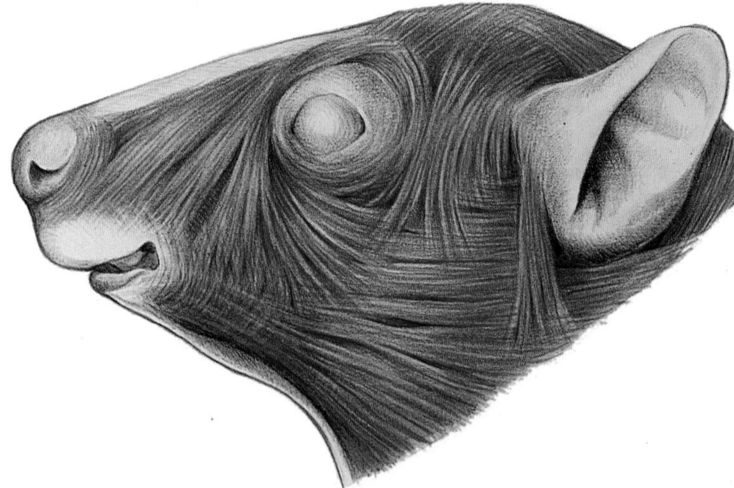

A reconstruction series of Riversleigh's *Bettongia moyesi*, an Oligo-Miocene bettong that appears to have been ancestral to or at least more primitive than all living species of this group.

RECONSTRUCTIONS BY P. MURRAY

Australia. However, there is no doubt in our mind that it is directly ancestral to the species of *Propleopus*, similarly carnivorous roos from the Pliocene and Pleistocene of many areas of at least eastern Australia. It would be interesting to know whether or not this lineage of carnivores survived long enough to gobble up any of Australia's first humans. Perhaps the last late Pleistocene survivor of the lineage *Propleopus oscillans*, which stood about 2 metres tall, was trained as an 'attack roo'?

Equally interesting from an evolutionary point of view are the balungamayines. These have been controversial from the moment they first emerged from Riversleigh's System A sediments. The problem is that, although their overall anatomy most resembles that of rat-kangaroos, their molar teeth are lophodont like those of the macropodid kangaroos. When we described the first known example of this group (*Balungamaya delicata*), we suggested it was a potoroid that, in molar shape, had converged on the ordinary kangaroos—in other words, they were an early rat-kangaroo experiment that failed to hold their own in the face of the rapidly diversifying macropodids. We were challenged by Mike Woodburne and Judd Case who argued exactly the opposite case: balungamayines were ordinary kangaroos, as indicated by their molar teeth, and their potoroid-like features were convergent on potoroids. We read and considered their arguments, then examined the more complete

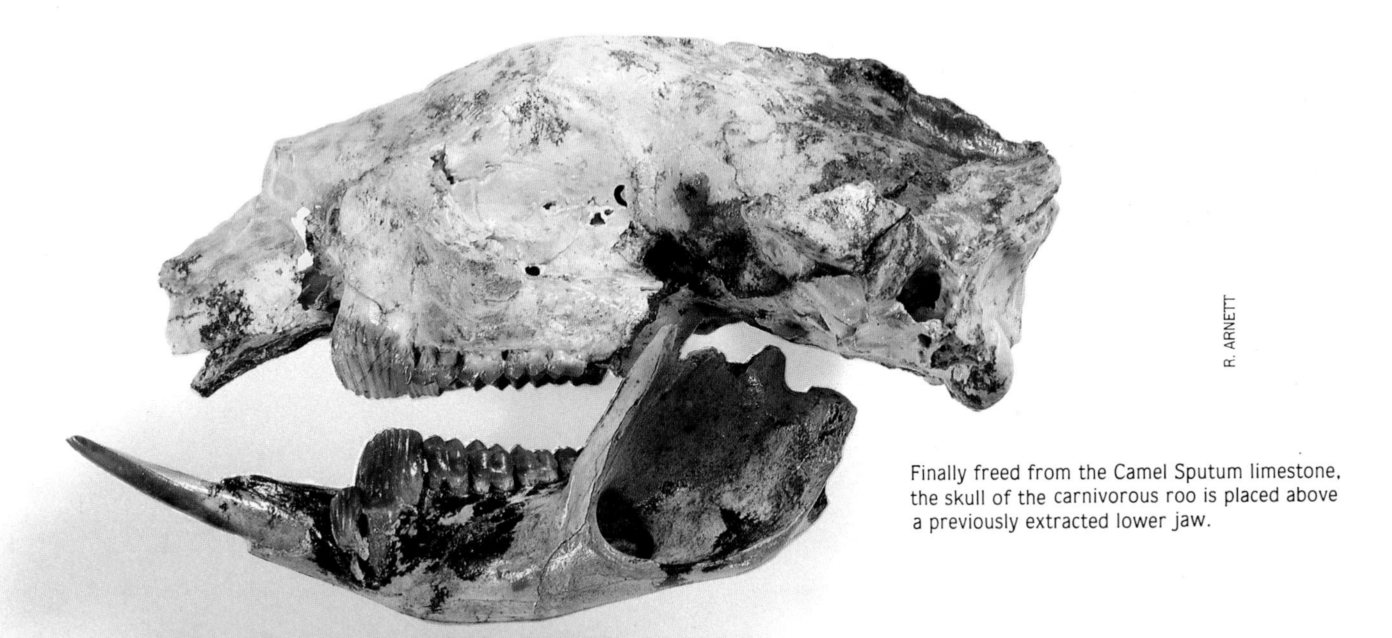

R. ARNETT

Finally freed from the Camel Sputum limestone, the skull of the carnivorous roo is placed above a previously extracted lower jaw.

material that had accumulated since we wrote the first paper. This examination urged us to restate our original case but now with even better reasons for concluding these Riversleigh roos to be potoroids.

Scientific debates, such as this about the evolutionary significance of particular fossils, are an essential part of palaeontology because they serve as tests of hypotheses that might otherwise be accepted without considering other equally plausible interpretations. While we originally considered, and abandoned as less parsimonious the hypothesis that balungamayines were macropodids, the challenge from Woodburne and Case forced us to more thoroughly examine this alternative using more complete material, something we otherwise might not have done.

The other family of kangaroos, the Macropodidae, has members which dominate Australia's modern grasslands wherever they occur, as well as a few browsers that are at home on the rainforest floor (e.g. the forest wallabies of New Guinea) or clambering in the tops of the trees (tree kangaroos). Three subfamilies are recognised: the macropodines, which include most living kangaroos but so far none from Riversleigh's Oligo-Miocene local faunas; the sthenurines, which include many Pliocene and Pleistocene giant kangaroos, as well as one living species, the Banded Hare-wallaby which some would argue should be placed, with the other sthenurines, in a distinct family; and the balbarines, which include only extinct kangaroos first recognised from Riversleigh and Bullock Creek.

Two halves of a block of Camel Sputum limestone, each containing one half of the skull of the carnivorous kangaroo. *Ekaltadeta ima.*

R. ARNETT

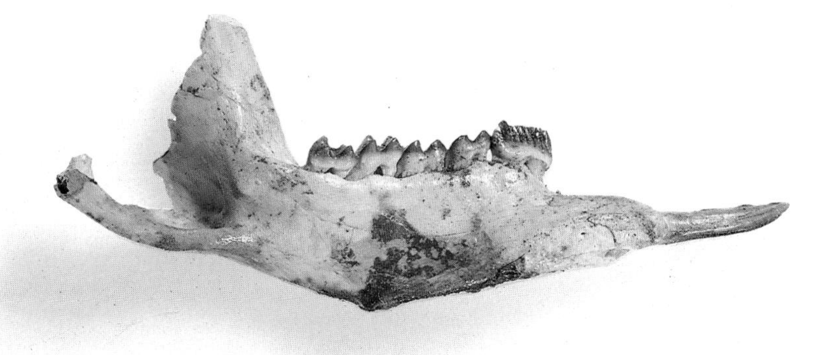

The lower jaw of a Riversleigh balbarine, a primitive member of the living kangaroo family Macropodidae.

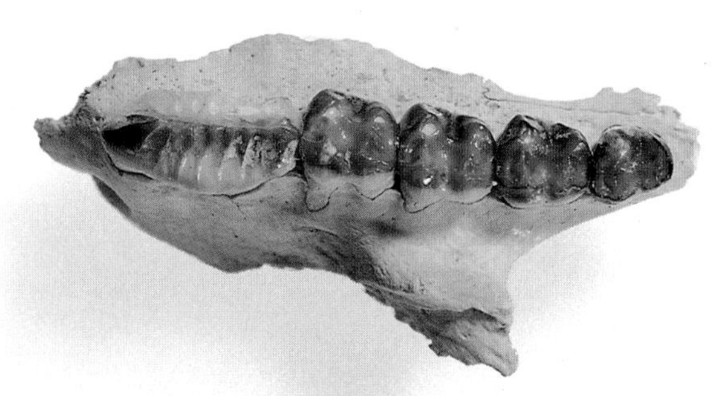

The upper tooth row of a more 'conventional' rat-kangaroo from Riversleigh.

A fortuitous blow with a sledge hammer split a block of Last Minute Site limestone to reveal this perfect lower jaw of a balbarine kangaroo.

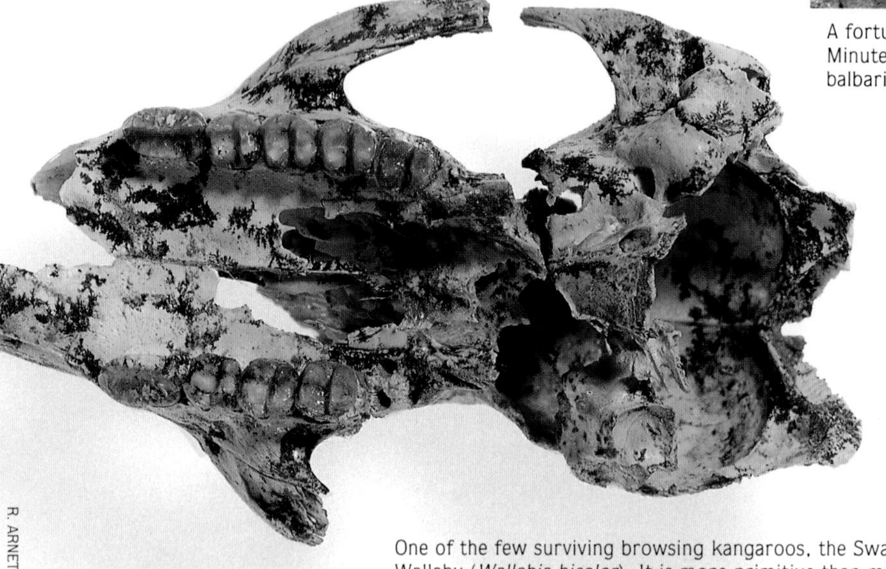

A partial skull of a balungamayine rat-kangaroo from Riversleigh. These strange potoroids converged in aspects of tooth shape on 'ordinary' roos (in the family Macropodidae).

R. ARNETT

One of the few surviving browsing kangaroos, the Swamp Wallaby (*Wallabia bicolor*). It is more primitive than modern grazers such as the Red Kangaroo but still far more specialised than any of the roos of Riversleigh.

R.&A. WILLIAMS

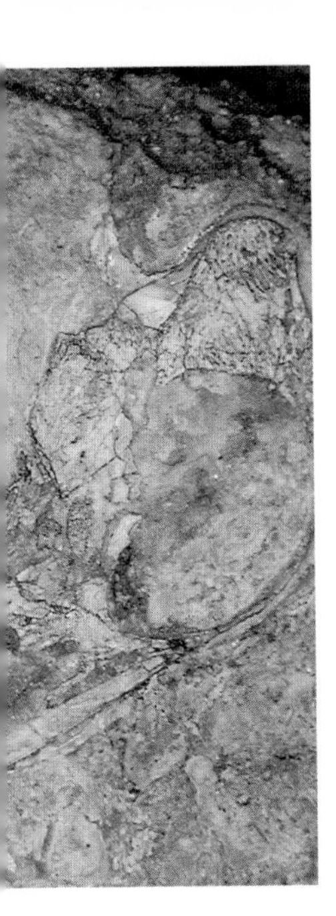

A kangaroo skull and lower jaw partially revealed in a block of Outasite limestone.

At Queensland University of Technology Bernie Cooke's studies of Riversleigh's kangaroos have so far focused on the balbarines. His initial studies of the first-known skull of a balbarine are providing a very detailed view of the structure of primitive macropodids.

By Rackham's Roost time, Riversleigh's balbarines had been replaced mainly by macropodines and probably sthenurines although so far we have found none of the latter. By the time Terrace Site accumulated, at least one modern macropodine species, *Macropus agilis*, the Agile Wallaby, had made its appearance, although this species had a long time range from the early Pliocene to the present.

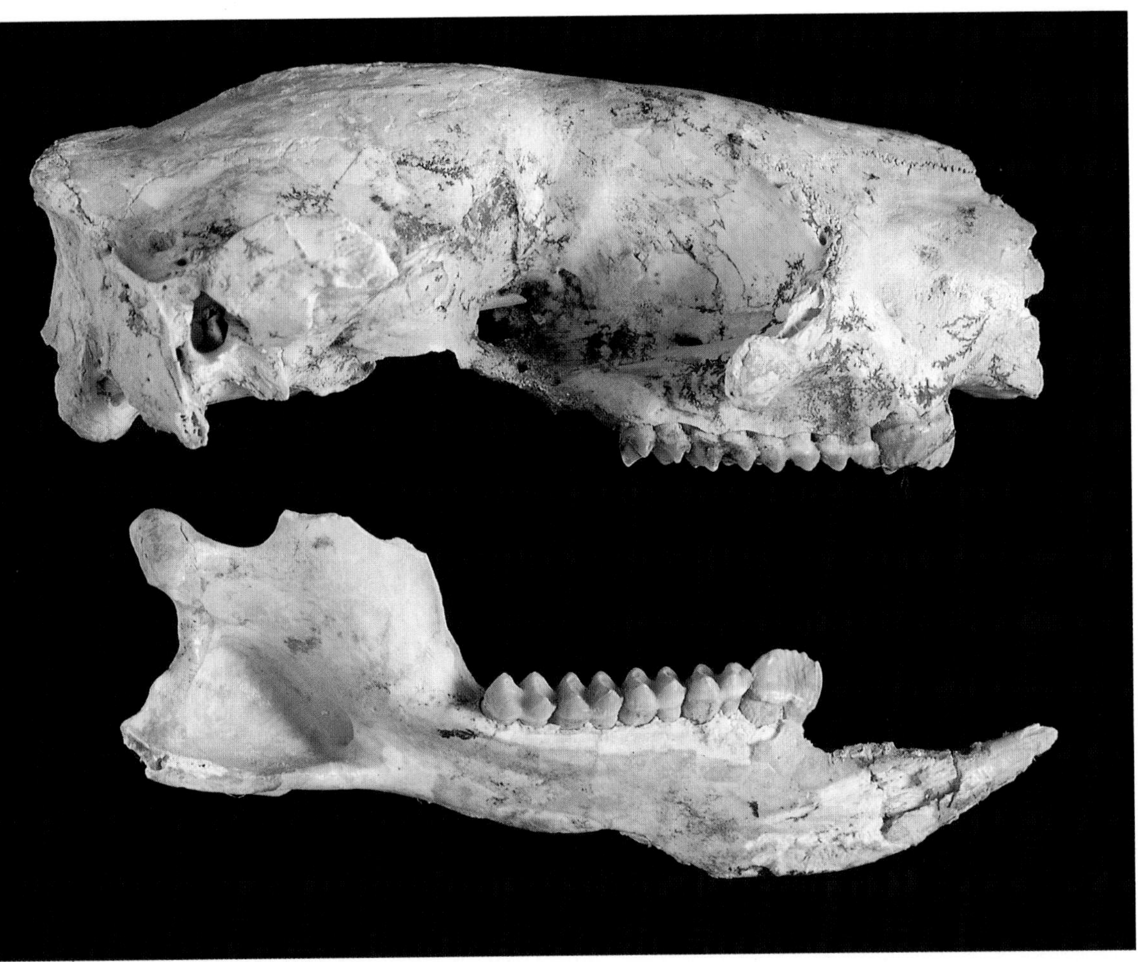

After preparation, the Outasite skull and lower jaw proved to be a very distinctive kind of balbarine kangaroo now being described by Bernie Cooke.

QUEENSLAND MUSEUM

PYGMY-POSSUMS

fur balls with giant histories

Today there are two basic kinds of burramyids or 'pygmy-possums': the ordinary large-mouse-size pygmy-possums in the genus *Cercartetus*; and the slightly larger Mountain Pygmy-possum, *Burramys parvus*. The Mountain Pygmy possum is a highly specialised inhabitant of the alps in south-eastern New South Wales and northern Victoria. It was discovered as a Pleistocene fossil long before it turned up alive in 1966, in a woodpile in a Mount Kosciusko ski lodge.

But that was just the beginning of the distinctive *Burramys* story. Isolated fossil teeth and later a jaw (later named *B. triradiatus*) were discovered in the early Pliocene Hamilton LF which represents a lowland rainforest habitat in western Victoria. Then Riversleigh began to kick in with hundreds of Oligo-Miocene specimens referable to a new species of *Burramys* now under study by Honours student Jenni Brammall, that must have been relatively common in the lowland rainforests of north-western Queensland. Finally, a third species (*B. wakefieldi*), based on a single dentary, turned up in the Miocene Ngama LF from central Australia.

What was surprising about these finds was the fact that of the two kinds of pygmy-possums known, *Burramys* was regarded as the more specialised and hence was not expected to turn up in the older deposits ahead of *Cercartetus*-type burramyids. To confuse the issue, a putative 'giant' *Cercartetus*-type pygmy-possum from the Oligo-Miocene deposits in central Australia turned out, upon detailed examination, to represent a new family of possums, the Pilkipildridae, with nothing whatsoever to do with pygmy-possums.

Clearly, without a fossil record, any attempt to explain why the specialised living Mountain Pygmy-possum is where it is today would be doomed to failure. Although physiologists and ecologists can establish *how* it manages to survive now in those regions, they cannot determine how it came to be there in the first place. What does the fossil record suggest about how this diminutive possum's situation has been changing?

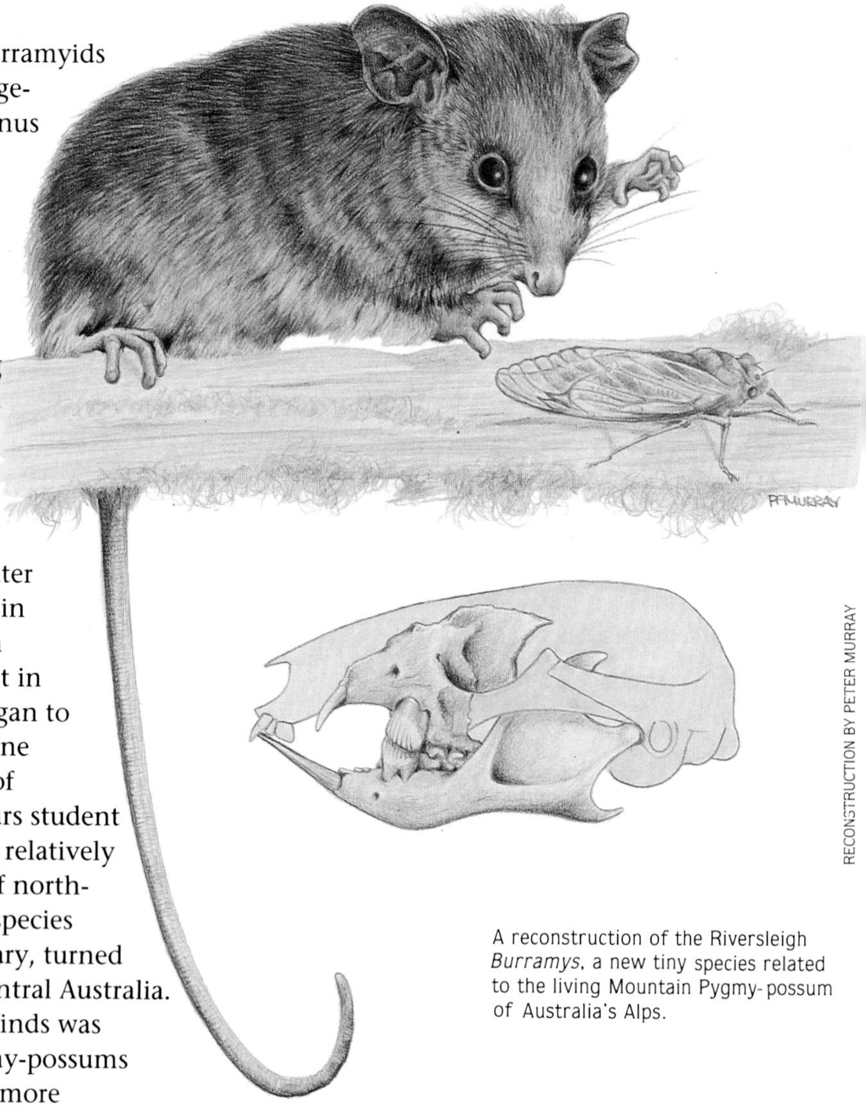

We have previously suggested that 20 million years ago, this lineage was widespread in the lowland rainforests of at least the eastern half of the continent. In these complex communities they may have specialised on hard seeds and nuts that accumulated on the forest floors, a diet for which their teeth seem singularly well adapted. As the dry, colder climates of the late Cainozoic approached, the vast forests began to dry and wither, first from the central and northern central portions of the continent. The *Burramys* populations that had once abounded in the Riversleigh region and

PF MURRAY

RECONSTRUCTION BY PETER MURRAY

A reconstruction of the Riversleigh *Burramys*, a new tiny species related to the living Mountain Pygmy-possum of Australia's Alps.

R. ARNETT

The skull of a Riversleigh 'cuscus', *Strigocuscus reidi*. This species appears to be related to the ground cuscuses of New Guinea.

South Australia would have been among the first to vanish. Populations in western Victoria (e.g. Hamilton) and probably in much of eastern Australia hung on at least until 4.5 million years ago. As the drying became more severe in the Pleistocene and the rainforests contracted even further, all *Burramys* populations, other than those in islands of rainforest dotting the tops of the Great Divide, would have perished. Sometime during this interval, the ancestors of the living species must have begun to depend more on the rocky substrate of the alpine rainforests than on the plant species themselves. Perhaps this was a necessary response to the marked seasonality that would have characterised much of the late Pliocene and early Pleistocene. Certainly it would have encouraged food hoarding and storing, a behavioural trait characteristic of the living species which spends months each year within rock piles covered by snow. By the time the alpine climates became too severe to support rainforest and it gave way to sclerophyll woodland, the ancestors of the living species had clearly crossed the threshold that once bound their ancestors to rainforest.

As a footnote to this overview, after we had finally begun to accept the fact that *Burramys*-type pygmy-possums were all the record had to reveal, a genuine *Cercartetus*-type taxon turned up at Riversleigh. From a preliminary look, it appears to be at least distantly similar to *C. caudatus*, a species that today inhabits the rainforests of north-eastern Queensland and New Guinea.

PHALANGERIDS

three success stories

The Common Brushtail Possum (*Trichosurus vulpecula*) is one of Australia's most frequently seen marsupials. Brushtails as a group occur in almost all Australian habitats, until recently even being common in the gums along the river banks of central Australia. Their relatives include the Scaley-tailed Possum (*Wyulda squamicaudata*) and the many cuscuses (species of *Phalanger, Strigocuscus, Spilocuscus, Ailurops* and *Phalanger*). All phalangerids are cat-size or larger and more or less omnivorous, which may be one of the clues to the ability of at least the Common Brushtail Possum to survive despite widespread destruction of their original habitat. It is not that unusual to find one foraging through a city rubbish bin at night while standing off an alley cat or two before retreating to a 'den' in the roof of a house. These possums are survivors.

The fossil record for the group has been slow to appear but is now steadily accumulating. Among the oldest representatives is a small jaw fragment from the early Miocene Geilston Bay LF of Tasmania. Without the record from

P. GERMAN

Modern cuscuses abound in New Guinea although only two species occur in Australia.

Riversleigh, the next oldest phalangerids would be two recently described from the early Pliocene Hamilton LF: *Trichosurus hamiltonensis* and *Strigocuscus notialis*. But, fortunately, there are also Oligo-Miocene phalangerids from Riversleigh. Three are so far recognised: *T. dicksoni, S. reidi* and an as yet unnamed species of *Wyulda*. Hence, all three modern groups were represented in the early to middle Miocene rainforests.

It is interesting to consider the trends through time for these three groups. The brushtail possum *T. dicksoni* is the most common of the Riversleigh phalangerids. Although it is more primitive in a number of features than any living species, it is about as common in the Riversleigh assemblages as the only rainforest species, *T. caninus*, is today. The same is true of the Hamilton *Trichosurus* which appears to be a 'missing link' between the relatively more primitive Riversleigh species and the more derived modern species. Hence, we see no significant change in the diversity or abundance of brushtail possums over the last 20 million years. The same could be said for the *Wyulda* lineage. Today there is a single, relatively rare species. In the Oligo-Miocene there was, as far as we have found, just one species and it, too, was rare, hence there is no evidence here for a significant change in the 'evolutionary health' of this lineage.

In distinct contrast is the change in fortunes for cuscuses. In Riversleigh's Oligo-Miocene rainforests, we have no evidence for more than a single, relatively uncommon species,

The lower jaw of the Riversleigh cuscus, *Strigocuscus reidi*, showing the powerful third premolar useful for cutting seeds, fruits or other creatures.

R. ARNETT

R. ARNETT

The skull of a Riversleigh brushtail possum, *Trichosurus dicksoni*. Brushtail possums were not uncommon in Riversleigh's rainforest biotas.

A lower jaw of the Riversleigh brushtail possum.

Strigocuscus reidi. The same low diversity and scarcity characterises the cuscuses of the Hamilton LF. In contrast, there are now two reasonably common species in the small rainforest remnants of north-eastern Queensland and at least 13 species in the rainforests of New Guinea and surrounding islands. Clearly this is a group upon whom fortune has smiled. It may be that its post-Oligocene success has been facilitated by a semi-isolated centre for diversification in the rainforests of New Guinea and its adjacent islands. Whatever the reason, the fossil record rings no alarm bells about the future of this group.

The Hamilton and Riversleigh phalangerid suites are significant in another aspect: both are the only places where species of the genera *Trichosurus* and *Strigocuscus* are sympatric. This could either mean that: *Trichosurus* evolved in Australia before or after it was possible for members of its lineage to disperse to New Guinea; or it reached New Guinea but subsequently went extinct, perhaps through competition with smaller species of the diversifying genera *Phalanger* or *Strigocuscus*.

PETAURIDS

a complex array of enigmatic experiments

Today petaurids comprise two subfamilies: the petaurines, containing five gliding possums (species of *Petaurus*) and Leadbeater's Possum (*Gymnobelideus leadbeateri*); and the striped possums (species of *Dactylopsila*).

The fossil record for the insectivorous/omnivorous striped possums is a complete mystery. Despite the fact that all living species are inhabitants of rainforests, we have not had a 'smell' of this group (modern striped possums are highly odoriferous) from Riversleigh. Perhaps the group had not differentiated by Oligo-Miocene time.

The history of petaurines was also, until recently, a mystery. In 1987, two were reported from the early Pliocene Hamilton LF, both being similar to, but different from, the living Squirrel Glider (*P. norfolcensis*) and Sugar Glider (*P. breviceps*). Were these inhabitants of that Victorian rainforest gliders? While their close resemblance to living species supports the possibility, it is difficult to conceive of a reason why arboreal possums would need to glide in a rainforest where all the crowns are touching. However, there is also a pre-Greater Glider in the Hamilton LF, so the idea of rainforest gliders should not be too quickly abandoned. Perhaps, alternatively, the source of the Hamilton fossils included more than a rainforest biota. Some of the animals may have been collected by owls that had roosted and regurgitated bone-rich pellets onto the rainforest floor but collected some of their prey from open forest, beyond the rainforest's boundaries.

In Riversleigh, there is an embarrassingly rich pile of petaurid-like animals which have yet to be examined in detail. Some of the taxa occur in many of the Oligo-Miocene local faunas. None of them, however, seem particularly close to any living petaurid genus. Although many have molars that closely resemble the kinds displayed by living species, they have, in contrast, a very large posterior premolar. One taxon we have temporarily dubbed 'Pre-petaurus' looks like it was heading towards the *Petaurus/Gymnobelideus*-type morphology. Other recently collected isolated teeth look even more like the species of this modern group but as yet they are too poorly known to confidently assess their relationships.

Whatever the evolutionary relationships of these petaurid possums turn out to be, they were abundant in the Oligo-Miocene rainforests of Riversleigh and a significant part of Riversleigh's arboreal possum communities.

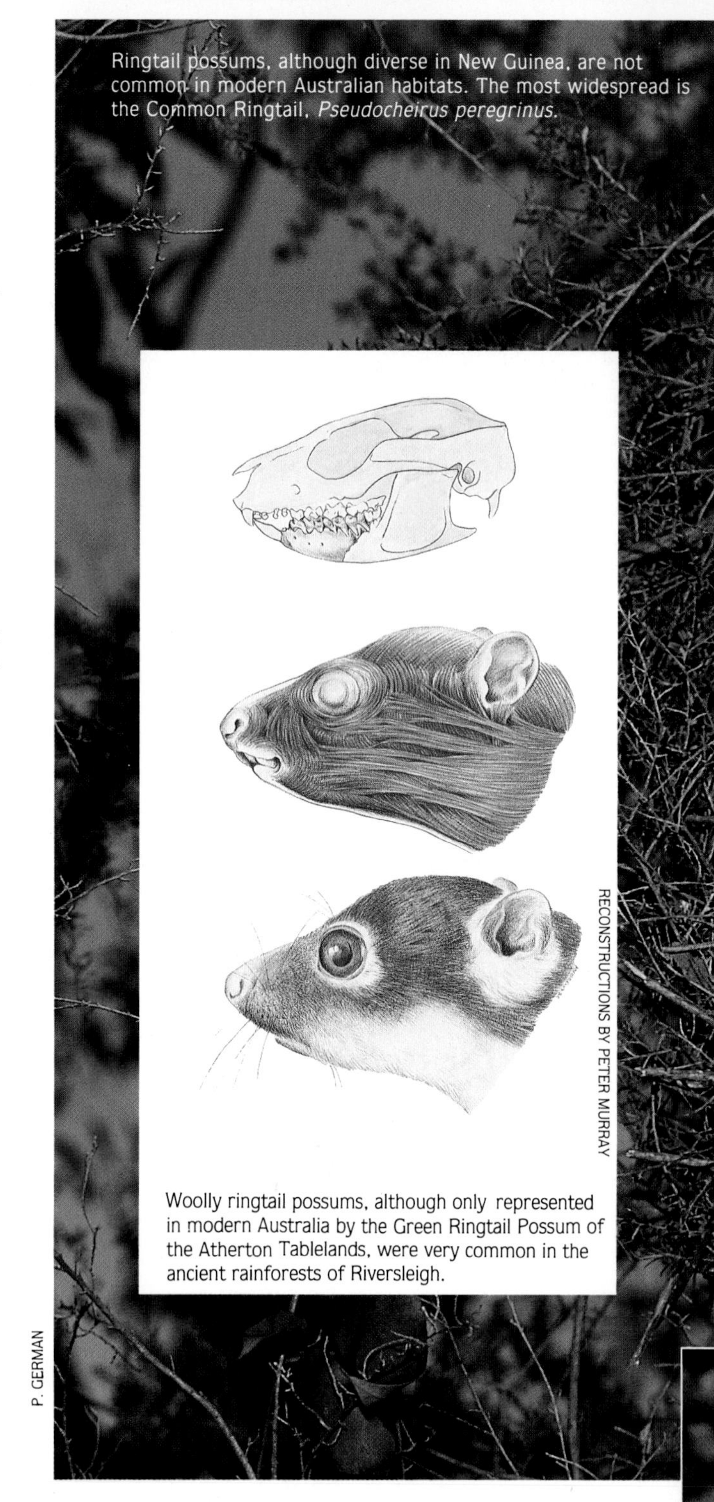

Ringtail possums, although diverse in New Guinea, are not common in modern Australian habitats. The most widespread is the Common Ringtail, *Pseudocheirus peregrinus*.

Woolly ringtail possums, although only represented in modern Australia by the Green Ringtail Possum of the Atherton Tablelands, were very common in the ancient rainforests of Riversleigh.

RECONSTRUCTIONS BY PETER MURRAY

P. GERMAN

RINGTAIL POSSUMS

gentle gardeners in the roof of the forest

All living ringtail possums (pseudocherids), whose diets are known, are tree-leaf-eaters—almost to the point of obsession. Most of the other possum groups will not turn up their pink noses if a moth presents itself as protein on their salad but ringtails will normally turn away in disgust. These gentle possums are a herbivore's herbivore. The only possible exception may be the Pygmy Ringtail of New Guinea (*Pseudocheirus mayeri*), one captive individual of which refused leaves but took sugar water.

There are three groups of living ringtails: the 'ordinary' ringtails (species of *Pseudocheirus*), the pseudochirops ringtails (species of *Pseudochirops*), and the Greater Glider and Lemuroid Possum (*Petauroides volans* and *P. hemibelideus* respectively). The fossil histories of these groups are quite different. The earliest *Petauroides*-like fossils are from the early Pliocene Hamilton LF; none are known from Riversleigh. The evolution of this lineage could have started in rainforest, where the Lemuroid Possum still persists, but the Greater Glider's

RAY WILLIAMS

The Green Ringtail Possum *Pseudochirops archeri* , confined today to the rainforests of north-eastern Queensland, is a close relative of several of Riversleigh's Oligo-Miocene ringtails.

origin is likely to have been in the more open forests of the kind that it occupies today. Here it puts its extraordinary gliding skills to full advantage.

The evolutionary significance of the pseudochirops possums has been debated for some time. Oddly, while so far we have not recognised among the mass of different Oligo-Miocene ringtails from Riversleigh any

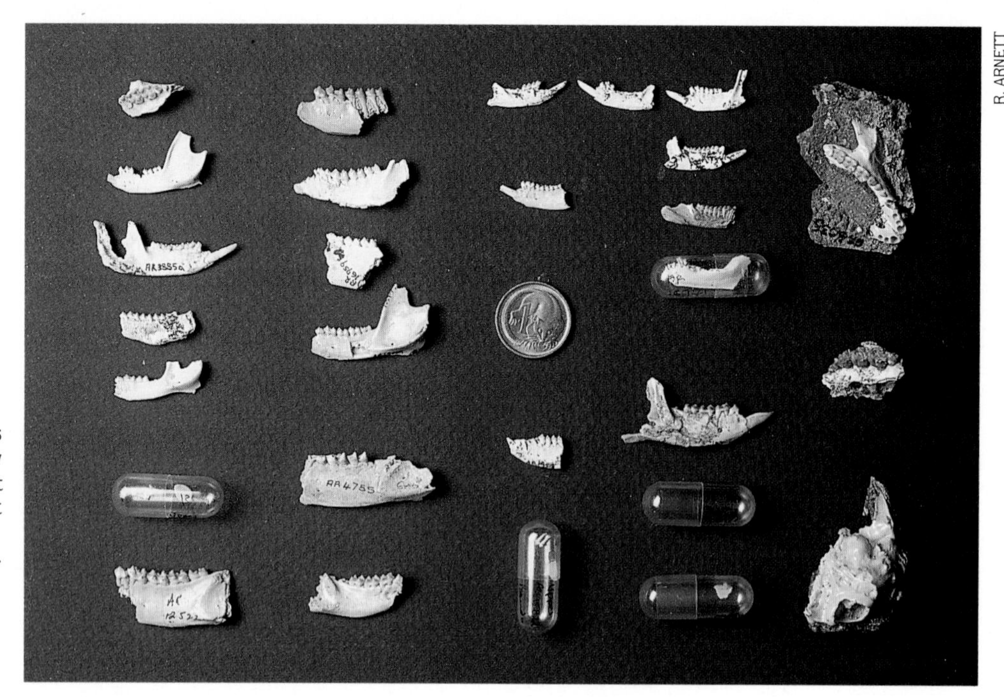

The fossil ringtail possums from Riversleigh are very diverse, there being at least 18 different kinds, most from the Oligo-Miocene rainforest biotas.

representative of the 'ordinary' ringtails, there are at least three species of *Pseudochirops*. All living members of this genus are confined to rainforest in north-eastern Australia (the Green Ringtail Possum which delights in eating, among others, the leaves of the Stinging Nettle Tree!) and New Guinea (the Coppery, Plush-coated and D'Albertis Ringtails).

The majority of Riversleigh's Oligo-Miocene ringtails represent bizarre and sometimes very primitive groups that appear to have left no living descendants. Three of the genera, *Pildra, Marlu* and *Paljara*, also had species in the Oligo-Miocene deposits of central Australia. Others appear to represent groups unique to the Riversleigh region.

When all of Riversleigh's ringtails are studied in detail, they should reveal a great deal about how this major group of arboreal folivores divided up their forest resources. Considering that the highest diversity of sympatric ringtails today is four species in some of the mid-montane rainforests of New Guinea, Riversleigh's Dwornamor LF, with nine sympatric species, suggests that something quite different was going on in the Oligo-Miocene rainforests of northern Australia. The wide palaeogeographic spread of the Oligo-Miocene species of *Pildra, Paljara* and *Marlu* (all of which are, for example, represented in the Kutjamarpu LF of central Australia) should also provide a powerful tool for refining current understanding about the biostratigraphic relationships of the sediments in central and northern Australia.

One other extinct ringtail deserves mention in connection with Riversleigh's rainforest communities: *Pseudokoala erlita*. This relatively large folivore was a rare inhabitant of the early Pliocene rainforests of Hamilton. When first found as isolated molars, it superficially resembled koalas until its diagnostic ringtail-like features were recognised. This made us wonder about a possible relationship between the presence of this animal in late Tertiary rainforests and koalas in the Oligo-Miocene rainforests. Is it possible that ringtails produced a large koala-like possum to fit into the niche in the eastern Australian rainforests abandoned by koalas when they followed at least one species of *Eucalyptus* into the new drier forests of the late Tertiary and Quaternary?

FEATHERTAIL POSSUMS

tiny but definitely not pygmy-possums

There are two feathertail possums living today: the Australian Feathertail Glider (*Acrobates pygmaeus*) and the New Guinean Pen-tail Possum (*Distoechurus pennatus*). Until very recently they were usually regarded to represent a very peculiar subgroup of the burramyids

The living tiny Feather-tailed Glider (*Acrobates pygmaeus*) is closely related to the Pen-tailed Possum of New Guinea, both being in the family Acrobatidae. Riversleigh's Oligo-Miocene rainforest deposits have produced at least two acrobatids.

(pygmy-possums). Then studies of the structure of the ear regions and relationships of blood proteins clearly indicated that they had nothing to do with the pygmy-possums. But what, then, *were* they? As an interim solution to the puzzle, they have been separated by Ken Aplin and Mike Archer as a distinct family of their own—the acrobatids. At this stage in the game, we are reasonably convinced that this family is most closely related to the very different-looking Honey-possums or tarsipedids.

Prior to the Riversleigh discoveries, there was no pre-Pleistocene fossil record for this family. Then, in 1985, ridiculously tiny possum teeth began to drop out of the concentrates from Gotham City Site, a deposit otherwise best known for its diverse bats. Although these teeth (and now whole jaws and partial skulls from this and other Oligo-Miocene sites at Riversleigh) have yet to be studied in detail, it is clear that they represent acrobatids. What we have yet to establish is whether this extinct taxon is a species of *Acrobates*, *Distoechurus* or a new taxon structurally between the two living forms.

Since we have mentioned the Honey-possum (*Tarsipes rostratum*, the living and only known member of the family), it is irksome to report that this is one of the very few family-level groups of Australian mammals that has not, so far, been recognised in any of Riversleigh's Oligo-Miocene assemblages—unless, of course, it is represented by one of the many enigmatic forms whose family relationships are as yet unclear. The Honey-possum has degenerate, spicule-like cheekteeth. Hence fossil teeth that represented a more 'primitive', less specialised ancestor might not be recognised as a tarsipedid. This same proviso applies to the apparent absences from the completely unknown, and numbats, whose teeth are degenerate. Perhaps we *do* have ancestors for these animals but have failed to recognise them for what they really are?

EKTOPODONTIDS

strangers in the forest

When the first ektopodontid molar turned up in the Leaf Locality of central Australia, Reuben Stirton, Dick Tedford and Mike Woodburne were utterly confused. It was, like when Thingodonta first appeared to us, a most unexpected and incomprehensible thing. Their field notes for that day read like a table-tennis ball, bouncing from one mad idea to another! Their first thought was very exciting—that it was probably a long-surviving multituberculate mammal, a group that had vanished from the rest of the world millions of years before the age of the Kutjamarpu LF. But as quickly, the idea was abandoned when it was realised that the orientation of the structures that made it resemble multituberculate teeth were in fact running in the opposite direction. Soon they found themselves in a corner. Having eliminated all other possibilities, they concluded that its high coefficient of weirdity made it a likely candidate for the oldest and first known Miocene monotreme. And so it was initially described—*Ektopodon serratus*, a possible monotreme.

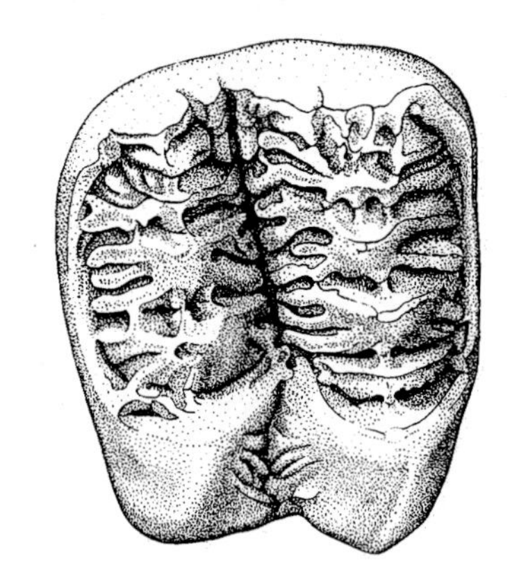

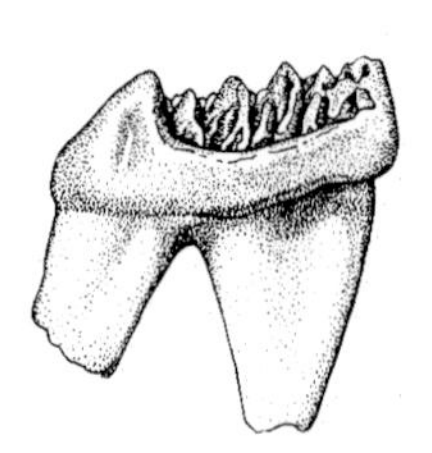

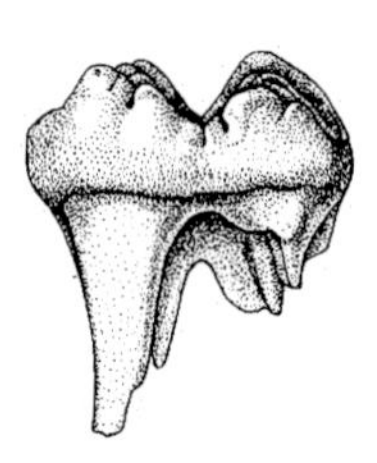

The distinctive teeth of ektopodontids are unmistakable. Riversleigh has produced two different kinds, both of which are similar to forms previously known from central Australia.

Much later, in 1972, Mike Archer and Mike Woodburne were sorting concentrates from late Oligocene deposits at Lake Palankarinna when yet more ektopodontid teeth turned up. But these, subsequently named *Chunia illuminata*, were much simpler in construction. Having recently examined a variety of possum molars, Mike Archer was struck by the similarity in a number of key features to the complex teeth of cuscuses. So when Neville Pledge finally discovered a nearly complete lower jaw from the Ngama LF of central Australia (*E. stirtoni*), it came as no surprise to discover that these animals were indeed possums that were at least distantly related to phalangerids.

Subsequent discoveries by Tom Rich and his

RECONSTRUCTION BY PETER MURRAY

The strangest possums of all are the brushtail possum-size ektopodontids. Their diet is a mystery. They barely survived into the early Pleistocene and then utterly vanished.

colleagues of ektopodontids from the Lake Frome basin Oligo-Miocene sediments (species of *Chunia* and *Ektopodon*), the early Pliocene Hamilton LF (*Darcius duggani*) and, most recently, from the early Pleistocene Portland LF (an unnamed form) of Victoria, added considerably to our understanding of the time and geographic range of these probably seed-eating possums.

So we were naturally puzzled about why the group seemed to be absent from the otherwise diverse Riversleigh local faunas of the same approximate ages. But before we could attempt to rationalise this absence, the first half tooth of an ektopodontid turned up from Wayne's Wok LF. It represented a species of *Ektopodon* at least superficially similar to *E. serratus*. Once the drought of non-representation broke, it tore loose! For just last year, from the newly discovered and stratigraphically low Dirk's Towers Site, we recovered a molar of a species of *Chunia*. Both genera are otherwise only known from the Oligo-Miocene sediments of central Australia. Now we await the first complete skull!

YINGABALANARIDS: 'WEIRDODONTA'!

When the mystery of Thingodonta became a little less of a mystery following discovery of the nearly complete skull, something else tantalising had to fill Thingodonta's place as a mystery. That something was not long in coming. From Upper Site came a single lower molar so odd that once again we could not convince ourselves that it was a marsupial, placental, or even a mammal as distinct as either of these two great groups. Although it has now been described as *Yingabalanara richardsoni*, the only member of the family Yingabalanaridae, we really have no idea what this thing is!

In the scientific paper describing the animal, we pointed out that it could represent one of six quite different kinds of creatures, from a long-surviving member of a group of Mesozoic mammals to a phyllostomid bat. But without more material, an upper molar, a jaw or partial skull, this one will remain a tantalising mystery.

A single, tiny tooth known from one of Riversleigh's most puzzling group of mammals, the yingabalanarids.

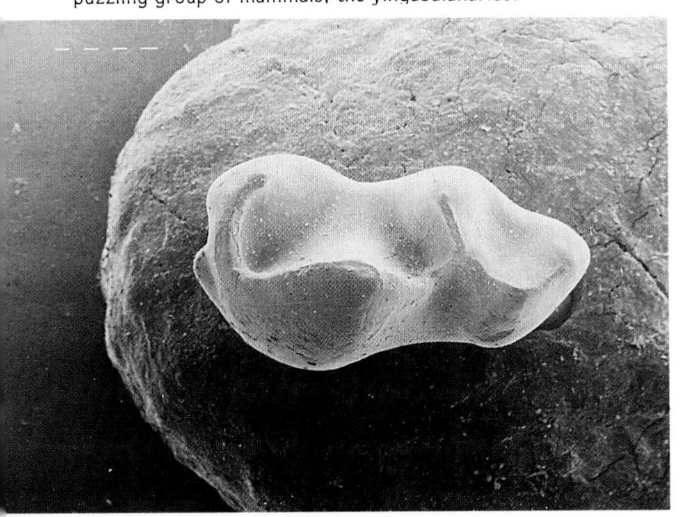

ETCETERA

more strange marsupials from a strange land

There are also a host of other kinds of marsupials in the Oligo-Miocene rainforest assemblages. Among the ones that have already been named are the pilkipildrids, a family of small possums whose relationships appear to lie with the petaurids or, perhaps, the phalangerids. A skull is required to settle the dispute. Members of this group from late Oligocene central Australia were previously misconstrued as 'giant pygmy-possums'. With discovery of the first complete jaw from Riversleigh's Last Minute Site (*Djilgaringa gillespieae*), their distinctive, non-pygmy-possum nature was instantly clear. Other new and even smaller Riversleigh pilkipildrids await description.

Another new family of possums is represented by distinctive jaws and numerous isolated teeth. This family appears to be distantly related to phalangerids. Many other previously unknown distinctive groups of possums also await study. Some appear to be distantly related to pseudocheirids and others to petaurids. Clearly, the Oligo-Miocene rainforests of Riversleigh teemed with an extraordinary variety of possums.

Also new and awaiting description are tiny ilariid marsupials, relatives of species of *Ilaria* and *Kuterintja* previously known from Oligo-Miocene sediments in central Australia. While the Riversleigh animal was probably no larger than a dog, *Ilaria illumidens* was the size of a small cow. These selenodont herbivores

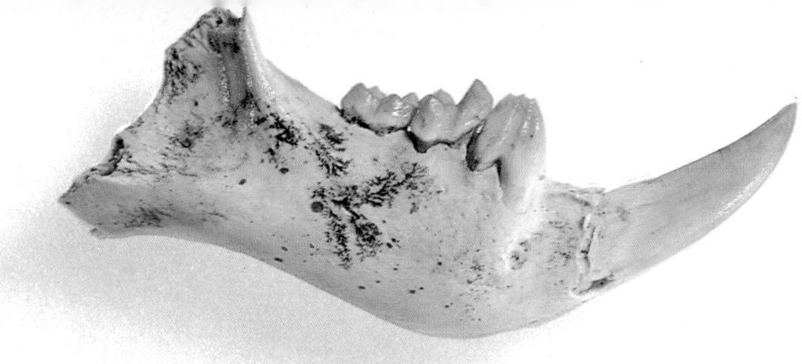

Pilkipildrids are yet another extinct family of bizarre possums. Riversleigh has produced one kind (a lower jaw is shown here); three others are known from different areas of central Australia.

R. ARNETT

probably ate the leaves of trees and it is not impossible that the Riversleigh species was arboreal although it would take a mighty tree to support *I. illumidens*! Before the lifestyles of these animals can be confidently interpreted, more information about their skeleton is needed. The relationships of the group are also

A reconstruction sequence of the Riversleigh pilkipildrid from Last Minute Site, *Djilgaringa gillespieae*, a small possum about the size of a Squirrel Glider.

A reconstruction sequence of an ilariid marsupial. This particular taxon found in central Australia was about the size of a cow. The Riversleigh ilariid is much smaller, about the size of a medium-size kangaroo.

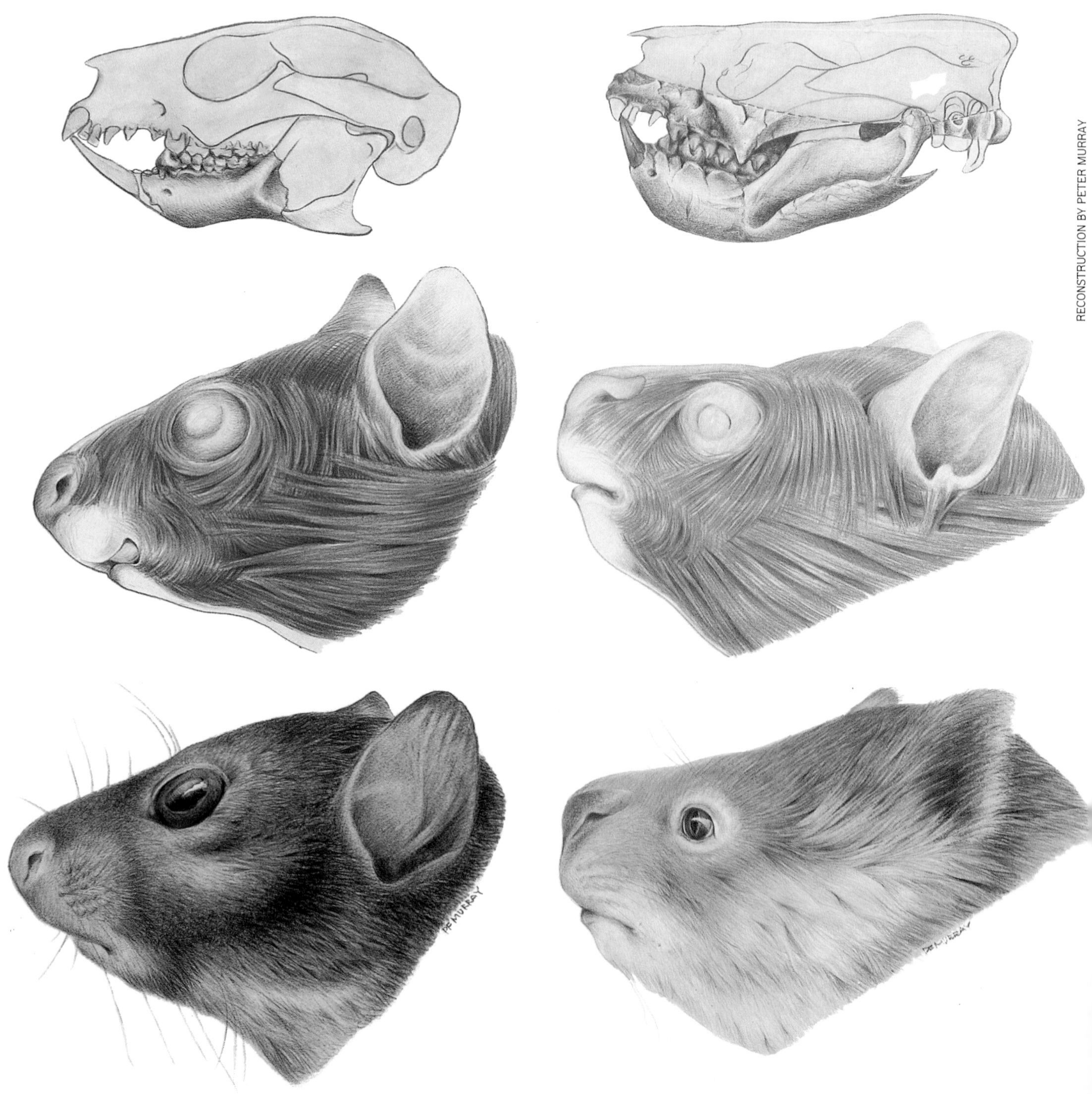

RECONSTRUCTION BY PETER MURRAY

obscure but they show distant similarities to wombats, as well as to koalas and other vombatiform marsupials.

Other new families of marsupials are known from Riversleigh, including a very strange, cow-sized herbivore from Hiatus A Site. So far we only have a partial palate with one tooth row. Although the teeth are very worn and difficult to compare with those of other kinds of marsupials, this beast is clearly distinct at a high taxonomic level from anything else we have seen. Unfortunately, the level that produced this specimen goes straight into an 8 metre-high cliff of extremely dense limestone—which has temporarily frustrated our efforts to obtain additional specimens.

BATS

rushes and whispers in the air of night

Millions of bat bones have been recovered from the Riversleigh fossil deposits. The tiny, delicate bones and teeth of bats are among the most common fossil remains found in Riversleigh's 100-plus fossil assemblages. The remains have been identified as belonging to more than 35 different kinds of bats; 23 of these have come from Riversleigh's many Oligo-Miocene deposits and 12 species from the Pliocene-aged Rackham's Roost deposit.

Many of these bats are represented by literally thousands of specimens each. The fossil material includes perfectly preserved skulls, jaws, and even natural casts of little bat brains. The fossil record for bats at Riversleigh is one of the world's richest, both in terms of number of taxa and individuals. The only other fossil deposits in the world that have produced so many extinct kinds of bats are the Tertiary cave and fissure fills of France, Germany and Austria.

Under certain circumstances, such as in caves and fissure fills, bats fossilise extremely well, but more typically bats are not likely candidates for fossilisation. Their bones are fine and fragile and tend not to survive transportation after death. Before Riversleigh's bat-rich limestone deposits were discovered in 1978, the Tertiary fossil bat record for Australia consisted of two isolated, unidentified teeth — one from late Oligocene lake deposits in the Tirari Desert of South Australia and the other from an early Pliocene soil near Hamilton in western Victoria. In 1988, another extinct bat, from the Pliocene-aged Big Sink carbonate deposit of Wellington Caves in New South Wales, was described.

Today, Australia's 65 species of bats make up a quarter of our native, non-marine mammal fauna. Eleven of the 65 species are fruit-, flower- and nectar-feeders, which include the flying-foxes and blossom-bats (megachiropterans). The other 54 are generally smaller, less conspicuous insect-eating bats (microchiropterans) although one of these, the Ghost Bat of northern Australia, also has a predilection for flesh.

All the fossil bats found so far in the Riversleigh deposits are microchiropterans. Most are what are commonly referred to as 'leaf-nosed' bats, named for the complicated flaps of

Twenty million-year-old skulls recovered from the Bitesantennary Site represent three different species of leaf-nosed or hipposiderid bat.

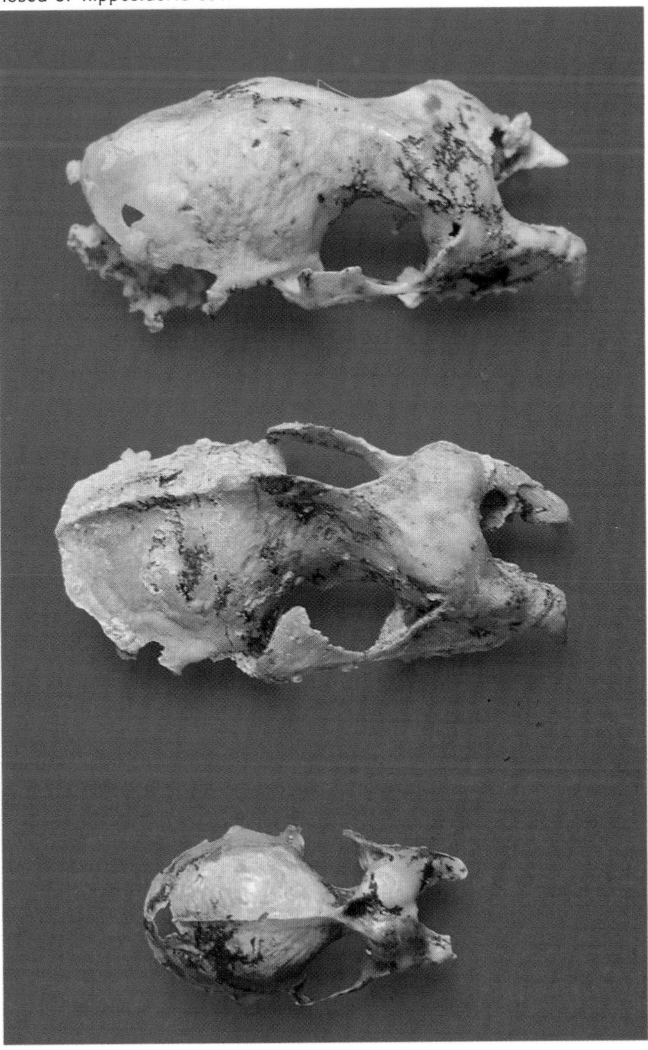

R. ARNETT

R. WILLIAMS

Dusky Horseshoe Bats hang from the roof of a warm, humid cave. Many such caves are developed in the limestone cliffs fringing the Gregory River.

skin that surround the nasal area. The fleshy 'nose leaves' help direct the pulses of ultra-high frequency sound that these bats use to detect objects and prey into a narrow beam or wide arc. As many as 100 pulses of sound per second are produced in the bat's larynx and, in this group, emitted through the nose (rather than the mouth). Special grooves in the leaf-nosed bats' large ears help channel the returning echoes that bounce off targeted objects. From this process of natural sonar or 'echolocation', information is gained about the size, shape, density, direction and speed of insect prey, objects and obstacles.

Hipposiderids. Riversleigh's Old World leaf-nosed bats, or hipposiderids, should not be confused with the only very distantly related New World or American leaf-nosed bats, the phyllostomids. Today about 60 species of Old World leaf-nosed bats, ranging in length (head and body) from between about 4 and 12 centimetres, live in caves and mines in tropical to subtropical areas of Africa, Asia, Australia and in the Pacific islands west to Vanuatu. They sometimes roost in very large numbers (in northern Australia colonies of over 40 000 have been recorded) or separately as solitary individuals.

Probably one of the greatest surprises about Riversleigh's leaf-nosed bats is the close relationship of a number of them to bats from French fossil deposits —'the French connection' (see Chapters 2 & 13). Six species of *Brachipposideros* are known from late Oligocene to middle Miocene fossil deposits in France; a further nine species have now been identified from Oligo-Miocene and Pliocene Riversleigh deposits. It is assumed that the distribution of the *Brachipposideros* group once included areas of Asia but remains from those regions are yet to be found.

A direct, living descendant of the Australian *Brachipposideros* bats is one of northern Australia's six living species of leaf-nosed bats, the brilliantly coloured Orange Horseshoe Bat (*Rhinonicteris aurantius*). Today, this rare,

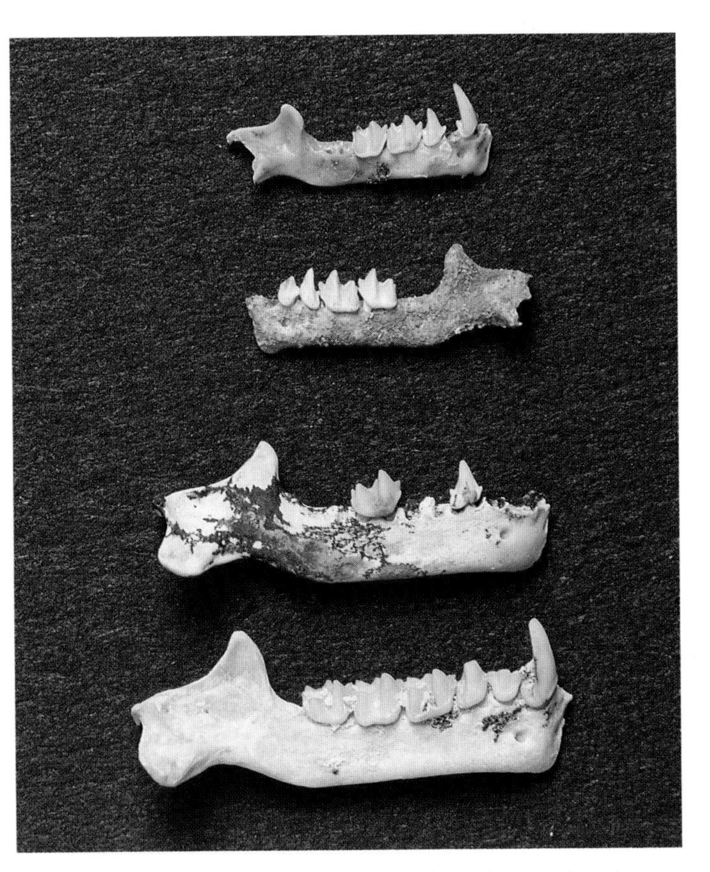

The size range of co-existing leaf-nosed bats in some deposits presumably enables them to specialise on different prey within the ancient Riversleigh rainforests.

The bat- and rodent-rich Rackham's Roost Pliocene deposit occurs as a pale fossilised guano pile (left) lapping against the remnants of the original cave walls (right).

endemic Australian bat lives only in extremely hot, humid caves, in which temperatures reach approximately 32°C and relative humidity 100%. Ancestors of the Orange Horseshoe Bat dominated Riversleigh's Oligo-Miocene bat faunas, evidently roosting in large numbers in the warm, humid limestone caves of the lush Riversleigh rainforests.

Sites spectacularly rich with the bones of *Brachipposideros* include Microsite—most of which belong to a single species, *Brachipposideros nooraleebus*, and Bitesantennary Site—the remnants of an old cave in which as many as five species of this group lived together. In many of the older Riversleigh sites, up to five species of *Brachipposideros* appear to have co-existed. By Pliocene time, as evidenced by the Rackham's Roost deposit, the numbers had dropped to four species. By recent times only the Orange Horseshoe Bat had survived the increased aridity that accompanied the later Pliocene and Pleistocene and it is now rare.

Megadermatids. The rare, flesh-eating Ghost Bat (*Macroderma gigas*) is another endemic Australian bat whose ancestors are well represented in the Riversleigh fossil deposits.

A piece of rain-etched Rackham's Roost limestone showing limbs and a jaw of Pliocene bats. The fragmentary nature of much of the material reflects the fact that it was chewed by carnivorous Ghost Bats.

With a wingspan of 60 centimetres and head-body length of 13 centimetres, the living Ghost Bat is Australia's largest microchiropteran. It roosts in caves and mines across northern Australia, emerging in the evenings to hunt for

G. HOYE

The flesh-eating Ghost Bat which today inhabits areas of northern Australia has a variety of ancestors in the fossil deposits at Riversleigh.

insects, other invertebrates and small vertebrates, such as fish, frogs, lizards, birds and small mammals, including other bats. It has been reported to take prey up to the size of a pigeon. Today, five species of ghost bats or false vampires (family Megadermatidae) live in tropical and subtropical areas of the Old World: two in Africa, two in Asia and one, the largest, in Australia.

At least five new kinds of false vampires have been identified from among the Riversleigh remains. In many Riversleigh deposits, at least one species is represented but in some there are two species, one normally smaller than the other. Presumably this difference in size between co-existing carnivorous ghost bats reduces competition for resources. Living Asian and African false vampires that co-exist over part of their range are either significantly different in size (e.g. the Asian *Megaderma spasma* and *Lyroderma lyra*) or represent distinctly different lineages (e.g. the African *Lavia frons* and *Cardioderma cor*). The latter in each of these species pairs consumes significantly more vertebrate prey than the other. In the Riversleigh Dwornamor Local Fauna, the two false vampires, *M. godthelpi* and the 'Dwornamor Variant', are similarly-sized but belong to quite different lineages.

For the most part, the Riversleigh fossil species are slightly smaller than the living *Macroderma gigas* but one, from the middle Miocene Gotham City Site on the Gag Plateau, is around the same size and is in fact the largest fossil bat yet recorded from Riversleigh. Gotham City Site appears to be the indurated floor of an ancient cave. It contains the remains of prey collected, consumed and discarded by resident ghost bats. These remains represent a wide range of prey items and include frogs, fish, skinks, birds, dasyurids, bandicoots, acrobatids and even a tiny koala. Like that of ghost bats living in eastern Australian rainforests today, the prey of the Gotham City ghost bat included a large number of arboreal animals. At Rackham's Roost, however, ghost bat prey is dominated by terrestrial rather than arboreal small vertebrates, such as small dasyurids and rodents in particular. The Rackham's Roost deposit dates from a time when the ubiquitous rainforests of the early and middle Tertiary had already retreated to the edges of the continent and woodlands covered great tracts of northern Australia.

Leaf-nosed bats and false vampires are the most common of all Riversleigh fossil bats. They are also common in European fossil deposits of similar age, particularly between Eocene and middle Miocene times. During this period, climates in at least southern Europe were warmer than today and forests covered large areas of the continent. In France, where six species of *Brachipposideros* are recorded from late Oligocene to middle Miocene sediments, the disappearance of the group coincided with the cooling and drying of southern Europe. False vampire bats appear to have disappeared from

Bats leave a cave at dusk to forage for insects as they have done throughout Riversleigh's ancient history.

The endemic Orange Horseshoe Bat has a long Australian pedigree as evidenced by many closely related fossil bats found at Riversleigh. The late Cainozoic decline in diversity of this kind of bat is cause for concern.

Europe a little later, at the end of the Tertiary.

In Australia, there has been no dramatic change in ghost bat diversity in contrast to the decline evident in the Orange Horseshoe Bat's lineage but the modern Ghost Bat's range has contracted alarmingly since human settlement of Australia, particularly since the arrival of Europeans and the many changes in Australia's natural habitats precipitated by their activities.

The rarer bats. Less common types of bats encountered in the Riversleigh deposits are molossids (free-tailed bats, also commonly called mastiff bats), vespertilionids (plain-faced bats) and emballonurids (sheath-tailed bats). Unlike the Old World leaf-nosed bats and false vampires, members of these three bat families are today found in the New World, as well as the Old.

Molossids. At least three extinct kinds of free-tailed bats have been found in Riversleigh's Oligo-Miocene deposits. The relationships of two of these are yet to be determined but those of the third species, *Petramops creaseri*, are better understood. Surprisingly, *P. creaseri* does not appear to be closely related to any of Australia's six living free-tailed species, all of which belong either to the cosmopolitan genus *Nyctinomus* or the Old World genera *Chaerephon* and *Mormopterus*. In contrast, it appears to have been part of a widespread more 'archaic' bat fauna that, as far as we know, left no living Australian descendants. Other members of this archaic Old World bat fauna would have included the false vampires and species of *Brachipposideros*.

Petramops creaseri seems to be most closely related to primitive free-tailed bats recovered

G. HOYE

from late Eocene-early Oligocene deposits in Europe (species of *Cuvierimops* and *Rhizomops*). Some palaeontologists believe that this lineage may still survive, represented by the living Mexican Free-tailed Bat (*Nyctinomus brasiliensis*) which is arguably the most primitive of all free-tailed bats. Today, Mexican Free-tailed Bats congregate in colonies of millions in the warm, low-latitude caves of Mexico. Like many free-tailed bats, they are exceptionally good long-distance fliers, some individuals seasonally migrating from Canada to Mexico, a distance of over 1300 kilometres.

Free-tailed bats are also direct, fast fliers that are less capable of manoeuvring in confined spaces than, for example, leaf-nosed bats. By modern analogy, *P. creaseri* probably would have hunted its flying insect prey primarily in open areas above or at the edges of the canopy rather than within the lower, more cluttered levels of Riversleigh's rainforests.

Vespertilionids. In Riversleigh's Oligo-Miocene deposits plain-faced or vespertilionid bats are so far represented by little more than a few isolated, unidentified teeth. In the Pliocene Rackham's Roost deposit, they were far more common, although they were still outnumbered by the 'archaic' hipposiderids. Today vespertilionid bats outnumber all others in the Australian bat fauna (more than 30 species) and usually comprise about half of the cave-dwelling species of northern Australia. Although yet to be positively identified (hopefully on the basis of more complete material yet to be recovered), the Rackham's Roost fossil vespertilionids appear to be most similar to species of *Scotorepens* and *Chalinolobus*, others of which roost today with leaf-nosed bats and the Ghost Bat in caves of north-western Queensland.

Emballonurids. One of the most common inhabitants of limestone caves in the Riversleigh area today are sheath-tailed bats. These large, very distinctive bats with smooth faces and big eyes, are often found spread-eagled against rock walls just beyond the daylit zone of caves, fissures and rock shelters. Sheath-tailed bats don't make their appearance at Riversleigh (or in the Australian fossil bat record) until Rackham's Roost time, around 4 to 5 million years ago, but appear in the world fossil record, in Europe, some 42 million years ago. In the Rackham's Roost deposit, two species of *Taphozous* are represented by many specimens, suggesting they were probably as common as the single species now found in the Riversleigh area, *Taphozous georgianus*. **But no Riversleigh rhinolophids.** Horseshoe bats or rhinolophids are close relatives of the Old World leaf-nosed bats and false vampires and are placed with them in the superfamily Rhinolophoidea. The horseshoe bats are characterised by very large ears and complex, horseshoe-shaped nose leaves. This group does not appear to be represented at Riversleigh in either Oligo-Miocene or Pliocene sediments. However, Australia's oldest bat fossil, the isolated bat tooth from late Oligocene deposits in the Tirari Desert of South Australia, may be a representative of this family. Today, Australia's only two species of horseshoe bats are restricted to forested areas of Australia's east coast. The diversity of horseshoe bats is very much higher in South-east Asia (e.g. the island of Borneo has ten species) and New Guinea (which has four) than in Australia. Thirteen species are known from European Oligo-Miocene deposits.

And no pteropodids! The only modern Australian group not yet known from the pre-Pleistocene Australian fossil record are flying-foxes, blossom-bats and other fruit-bats (pteropodids). The absence of this group is a surprise. After all, many of these are large and conspicuous elements of modern tropical to subtropical rainforest faunas (including those of Australia). They are present in fossil deposits of similar age in Eurasia and Africa. Further, many living flying-foxes readily fly long distances nightly to reach suitable food trees and they have radiated widely in the islands north of Australia. In fact, on at least some of these islands they are the *only* bat colonists. Today, in the not too distant rainforests of south-eastern Asia, small fruit-, blossom- and nectar-feeding bats (many of which live in the entrances of caves) play a major role in the pollination and dispersal of plants. If fruit-bats *were* living in the Riversleigh forests, we would expect them to be preserved in the fossil deposits because cave-dwelling has not been a prerequisite for fossilisation at Riversleigh—as evidenced, for example, by the thousands of Riversleigh possums, crocodiles, insects and birds now

G. HOYE

Fruit-, flower- and nectar-feeding bats, like this Spectacled Flying-fox of north-eastern Queensland, are conspicuously—and surprisingly—absent from the Riversleigh fossil deposits.

found as fossils.

It is interesting to note, however, that the modern Australian fruit-bat fauna, which is comprised of eleven pteropodid species (only one of which is endemic), is markedly depauperate compared with those of many of the islands north of Australia. The modern New Guinea bat fauna, for example, includes 25 pteropodids, 18 of which are endemic. The relative success of fruit-bats in New Guinea may be at least in part due to historical events. When pteropodids reached the islands that formed New Guinea in the early to middle Tertiary, fruit-, blossom- and nectar-feeding possums were probably also colonising from the south. On the other hand, when pteropodids arrived in Australia, they would have encountered a long-established and very diverse herbivorous, arboreal marsupial fauna. Today, roughly equal numbers of pteropodids and possums cohabit New Guinea rainforests, while in Australia possums far outnumber fruit-bats.

RODENTS

resilient invaders of the grasslands

In Australia today, 59 species of rodents comprise nearly a quarter of our native, terrestrial mammals. The world over, rodents are a spectacularly successful group of mammals—in species, they outnumber all other mammal groups combined. They include the squirrels, guinea pigs, porcupines, beavers, rats and mice.

All of Australia's rodents belong to the family of rats and mice, the Muridae. The other rodent groups apparently never made it to Australia, despite the fact that they have a much longer fossil history. Murids are native to the eastern hemisphere but a few species, i.e. the House Mouse (*Mus musculus*) and Black and Brown Rats (*Rattus rattus* and *R. norvegicus*) have been distributed worldwide by the activities of humans.

Australian murids are commonly split into two groups: the 'old endemics' (hydromyines) and 'new endemics' (murines). The 'old endemics' are almost exclusively confined to continental Australia and are believed to have evolved in the region. They include the false mice (species of *Pseudomys*), water rats (species of *Hydromys* and *Xeromys*) and rock-rats (*Zyzomys* spp.), among many others. The 'new endemics', which include the 'true rats' (species of *Rattus*), appear to have entered the region probably less than a million years ago.

Until the day Alan Rackham found the Rackham's Roost deposit perched on a cliff high above the Gregory River, only three Tertiary rodent fossils were known from Australian fossil deposits. These fossils were distributed in such a way that palaeontologists were led to believe that murids probably entered Australia just over 4.5 million years ago. The oldest rodent fossil was an incisor recovered from the radiometrically dated 4.5 million-year-old Bluff Downs deposit in north-eastern Queensland. From Chinchilla in south-eastern Queensland, two fossil rodent teeth had been found that were judged to be approximately 4 million years old. However, at Hamilton in western Victoria, a fossil deposit radiometrically dated at 4.5 million years, and rich in small mammal remains, has failed to produce any sign of rodents. On the basis of the available fossil evidence, it was surmised that murids colonised Australia from South-east Asia not long before the Bluff Downs deposit accumulated 4.5 million years ago and did not reach the southern part of the continent for at least another 40 000 years.

More recently, molecular data gathered in the course of studying modern Australian murid relationships has inspired another hypothesis to explain the origin and affinities of the Australian rodent radiation. Molecular evidence suggests that Australia's Old Endemic group had evolved by 15 million years ago and that by 10 million years ago representatives of modern genera were alive and well. Exactly where they were living — i.e. in Australia or on the islands to the north — was another question.

It is of interest to note here that the world's oldest murid, *Antemus chinjiensis* from Pakistan, which is considered very primitive and probably basal to the whole murid radiation, is only 15 million years old.

Enter Alan Rackham with a fist-sized lump of limestone studded with the teeth of seven different kinds of rodents — and, in another block, teeth of the extinct kangaroo *Protemnodon snewini*, previously only known from the 4.5 million-year-old Bluff Downs Local Fauna. The tally of new kinds of rodents in the Rackham's Roost deposit now stands at thirteen. Such diversity strongly suggests that murids had been residents of Australia for a long time before the Rackham's Roost deposit accumulated. However, we also know, from Oligo-Miocene deposits, that rodents were not in the Riversleigh area before 12-15 million years ago. Nor are they known from the middle Miocene Bullock Creek or late Miocene Alcoota Local Faunas of the Northern Territory or the late Miocene Beaumaris Local Fauna of Victoria. It seems likely that murids entered Australia in the latest Miocene, about 6-7 million years ago.

All of the Rackham's Roost rodents are now extinct but among them are ancestors of living taxa. The first of these to be recognised was a new species of rock-rat (*Zyzomys* sp.). Today, rock-rats are found across northern Australia. The Common Rock-rat (*Zyzomys argurus*) has the widest distribution of the five living species. It is always found in habitats associated with rocky outcrops, including the dry, rocky hillsides around Riversleigh Station. The remains of the fossil *Zyzomys* are the most abundant of all the

From a cliff 20 metres above the Gregory River, Alan Rackham Snr lowers a drum of fossil-rich limestone from Rackham's Roost to those waiting below.

Sledge-hammers and light explosives are used to collect the pink-coloured, tooth-filled limestone floor of the eroded cave at Rackham's Roost.

The Common Rock Rat is a specialised inhabitant of semi-arid areas in the northern tropics. A related species lived in the Riversleigh area long after the ancient rainforests had vanished and was found in the Pliocene Rackham's Roost deposit.

mammals recovered from Rackham's Roost and it is likely that these animals occupied the same hillsides their cousins occupy today.

In some of the many caves that occur in the cliffs of the Gregory River, the prey remains of modern Ghost Bats can be found. That the hapless rock-rat is still a favourite on the menu is evident from the large numbers represented in these deposits. It seems that habitats and habits along the Gregory River have not changed greatly over the last 4 million years.

In fact, habitat preferences for many of the Old Endemics may have been established well before the first murids entered Australia, and perhaps have hardly altered since. The ancient Eurasian *Antemus* and its relatives were mostly inhabitants of open woodlands and grasslands. Towards the end of the Miocene, when murids probably entered Australia, the continent was drying and the rainforests receding. Open forests and grasslands were becoming dominant, providing the kind of habitat early murids were used to. There was probably little competition from marsupials, which were just beginning to take full advantage of this relatively new habitat.

Among the other murids found in the Rackham's Roost deposit are at least six species of false mice (*Pseudomys* spp.). As in Australia today, the genus *Pseudomys* was the most diverse group represented. Also found have been remains belonging to a tiny species of the closely related genus *Leggadina*.

Not all the rodents present in Rackham's Roost belong to extant genera and these may yet turn out to be the most informative. It is possible that the extinct genera might represent the last of Australia's earliest rodent invaders and hence may be key to understanding the origin of Australian murids. So far, no close relatives of Australia's Old Endemics have been found in the regions surrounding Australia. This could be because the original stock that gave rise to the Australian radiation has been displaced by a second invasion of more advanced rodents. Alternatively, those first murids may have radiated rapidly in Asia and their living descendants are now unrecognisable as close relatives of the Old Endemics.

Rackham's Roost does not hold a monopoly on fossil rodents at Riversleigh. There are many caves in the limestone cliffs flanking the

The Forest Mouse, whose relatives occur in the Rackham's Roost deposit, is today an inhabitant of arid habitats in north-western and central Australia.

Gregory River. Some of these are mere overhangs but others are probably not unlike Rackham's Roost might have been in the Pliocene. In these caves, breccias preserving late Pleistocene and Quaternary mammals have been found. These represent essentially modern species, most of which are still present in the area today. Commonly represented in these deposits are 'new endemics', such as *Rattus* species, one of which, the Long-haired Rat (*Rattus villosissimus*), we often encounter live near our camp site on the Gregory River.

HUMANS

last but definitely not least of Riversleigh's native fauna

So far we have not encountered pre-'modern' human remains in any of the fossil deposits at Riversleigh. But, that may change. There are two deposits from which we may fairly expect to find indirect, if not direct, evidence of humans: Terrace Site and Carrington's Cave.

Sue Solomon, an anthropologist who visited and helped excavate Terrace Site in 1987, suspected the possibility that some of the small chipped chert pebbles forming part of the bed of river sediment that contained the bones of *Diprotodon optatum* and other Pleistocene animals had been 'worked'. The flaked edges of these pebbles could be the result of deliberate pressure flaking by humans intent on making stone tools. However, flaked stones can also be produced by other means such as large animals stepping on them while they are resting on other larger stones or by pebbles being naturally tumbled against each other as racing water rolls them along. Perhaps it will require the discovery of actual human remains in this deposit to make the pebble tools appear more plausible.

The second place where we are likely to encounter old humans is Carrington's Cave but work here has not yet begun. Therefore, the possibility of human remains being discovered here is just reasonable speculation. In any event, when excavation of the deposits in Carrington's Cave begins, it will be by a team of diverse specialists experienced in the recovery of archaeological and anthropological, as well as palaeontological remains.

WHEN, IF EVER, WILL IT ALL END?

And so it goes, every year a new beast or five, more informative specimens of creatures already known and new localities for taxa previously thought confined to a narrower suite of sites. Certainly it is true that the number of new species found each year has never been as high as it was in the first three 'great' years of 1983 to 1985. But new creatures *do* drop out of the blocks at a continually exciting pace. Further, having developed ideas about the nature of the new creatures encountered, when more complete specimens turn up, such as the skull of the carnivorous kangaroo *Ekaltadeta ima*, opportunities arise to test initial interpretations which may have been developed using little more than a single jaw.

The process of documenting and understanding the diversity of creatures present in Riversleigh's fossil deposits will go on for centuries, just as it has done in the major fossil fields of other continents. What we have so far discovered about prehistoric biodiversity at Riversleigh is only the beginning.

CHAPTER 5

Why Riversleigh is significant

Ten Good Reasons

While it is certainly true that much of Australia's vertebrate story must still be composed from widely separated fossil deposits, Riversleigh's sediments are providing a lot of information about its last 25 million years. If we were to summarise their main value in an Australian-wide context, at least ten points should be made.

1. The problem with constructing hybrid puzzles

Although there are many different-aged faunal assemblages from Australia, very few occur within the *same* area. Bluff Downs in north-eastern Queensland produced an important Pliocene local fauna but so far has revealed nothing about the faunas that led up to or followed it. Hence we cannot tell if Bluff Downs' Pliocene fauna is a 'relict' for its time, persisting in the area because it retained a patch of lusher habitat, or whether it is characteristic of a widespread and 'typical' Pliocene biota. Similarly, Geilston Bay produced a tantalising pile of early Miocene bits but nothing preceding or following this assemblage.

Problems of this kind forced palaeontologists to compile a continent-wide picture of faunal evolution by bringing to the puzzle pieces that were often gathered from widely separated areas. To do so involves making the potentially wrong assumption that contemporaneous faunal assemblages were similar across the continent. Modern Australia provides a stark demonstration that regional but contemporary differences can be profound. For example, the modern rainforests of the Atherton Tableland contain a host of relatively 'primitive' animals, such as the Musky Rat-kangaroo, whereas south-western Victoria is today inhabited by relatively derived, or specialised rat-kangaroos such as the Long-nosed Potoroo.

In Riversleigh, fossil-bearing sediments of many different ages are superimposed, one upon the other spanning the period from late Oligocene to Holocene times. In this way it is obviously easier and more meaningful to reconstruct changes in terrestrial ecosystems within single regions. Unfortunately, some pieces of Riversleigh's regional puzzle must, nevertheless, be imported, such as our understanding of the late Miocene and probably the late Pliocene which are at present missing from these sequences. Future discoveries at Riversleigh may fill in these gaps.

A similar situation exists in central Australia where late Oligocene sediments underlie Plio-Pleistocene sediments which in turn underlie Pleistocene sediments.

BITESANTENNARY SITE

For many of the creatures living in the Riversleigh rainforests during the early Miocene—perhaps 20 million years ago—holes in the forest floor became passageways to eternity. The limestones needed only the trickle of water, charged with the acids of decaying leaves, to erode into caverns beneath the mighty forests. Within days of the surface collapsing under its own weight, insect and flesh-eating bats colonised a newly exposed cave's walls and ceiling. Soon pythons came, finding easy prey in the hungry bats that emerged from the cave in a rush to feed at night. Water trickled over the rims of the cave creating crystalline decorations and accumulating below as pools which nourished vast colonies of snails and algae. As hundreds of generations of bats were born and died, they produced their own kind of steady 'rain' which accumulated in the pools below as dark domes of bone-rich guano. Momentarily distracted bandicoots and other creatures of the forest swan-dived into the pool to add their bones to what was to become the spectacularly rich Bitesantennary fossil deposit.

1 Pythonid snake.
2, 5, 16-17 Leaf-nosed bat (Hipposideros sp. 1)
3-4, 7, 10 Carnivorous ghost bat (*Macroderma* sp.)
6, 24, 27-30 Leaf-nosed bat (Hipposideros sp. 2)
8 Cicada (insect).
9 Marsh frog (*Lymnodynastes* sp.).
11 Gekkonid lizard.
12 Insectivorous dasyurid marsupial.
13 Threads of 'glow worms' (beetle larvae)
14 Collapsed blocks of what was previously the roof of the cave.
15 Guano piles, filled with bones and snail shells, accumulating beneath bat roosts—the eventual fossil deposit of Bitesantennary Site.
18 Moth.
19 Small bandicoot marsupial headed for immortality.
20 Leaf-nosed bat (Hipposideros sp. 3)
21 Leaf-nosed bat (Hipposideros sp. 4)
22 Stalagmites forming under lip of overhung edge of collapsed cave.
23 Lime-rich freshwater pool that filled Bitesantennary Cave.
25 Stork (ciconid).
26 Roots of trees that have eroded their way down through the limestone.
31 Tunnel possibly linking Bitesantennary Cave to Neville's Garden.
32 Massive System A limestone walls of cave.

Dunphy

2. Crossroads in time

The Oligo-Miocene deposits of central Australia contain a number of groups of vertebrates that could be described as relatively archaic. That is, they appear to be Gondwanan groups that either failed to persist into the modern world or are forms that were subsequently replaced by relatively specialised members of those groups. Examples of extinct groups include: meiolaniid turtles, ilariids and wynyardiids; examples of replaced groups include balbarine kangaroos and 'VD' bandicoots.

In most of the central Australian deposits these Gondwanan groups rarely occurred alongside the more specialised forms. In contrast, the oldest Oligo-Miocene Riversleigh local faunas (Systems A and B) contain the old as well as the new. Here, for example, are the last known ilariids and wynyardiids in the same faunas as pre-wombats. Archaic ringtail possum genera, such as *Pildra* and *Paljara*, occur here with the first known representatives of modern ringtail groups, such as *Pseudochirops*. Similarly, 'primitive' dasyurids overlap with the first representatives of more modern *Murexia/Antechinus*-type dasyurids.

By the time Riversleigh's System C local faunas were being accumulated, many of the more 'primitive' groups (e.g. ilariids, wynyardiids and *Ngapakaldia*-like palorchestids) had disappeared from the region. And by the time its Pliocene faunas had accumulated, almost all of the archaic groups had vanished.

As a result, Riversleigh's parade of faunal assemblages enables us to watch the last of the 'old regime' phasing out as it was gradually replaced by the new—the inevitable consequence of evolutionary processes that never-endingly convert the past into the present, on its way to the future.

3. Refining immigration schedules

Throughout the last century, Australian biogeographic history has appeared at any one time to be a relatively simple matter. Before the 1960s, when most geologists and biologists were convinced that the world's continents were all more or less as we find them now, everyone presumed that Australia received its immigrants via south-eastern Asia. Marsupials were presumed to have come in from the north despite the fact that at the time not a single fossil, let alone living marsupial, was known from the whole of the Asian subcontinent. Then, as plate tectonics unfolded its mysteries and the world's surface became a kaleidoscope of shifting lands, Australia was understood to be an alien invader from the south, a seceded section of ancient Gondwana—and marsupials were presumed to have come from the south, dragged up through the Indian Ocean on a colossal ark that moved north at a rate just slower than the growth of human hair. As a result of the presumed structure of the Indian Ocean, we visualised Australia creeping up through a vast expanse of otherwise featureless southern seas for nearly 30 million years after its final separation from Antarctica. For this reason, Australia's biotic isolation, before its 'crash' into south-eastern Asia approximately 15 million years ago, was more or less regarded as an inviolable axiom. Accordingly, it was always presumed that bats and rodents dispersed into Australia sometime this side of the 'crash' date.

Now, as a result of recent studies of the island arcs of south-eastern Asia, it seems clear that the 'vacant ocean' concept was wrong. Recent palaeogeographic studies have suggested that an island arc (including Timor) dotted its way between Asia and the north-western corner of Australia well before the latter separated from Antarctica in the Eocene.

Riversleigh's fossil deposits demonstrate that at least four families of bats—

megadermatids, hipposiderids, molossids and vespertilionids—were in Australia by 20 to 25 million years ago. How much earlier than this bats reached Australia is still uncertain but it now seems highly likely that they were here even before Australia separated from Antarctica.

Unfortunately, Riversleigh's record does not give us as much useful information about the time of colonisation of rodents. Before the discovery of rodents in the Pliocene Rackham's Roost, teeth from Bluff Downs in north-eastern Queensland had demonstrated that they were in northern Australia by the early Pliocene. Although the Rackham's Roost rodent assemblage is highly diverse thus suggesting a considerable prior time to this for its diversification from a common stock (if such a thing actually occurred), we cannot as yet refine how long prior to the early Pliocene rodents first entered Australia because Riversleigh appears to lack a late Miocene record.

Riversleigh's late Oligocene to early Miocene fossil deposits provide us with Australia's earliest records for other 'northern' animal groups such as the poisonous elapid snakes which are otherwise widespread throughout the Old World as, for example, King Cobras and Banded Kraits. Similarly, Riversleigh's Quaternary history of vertebrates in northern Australia may also help to clarify the time of entry of Humans and Dingoes.

4. Heritage of a changing forest

Because Riversleigh has already produced more than 100 assemblages that accumulated in rainforests between approximately 15 to 25 million years, it is Australia's best record of sequential change within this ecotype. As this record unfolds, it has also become clear that the creatures of these lowland forests were the progenitors for almost all of Australia's living animals, from those that still persist in its remnant rainforests to those that dart through its grasslands and burrow through its deserts. This is the 'Green Cradle' concept. As palaeontologists who see the present as an evanescent horizon between the past and future, Riversleigh's parade of life tells us that Australia's surviving rainforests are more than just beautiful remnants of a once green continent. They contain many of the descendants of the 'seminal' creatures that spawned thousands of new creatures to rapidly fill a continent that had become 44% arid. These rainforests may be the best genetic resource we have left to do the same thing again if we lose our deadly game of balancing the world's ecosystems on the threshold of disaster.

5. Biodiversity's finest portrait in green

Because of Riversleigh's Oligo-Miocene cloak of rich lowland rainforest, its faunal assemblages of these times are the most diverse of any known from the continent. In fact, to match this diversity for some groups, it is necessary to prowl through the crowns and forest floors of some of the less devastated Indo-Malay lowland rainforests, such as parts of Sarawak or Kalimantan.

If we consider just family-level diversity of mammals, Riversleigh's older rainforest biotas contain almost twice as many families as *any* of Australia's living biotas. We interpret the high diversity of animals to be a reflection of high plant species diversity. For many species of animals to be able to live in the same place, they must have been able to specialise on different food resources or some other aspect of the shared environment. Because many of Riversleigh's fossil biotas come from very restricted deposits, it seems probable that their high diversity was characteristic of relatively small areas. In Malaysia's lowland rainforests, 835 distinct tree species have been recorded from a single 50 hectare patch of forest. This is the kind of plant diversity we presume characterised the ancient lowland rainforests of Riversleigh, a forest filled with distinctive 'patches' that invited and enabled the Kingdom Animalia to respond in kind.

6. Understanding the neighbourhood and community relations

In some fossil vertebrate sites, such as the Leaf Locality of South Australia—source of the important Kutjamarpu LF, where bones and teeth are stream-worn—it is clear that these were gathered by water from a much wider source area than the fossil deposit itself. In contrast, because most of Riversleigh's Oligo-Miocene faunal assemblages contain only delicate, unworn bones and teeth, it is clear that they accumulated in situ without significant transportation and are almost certainly fairly representative of the community immediately surrounding the fossil deposit itself. They may, therefore, provide one of the best opportunities to examine the evolution and structure of Australia's Cainozoic rainforest vertebrate communities.

Further, it should be possible, by comparing presences, absences and abundances in the more than 150 local faunas, to determine or clarify patterns of species- and generic-level interactions within Australia's rainforest communities. These patterns may still operate in Australia's living ecosystems but are difficult, if not impossible, to understand without experimentally interfering with these ecosystems. By understanding these probably very complex patterns, it should be possible to better manage and, if desirable, cleverly manipulate modern environments to sustain maximum levels of biodiversity.

7. 15 million years of natural deforestation

Riversleigh's deposits span and pass the period when Australia's great continental rainforests largely vanished and document the profound effects this change had on Australia's terrestrial faunas. They provide a graphic record of the often contradictory fates of particular groups of vertebrates as the closed forests gave way to more open forests and eventually grasslands. For example, a wide range of browsing and even carnivorous kangaroos in the older Riversleigh rainforests gradually gave way to an explosion of grazing, mainly non-rainforest kangaroos like those that dominate Australia today. Similarly, although rhinolophoid bats dominated the ancient rainforests, as this habitat declined the balance of bat power shifted to the vespertilionoid bats that now dominate the world as well as most of Australia.

8. Measuring life's rates of change

Recent evolutionary debates have focused on the *rates* at which evolutionary change takes place. Some biologists have suggested that these are reasonably constant through time; others that evolutionary change has been 'stochastic', occurring in leaps and bounds punctuated by long periods of relatively little change. Related debates focus on questions about group-specific differences in the rates of change. For example, do species of kangaroo 'turn over' or change at a faster rate than possums? Unfortunately, the antagonists in these debates often base their arguments upon depressingly little hard data.

Because Riversleigh's fossil deposits span, albeit unevenly, the last 25 million years and because they contain many groups that have long ranges through time it should be possible to determine rates of evolutionary change within these groups and shed light on what may be group-specific differences in these rates.

Where the fossil record begins to improve, we may also find support for or challenges to hypotheses about the antiquity of Australia's creatures as determined by the 'molecular clock' concept. This concept assumes that the number of structural differences between the same molecule (e.g. albumin—a protein molecule) in different creatures will be a measure of the amount of time elapsed since both creatures shared a common ancestor. Although a reasonable idea and one commonly supported by the fossil record (e.g. albumin suggests that humans and chimpanzees shared a common ancestor about 6 million years ago, a date which the fossil record so far supports), this 'clock' does not always appear to 'keep good time'. For example, although a 'molecular clock' date of 8 million years was suggested for the common ancestor of thylacine (*Thylacinus cynocephalus*) and dasyure albumin, Jeanette Muirhead is busy describing at least five *different* kinds of thylacinids from 15-25 million-year-old Riversleigh deposits—7 to 14 million years before there should have been *any* thylacinids.

A similar thing happened with discovery of elapid snakes in Riversleigh's Oligo-Miocene deposits. According to recent studies of the protein transferrin, the elapid snake radiation in Australia is no older than 10 million years old. Discovery of 15-25 million-year-old elapids at Riversleigh does not appear to fit comfortably with the 'molecular clock' date and suggests that this particular clock may be 'running too fast'.

9. Humans and their introduced commensal friends

Most of our understanding about the effects of the first human invaders on Australian environments comes from the southern half of the continent. What we know about these events in northern Australia, the supposed first stepping stone of 'man the destroyer', is by comparison negligible. The little that is known is largely confined to the very end of the Quaternary (the last 10 000 years). Part of the problem is the relative scarcity of suitable deposits. Caves and river basins in the seasonally wet tropical regions are often flushed clean of the previous year's deposits by monsoonal floods. In the drier regions of the north, there are few caves or other traps to accumulate fossiliferous sediments in the first place. The few exceptions include a handful of caves in north-eastern Queensland (e.g. Tea Tree Cave at Chillagoe), a fluviatile deposit in Windjana Gorge and shelters in Arnhem Land.

Riversleigh's Pleistocene and Holocene deposits will enable clarification of the initial and subsequent effects humans had on the 'virginal' biotas of northern Australia in three main ways. First, the Terrace Site and other fluviatile fossil deposits left by the ancestral Gregory River and its tributaries are distinctive Pleistocene sites at least some of which contain 'megafaunal' creatures such as *Diprotodon optatum*. With some of these are pieces of chert that resemble human artifacts, an association that, if confirmed, may elucidate early human/faunal relationships along the doorstep of the continent.

Second, Riversleigh's Holocene deposits include 'pre-' as well as 'post-Mouse' faunas which should help clarify the regional effects of the arrival of Europeans with their inevitable companion, the House Mice.

Third, because some of Riversleigh's Quaternary sites are complexes of sediments that accumulated in different ways, they may reveal different but contemporaneous views of the effect of humans on the biota of the region. Carrington's Cave, for example, has two quite distinct sedimentary deposits: an outer entrance cave chamber with clear evidence of human activity at the surface;

and an inner chamber with a thick bone-rich deposit representing the remains of prey accumulated over a long period of time by carnivorous Ghost Bats. In this case, the palaeontological/anthropological record stored in the outer chamber will hopefully reveal the history of interaction of humans and the 'megafauna' while the record in the inner chamber ought to reveal a temporally overlapping view of changes in the 'microfauna' of the same region.

10. Timing the events of Australia's marathon of life

Because Riversleigh contains many different-aged deposits with high biological diversity, its biostratigraphic sequence can be used as a framework of interrelationships linking many of Australia's otherwise poorly correlated faunal assemblages. For example, the Kutjamarpu LF of the Tirari Desert is *unlike* the vast majority of other central Australian assemblages to which it is geographically close and hence very difficult to place in a local stratigraphic context. In contrast, it shares very distinctive taxa, such as the rat-kangaroo *Wakiewakie lawsoni*, with Riversleigh's Upper Site LF from System B.

Geographic correlation of this kind forms much of the basis for temporally relating Australia's isolated Tertiary terrestrial assemblages within and beyond Riversleigh. Although there are circular-reasoning traps for the unwary here, the technique seems to work reasonably well. As a consequence, we are building a framework in space and time within which Australian biological events can be understood in terms of each other and in terms of developing knowledge about the history of this continent's geological and climatological trends.

Riversleigh has also demonstrated how the Cainozoic records of other continents can be woven together with those from Australia. Because of fortuitous overlaps in the kinds of bats represented in the Oligo-Miocene Riversleigh LFs and those in dated non-Australian assemblages, it is possible to extend understanding about the non-Australian deposits from those at Riversleigh and vice versa.

CHAPTER 6

The Australian Fossil Record

A Broader View

It is often said that Australia is 'the most ancient of continents'. While strictly speaking this is not true, all of the Earth's continental material being of roughly the same age, Australia has a fossil record of great antiquity. Although this book focuses on the fossil deposits of Riversleigh, it is important to place these into a broader, continent-wide context.

Some of the world's oldest life forms (3.5 billion years old), single-celled creatures known as cyanobacteria that built vast but simple reefs, have been found at North Pole in Western Australia. Younger but no less significant, the late Precambrian Ediacaran fauna from the Flinders Ranges in South Australia, approximately 650 million years old, is the best preserved and biologically most informative of any of the world's rare Precambrian faunas. Late Silurian plants including *Baragwanathia* from Victoria, approximately 415 million years old, were among the first to conquer the land and among the most diverse of any in the world. Spectacular late Devonian fish sites such as those on Gogo Station, about 370 million years old, in the Kimberley region of Western Australia are unsurpassed in the world in terms of diversity and preservation. The early Cretaceous vertebrate faunas from Cape Patterson and the Otways in southern Victoria, approximately 110 million years old, are among the most important in the world in terms of understanding how dinosaurs survived in extreme climates—in this case within the Antarctic Circle when mean annual sea temperatures were less than 5°C in this area. Similarly, the early Cretaceous opal deposits of Lightning Ridge have provided a wealth of dinosaur, crocodile and other vertebrate remains including the world's oldest known monotreme. The early Tertiary Murgon Local Fauna from south-eastern Queensland is the only terrestrial mammal community known that existed before Australia peeled away from Antarctica and, as such, is providing unique information about early intercontinental biogeographic relationships of Australian vertebrates.

Hundreds of other Australian fossil deposits have provided important information about either periods of time so far unrepresented at Riversleigh or contemporary but regionally or ecologically distinctive biotas. In particular, at Riversleigh we have so far failed to find deposits containing vertebrates of pre-Oligocene age, late Miocene or late Pliocene age. Riversleigh's primary contribution comes from its deposits of late Oligocene, early and middle Miocene, early to middle Pliocene, Pleistocene and Holocene age. If it has rocks containing vertebrates from other periods of time, we have yet to find (or recognise!) them.

Here we summarise some of the high as well as the low points of the Australian terrestrial vertebrate fossil record.

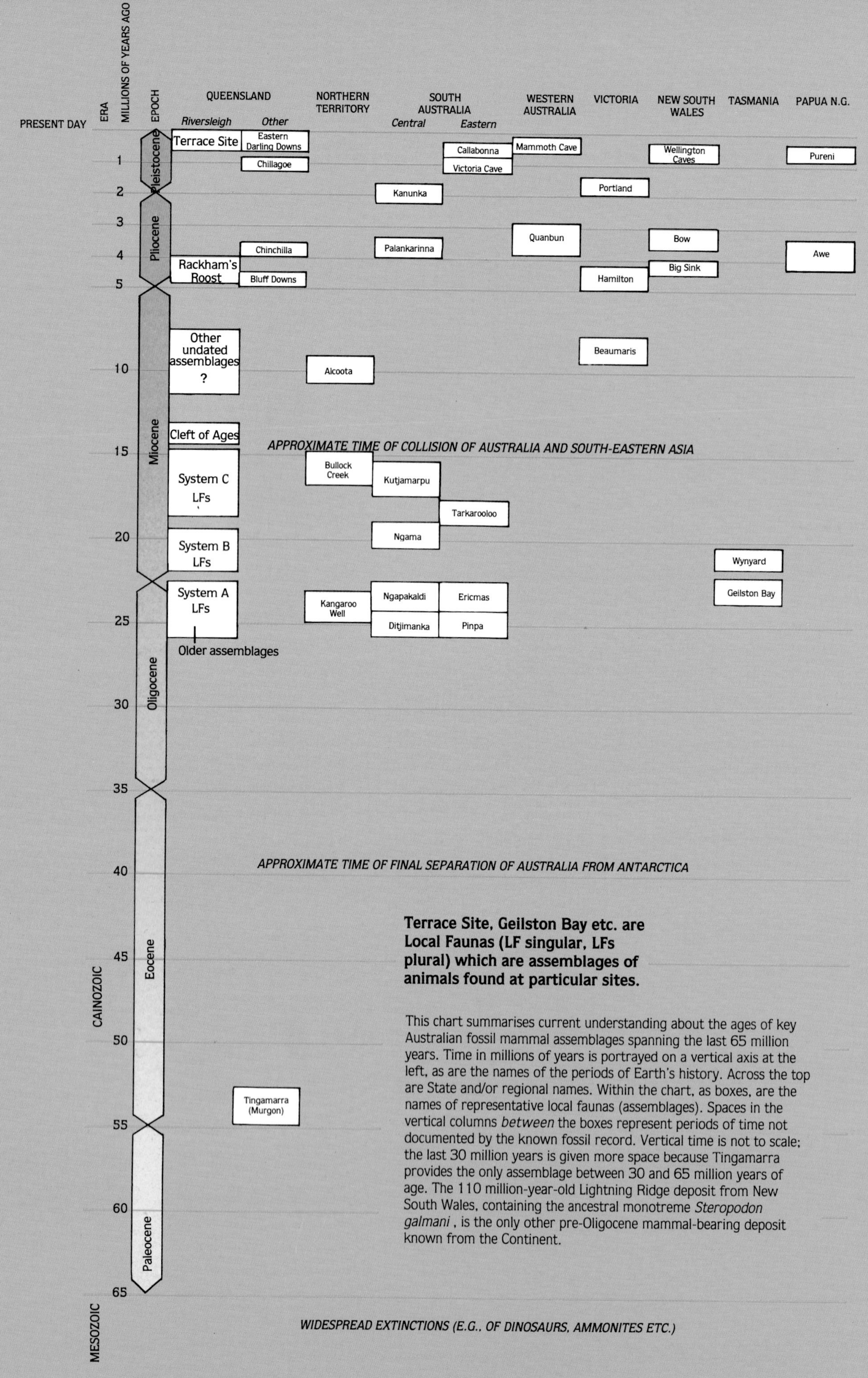

Terrace Site, Geilston Bay etc. are Local Faunas (LF singular, LFs plural) which are assemblages of animals found at particular sites.

This chart summarises current understanding about the ages of key Australian fossil mammal assemblages spanning the last 65 million years. Time in millions of years is portrayed on a vertical axis at the left, as are the names of the periods of Earth's history. Across the top are State and/or regional names. Within the chart, as boxes, are the names of representative local faunas (assemblages). Spaces in the vertical columns *between* the boxes represent periods of time not documented by the known fossil record. Vertical time is not to scale; the last 30 million years is given more space because Tingamarra provides the only assemblage between 30 and 65 million years of age. The 110 million-year-old Lightning Ridge deposit from New South Wales, containing the ancestral monotreme *Steropodon galmani*, is the only other pre-Oligocene mammal-bearing deposit known from the Continent.

· (e.g., Dredge's Ledge)

·· (e.g. Low Lion)

Amphibians. The history of Australia's amphibians and reptiles extends, albeit intermittently, back to the early Triassic. Spectacularly rich deposits, like those on Rewan Station, south-eastern Queensland, have produced a plethora of labyrinthodonts (extinct, vaguely 'salamander-like' amphibians), as well as a few small reptiles and even a mammal-like reptile, a member of the group from which all mammals subsequently evolved. Australia's frogs, however, lack a record prior to the 55 million-year-old Murgon deposits of south-eastern Queensland, despite having a patchy record in other parts of the world that extends back to the early Triassic (e.g. *Triadobatrachus* from Madagascar). The Oligo-Miocene deposits from the Lake Eyre Basin have provided representatives of the family Hylidae but it was not until the thousands of Oligo-Miocene frogs leapt from Riversleigh's rocks that representatives of many distinctive Australian frog groups made their first appearances.

A locality map showing the locations of the major fossil sites discussed in this book. Although most of the sites are in eastern Australia, this probably reflects the fact that the western half of the continent has been less thoroughly explored.

Early Cretaceous dinosaur deposits in Cape Otway, Victoria, accumulated at a time when this area of Australia was subject to near freezing temperatures.

Turtles. Turtles, which have a record back to the late (or possibly middle) Triassic of Europe, are unknown from the Australian record prior to the early Cretaceous. Further, these oldest Australian turtles are marine. The oldest freshwater turtle known from this continent is a single specimen of *Chelycarapookus arcuatus* from Cretaceous deposits in Victoria. The Cainozoic record of Australian turtles is even patchier until the early to middle Tertiary but it does include the now strikingly well represented 55 million-year-old trionychids from Murgon. Riversleigh's chelid ('ordinary') and meiolaniid (horned) turtles are about to provide a very good look at many of Australia's Oligo-Miocene turtles.

Crocodiles. Crocodiles are known in the rest of the world from deposits as old as the late Triassic but in Australia, their earliest appearance is in the 110 million-year-old early Cretaceous opal deposits of Lightning Ridge where *'Crocodylus' selaslophensis* occurs with *Steropodon galmani* (an ancestral platypus-like animal) and a variety of dinosaurs. However, the early and middle Tertiary record of crocodiles is rapidly improving with good material now known from Murgon and other early Tertiary deposits in south-eastern Queensland. Similarly, a number of excellent specimens have been recovered from the Oligo-Miocene deposits of Bullock Creek (e.g. *Baru darrowi*) in the Northern Territory and the Tirari Desert (e.g. *Australosuchus*) and Frome Basin of South Australia. But again, the *variety* and in many cases preservation of middle Tertiary crocodiles coming from the Riversleigh deposits is singular.

Lizards. The early Triassic Rewan *Kudnu mackinleyi* is Australia's and possibly the world's oldest lizard. Despite these early beginnings, the subsequent record of lizard history in Australia is relatively lousy—at least prior to the Cainozoic. There are a few possible early Cretaceous lizards and a marine mosasaurid from the late Cretaceous of Gingin, Western Australia. Cainozoic lizards did not pop up in the record until a handful of bits from the Oligo-Miocene deposits of central Australia were described as representatives of the modern scincid (skink) and varanid (goanna) families. When Riversleigh began to produce its wealth of Oligo-Miocene lizards a wide variety of modern, as well as extinct Cainozoic, lineages put in their first appearances. Plio-Pleistocene varanids in Australia included the largest land-dwelling lizard known, *Megalania prisca*, which reached lengths of at least 5 metres. So far, lizards of this size have not been found in any of Riversleigh's Cainozoic deposits, although we know they were present in northern Australia at the time because of occurrences in Pliocene and Pleistocene deposits in north-eastern Queensland (e.g. Bluff Downs LF and Wyandotte LF respectively).

Snakes. Australian snakes also have a lousy early Australian record. In the world they first appear in the Cretaceous but in Australia they first appear as madtsoiid pythons at Murgon. This group is unmistakably Gondwanan in origin with Cainozoic representatives in South America, as well as Australia. It persists through the Australian Cainozoic record into the Pleistocene, when it was represented by the giant *Wonambi naracoortensis* from, for example, Victoria Fossil Cave, South Australia. With discovery of the Oligo-Miocene Riversleigh deposits, the earliest Australian occurrences of boids (pythons) and elapids (poisonous snakes) turned up, as well as the earliest occurrences in the world of typhlopids (blind burrowing snakes). Of the other Australian snake groups, acrochordids (file snakes) make their first appearance in the early Pliocene Bluff Downs LF and colubrids ('ordinary' snakes) in the Pleistocene deposits of various caves and river sediments of eastern Australia.

Birds. The world's oldest undoubted bird is the late Jurassic *Archaeopteryx lithographica* from Germany, although there are older contenders for the throne among forms that have otherwise been interpreted as small dinosaurs. Australia's bird record is impressive not only for its variety but for its antiquity, with feathers coming from the early Cretaceous sediments at Koonwarra, southern Victoria and bones from western Queensland. Like all other southern and some northern continents, Australia's Cainozoic record is remarkable for its variety of giant flightless birds, although on this continent all were evidently herbivorous (unlike the ferocious, meat-eating phororhacoids of South America). The early to middle Tertiary record of Australian birds was, until Riversleigh's forest inhabitants began to appear, mainly made up of

Top: Cretaceous fossil deposits in Gingin, Western Australia, source of giant marine reptiles.

Above: Maureen and Jim Porter, owners of Tingamarra, one of the few early Tertiary vertebrate sites in Australia.

Top: At work in early Pliocene quarries in north-eastern Queensland, the source for the Bluff Downs LF.

Above: Lightning Ridge, N.S.W., source of Australia's oldest known mammal, the monotreme *Steropodon galmani*.

D. O'CARROLL

The late Pleistocene deposits of Victoria Fossil Cave, South Australia, are among the richest fossil deposits in the world. Work on the deposit has been continuous since its discovery in 1969.

penguins, dromornithids (large flightless ratites), emus and wading birds, such as flamingoes, ducks, rails, pelicans, waders, palaelodids (flamingo-like birds) and cormorants. These records were largely obtained from the Oligo-Miocene deposits of central Australia which must have accumulated in large lakes. In contrast, most of the Oligo-Miocene deposits of Riversleigh appear to have formed in small shallow pools within dense forest. As a consequence, Riversleigh's bird assemblages have few waders but abound in passerines and other less aquatic groups, as well as dromornithids and emus. As a result, the two records nicely complement each another and provide a more balanced view of the development of Australia's bird communities.

Mammals—an overview. In other parts of the world, mammals have a record that extends back to the late Triassic. For example, morganucodontids, among the most 'primitive' mammals known, come from 215 million-year-old deposits in England. Similar groups of primitive triconodont mammals are known from Africa, Asia and North America. Unfortunately, Australia's mammal record doesn't provide a hint of their nature until 110 million years ago, when a single jaw of the early Cretaceous platypus-like *Steropodon galmani* became entombed with the bones and shells of other Mesozoic animals in the opal deposits of Lightning Ridge. Apart from this minuscule glimpse, we lack information about the first 75% of what is probably a fascinatingly different history of mammals in Australia.

Not until the latest Palaeocene/earliest Eocene, do we get our next look at Australia's mammals with the 55 million-year-old vertebrate assemblage from Murgon. Here we have so far only identified a handful of different marsupials, as well as a primitive bat. But it is early days yet. Clearly we would not be surprised to find more marsupials, a monotreme, possibly a Gondwanan placental group and maybe a few representatives of mammalian groups present in the Cretaceous of South America. Then again, remembering that this is Australia, who knows what weirds might turn up as work progresses on this most tantalising of sites.

Following the Murgon window in time, our next view comes with a plethora of late Oligocene to middle Miocene sites in Tasmania (e.g. Geilston Bay and Wynyard), South Australia (e.g. Lake Palankarinna), the Northern Territory (e.g. Kangaroo Well) and Queensland (e.g. many sites at Riversleigh). Most of these appear to be 15-25 million years old. From this time on, the record is much better until the late Miocene, when only a few mammal-bearing deposits (e.g. Alcoota in the southern Northern Territory and Beaumaris in southern Victoria) fill in the vast gap between 15 and 5 million years. If there are sites of this age at Riversleigh, we have not, as yet, recognised them. But again, it is early days.

Considering all of Australia's late Oligocene to middle Miocene sites, most provide useful

Tingamarra has produced Australia's oldest marsupials and bats as well as the first indications of ground-dwelling placental mammals that competed, unsuccessfully, with marsupials.

Volunteers carefully quarry the green 55 million-year-old Tingamarra clays.

An early Miocene wynyardiid skeleton dropped out of marine sediments at Wynyard, Tasmania.

Late Oligocene clays and sands of Lake Palankarinna, South Australia, produced Australia's first-known Oligocene mammals.

Miocene limestones of Bullock Creek in the Northern Territory have produced a variety of creatures similar to others from Riversleigh.

Early Pliocene rainforest vertebrates from Hamilton, Victoria, were obtained from a soil just beneath the radiometrically dated basalt.

Pliocene vertebrates from Merriwa, N.S.W., were discovered when a road-cutting transected the fossil deposit.

Pleistocene deposits along Cooper Creek, South Australia, provided early explorers with the bones of flamingoes and other extinct creatures.

The historically important Pleistocene cave deposits in Wellington, N.S.W., contributed to Darwin's understanding of evolution.

information about the diversity and development of Australia's marsupials and a bit about its monotremes, but only Riversleigh's deposits are also packed with bats. Further, while the central Australian deposits contain the bones of many distinctive vertebrates, 99% of these are the disarticulated remains of teleost fish. In contrast, the fossil deposits at Bullock Creek and most of those at Riversleigh are dominated by terrestrial vertebrates making it a bit easier to interpret the actual structure of the terrestrial palaeocommunities of those areas. The lower latitude and higher biological diversity of Riversleigh's deposits compared with those from central Australia may also enable us to distinguish regional differences in the biotas across Oligo-Miocene Australia.

Australia's Pliocene mammal record demonstrates a number of widespread phenomena such as the rapid development of grass-eating kinds (e.g. kangaroos) and the development of gigantism such that many lineages are represented by animals much larger than those known from the Miocene. The Pliocene record is mainly documented from the eastern coastal area of the continent: the ?middle Pliocene Awe LF from Papua New Guinea; the early Pliocene Bluff Downs LF near Charters Towers, north-eastern Queensland; the middle Pliocene Chinchilla LF from south-eastern Queensland; early to middle Pliocene Merriwa LF from north-eastern New South Wales; the Big Sink LF from the Wellington Caves area of eastern New South Wales; and the early Pliocene Hamilton LF from south-western Victoria. The poorly known Quanbun LF from the Kimberley presents about all we know for the western half of the continent. Riversleigh's Rackham Roost LF represents the only Pliocene site for western Queensland and the only clear look at Australia's oldest rodent assemblages.

The Pleistocene record of Australian mammals similarly demonstrates a number of widespread phenomena including the increase in gigantism (with some lineages producing enormous representatives) and widespread late Pleistocene extinctions that resulted in the loss of almost a third of the kinds of animals present at the time. Pleistocene mammals are known from hundreds of sites, most of which are cave or lake deposits. These are known from all states but the vast majority come from the southern and eastern margins of the continent. The relatively rare Pleistocene sites from central Australia are mainly confined to fluviatile (river) deposits. Those from Riversleigh are cave as well as river deposits. However, none of Riversleigh's Pleistocene sites are particularly rich. Compared with the spectacularly rich southern sites such as Naracoorte (e.g. Victoria Fossil Cave), Wellington Caves, and the eastern Darling Downs, understanding of Pleistocene terrestrial vertebrate diversity in northern Australia has a very long way to go. Hopefully planned exploration of the newly discovered Quaternary sites in, for example, Carrington's Cave on Riversleigh will help to fill in a lot of the missing story for the northern half of the continent.

Monotremes. The egg-laying monotremes are so far known from nowhere else in the world other than Australia; so naturally the early Cretaceous *Steropodon galmani* is the oldest monotreme in the world! While Riversleigh has provided the only complete skull material of an Oligo-Miocene monotreme (*Obdurodon* sp.) and the only nearly complete dentition, late Oligocene deposits from central Australia provided our first glimpse of this kind of ancestral platypus (*Obdurodon insignis*). Now isolated teeth, a pelvic fragment and the rear of a jaw are known from late Oligocene to early Miocene central Australian localities.

Riversleigh has so far failed to provide any information about the early evolution of the echidnas, the oldest known representative being a late Miocene long-beaked echidna (*Zaglossus robustus)* from deep lead gold deposits in New South Wales. This lack in the Riversleigh deposits is a profound puzzle considering that *Zaglossus bruijnii* occupies rainforests in modern New Guinea and clearly representatives of the genus were in Australia by the late Miocene. Still, one serendipitous find could undermine whatever *ad hoc* explanation we might be tempted to offer for this curious absence!

Marsupials. There are interesting differences between the Oligo-Miocene marsupial assemblages of central and northern Australia. Those from central Australia were first explored by Reuben Stirton and his colleagues from the University of California in the 1950s and 1960s and we came to presume that they represented 'typical' Australian Oligo-Miocene vertebrate assemblages. In general, the first-known were characterised by low family-level diversity and, with the exception of koalas, dasyurids, bandicoots and kangaroos, most of the mammals recovered appeared to represent 'primitive' or at least highly distinctive groups that had failed to survive into the modern world.

Late Pleistocene fossil deposits of the Darling Downs, Queensland, provided some of the first-known marsupial lions and giant diprotodontids.

Errol Beutel prepares a section of bone bed from the richly fossiliferous late Miocene deposit at Alcoota, Northern Territory.

Although incredibly flat, the clays of Lake Pinpa contain some of the most extraordinary marsupial skeletons ever discovered.

Michael Woodburne excavates a 'wash pit' to field-process fossil sediments from the Miocene Leaf Locality in the Tirari Desert, South Australia.

An exception to this general picture came with the discovery by Stirton and his team, in 1962, of the early to middle Miocene Kutjamarpu LF from the Tirari Desert. Here biodiversity is much higher, with more still-living families represented, such as ringtail

Pliocene vertebrates from Chinchilla, Queensland, litter the ground near the banks of the Condamine River.

possums and wombats. It is probably a riparian accumulation gathered from a forest that surrounded a fast-moving stream and later a lake. What lay beyond the margins of the diverse Kutjamarpu forest that lined the river and lake is unclear but it may well have been less diverse communities of the kind that sired other relatively impoverished central Australian mammal assemblages.

When Riversleigh's System B and C deposits began to yield their goodies, our first impression was that these local faunas *most* closely resembled the Kutjamarpu LF of all other Australian faunal assemblages. Although the situation now seems more complicated than this, there is still an over-riding similarity between the marsupials from this central Australian assemblage and some of the Riversleigh assemblages such as the Upper Site LF. One reason may be the higher family-level diversity of the Kutjamarpu LF in contrast to that of most of the other central Australian assemblages. If so, this similarity between the Kutjamarpu LF and those from Riversleigh may be more an indication of comparable diversity rather than age. Detailed comparisons of the many kinds of marsupials in these faunas should resolve the nature of the similarity.

As time goes on, most of the marsupial groups that we once thought confined to the central Australian Oligo-Miocene deposits have turned up somewhere in Riversleigh's sequences. For example, the ektopodontid genus *Chunia*, recently discovered in Riversleigh's Systems A and B local faunas, was once thought to be restricted to the late Oligocene deposits of central Australia. Similarly, the bizarre family of central Australian leaf-eaters known as ilariids have now also turned up at Riversleigh. In contrast, many of the groups known as contemporaries in the older Riversleigh sediments are the first and sometimes only appearances of these groups in Australia (e.g. thylacinids, notoryctids, yingabalanarids, yalkaparidontids, petaurids, burramyids etc.), again perhaps a measure of the relatively high biotic diversity of this region in Oligo-Miocene time.

Despite the occurrence of most of central Australia's distinctive marsupial groups in Riversleigh's biotas, some of the central Australian sites have produced the best preserved examples of some of these groups. Specifically, although Riversleigh sports *Ngapakaldia* diprotodontoids, it has so far failed to produce skeletons of the kind that occur not infrequently at Lake Kanunka in the Tirari Desert. Similarly, while Riversleigh has produced the best preserved wynyardiid skulls (and the only skulls of *Namilamadeta*-type wynyardiids), the only complete wynyardiid skeletons have come from Lake Palankarinna in the Tirari Desert. Finally, although Riversleigh has produced ilariids, the only partial ilariid skeletons have come from Lake Pinpa in the Frome Basin.

So far, Riversleigh has also failed to produce undoubted late Miocene assemblages which is why most of our understanding about marsupial evolution during this interval is based on study of the Alcoota LF from the Northern Territory.

Similarly, although Riversleigh has Pliocene sites, none of those so far discovered have produced more than an inkling about the middle-to large-sized marsupials that roamed Northern Australia at that time. For this understanding, we must lean heavily on what has been learned about the Bluff Downs LF from north-eastern Queensland where giant diprotodontine *Euryzygoma* herbivores browsed among relatively modern-looking kangaroos. It is, nevertheless, infuriating to know that while megadermatid bats dropped the albeit fascinating half-eaten bones of tiny dasyurids, bats and rodents onto the floor of Rackham's Roost 3-5 million years ago, just outside the cave entrance were the middle-sized to large mammals that we desperately want to know so much more about—out of reach in space and time.

The Pleistocene marsupials of Riversleigh have also been slow to appear despite the fact that there are hundreds of Pleistocene marsupial sites from other areas of Australia, some having been known since the 1830s. Unfortunately, there are very few from northern Australia and even fewer that occur today within areas that are seasonally arid. Unravelling the history of the mammal faunas of these widespread habitats of northern Australia is bound to help us interpret the network of events that determined Australia's modern biotas. While tidbits keep turning up in the Terraces Site and some of Riversleigh's cave deposits (such as a palate of a species of *Palorchestes* in 1990), it will be many years before we have anything approaching the level of understanding about Riversleigh's Pleistocene mammals that we have for most areas of southern Australia.

Recent discoveries at Murgon have also provided surprises. For example, one marsupial appears to represent a group otherwise known only from the Eocene of Argentina, while another resembles South American Paleocene forms known from Argentina and Peru. But one of the most important Murgon discoveries, made by visiting Argentinian Dr Pancho Goin, was of a microbiotheriid marsupial, a group with one living representative (*Dromiciops* from the *Nothofagus* forests of southern Chile) but many representatives in the fossil record of South America. On the basis of studies of anklebone morphology and DNA, this may have been the stem group for all or part of the Australian marsupial radiation—so it is gratifying, and exciting(!), to find that it is represented in Australia's oldest assemblage of mammals.

?Condylarths. Murgon is so far alone in having produced what appears to be Australia's first and only condylarth-like placental mammals. This is a group that on other continents such as North America, Asia, Africa and South America, evolved an enormous diversity of mammal types including dogs, horses, cows and whales. Although it has always been presumed that marsupials dominated Australia because placental groups such as this never reached this continent, we may have to exchange this hypothesis for one that suggests marsupials dominated Australia because they outcompeted early placentals.

Bats. Before Riversleigh began to tip out millions of bat teeth and bones, like salt and pepper out of stone shakers, the whole of the older Australian bat record could be placed on the head of a nail. Riversleigh's fossil bat record has filled a significant world in our understanding about the history of this important group of mammals in Australia. In modern Australia, approximately 25% of the native mammal species are bats—something that is often overlooked by Europeans overawed by egg-laying monotremes and bizarre marsupials.

Before discovery of the Murgon bat (see below), the oldest Australian bat was a single tooth of an insectivorous bat from the late Oligocene Ditjimanka LF of South Australia. It appeared as a yellow speck in the bottom of a screen box used to wash the green Etadunna fish-rich clays of Lake Palankarinna. Although it was certainly exciting, that single tooth is not really enough to identify what kind of bat it is. However, because it looked at the time most like a rhinolophoid, that was the guess that went into the literature in 1978 with a portrait of the cherished object.

Apart from this central Australian tooth, the only other Tertiary bat known was a jaw fragment of an insectivorous bat from the early Pliocene Hamilton LF of Victoria. There is, however, some doubt about whether the fragment in fact came from the fossil soil or from the more modern soils that surround the fossil deposit. Again, although it was a tantalising bit, it has proved impossible to tie down its relationships to other bat groups because its morphology is simply too 'generalised'.

Riversleigh's millions of exquisitely preserved bats, including masses of whole jaws, skulls and even some semi-articulated skeletons have changed all that. But they are important for another quite different reason—they help reconstruct the timing of mammal 'invasions' into Australia from other lands.

It has long been assumed that Australia was biologically remote from south-eastern Asia until around 15 million years ago, when the Australian plate crashed into the island arcs of Indonesia. After breaking free from Antarctica 38-45 million years ago, Australia was supposedly isolated, her biota evolving in splendid isolation for some 30 million years. Accumulating evidence from the Australian fossil record, however, indicates that this was not the case. Long before the Australian ark docked —in fact perhaps before it even left Antarctica — it carried a biota that included Eurasian elements, as well as an older Gondwanan stocks.

From the Riversleigh deposits it is evident that 25 million years ago Australia already had a diverse bat fauna. Although there is at least one Riversleigh bat that is most closely related to a living American free-tail species, the affinities of the Riversleigh fossil bat fauna as a whole appear to lie primarily with Eurasian and

African bat faunas. Overall, there is little evidence of a Gondwanan or southern origin for Australia's bats. Migration routes for good dispersers, such as bats, evidently existed from the north well before 25 million years ago. Having arrived, the bats would have found the warm, humid, gallery rainforests of northern Australia in the early middle Tertiary to their liking.

Recent reconstructions of this region postulate that a complex of continental fragments existed between Australia and Asia during the Cretaceous and Tertiary. These fragments included what would later become South Tibet, Burma, the Thai-Malay peninsula and Sumatra. They are thought to have enabled exchange of plants and animals between Australia and Asia for much of the last 100 million years. Taking advantage of these routes were not only bats but also a number of other groups including varanids, some song-birds and rodents.

The recent discovery of a fossil bat in early Tertiary sediments at Murgon in south-eastern Queensland has really set the cat among the pigeons. The extraordinary Murgon find suggests that migration routes into Australia from landmasses to our north were available, for at least bats, much earlier than ever anticipated. At 55 million years old, the Murgon bat is one of the world's oldest. Primitive bats of this age have been previously found only in North America, Europe and Pakistan, areas that were once part of the northern supercontinent Laurasia. This information is helping to rewrite the history and development of Australia's ecosystems.

It is provocative and irritating that we have not yet found, in any of the many bat-rich sites at Riversleigh, a single tooth or toe of a fruit bat. This would not be so annoying if fruit bats were a relatively minor group in Australia today but there are two subfamilies, five genera and at least eleven species of Australian fruit bats—with not a trace of any member of the group in any Australian deposit older than 10 000 years BP!

Rodents. The record of Australia's fossil rodents is in many ways similar to that for its bats. Prior to the recent discovery of Riversleigh's fossil cave deposits, there was a minuscule sample of Australian Tertiary rodents to represent the early history of a group that again represents about a quarter of Australia's modern mammal species.

From the early Pliocene Bluff Downs LF of north-eastern Queensland two incisor fragments were collected during our work in 1975 and from the middle Pliocene Chinchilla LF of south-eastern Queensland we recovered two isolated molars—hardly a useful fossil record for 25% of Australia's native mammals but at least this time it was possible to determine the kind of rodent that the Chinchilla material represented, *Pseudomys vandycki*, a new 'false mouse'.

Riversleigh's diverse rodent assemblages from the Pliocene Rackham's Roost deposit provide a more balanced view of Australia's oldest rodents and indicate that many distinct lineages had already diversified by this time. This, in turn, suggests that rodents entered Australia well before the time of the Rackham's Roost deposit, probably in the late Miocene.

As with the bats, the nature of Riversleigh's assemblages help illuminate questions about the habitat preferences and entry route of Australia's first rodents—which now appears to have been the dry forest corridors of the Indo-Malayan islands, rather than the rainforests of New Guinea and Cape York.

NEVILLE'S GARDEN

Approximately 20 million years ago, in a luxurious corner of Riversleigh's lowland rainforest, perhaps near the edge of a vast wetland, there lived a very complex community of prehistoric creatures. Their existence was rediscovered in 1988 by Neville Whitworth who noticed a strange tooth projecting from a piece of limestone. Hours of hard work and excitement later, hundreds of clues about the nature of this vanished world lay around us. There were teeth of ancient platypuses and lungfish indicating that the main part of the deposit had accumulated in a pool. The bones of hundreds of forest creatures told us that these animals had either drowned in the pool or their delicate, sharp-edged bones and teeth had been gently washed in from the surrounding forest floor. The presence of a small cave at one corner of the pool was demonstrated by the finding of cave 'straws', 'travertine rim pool' ridges on a rock surface and even two small stalagmites found buried in place. Because Neville's Garden is on the other side of the ridge from Bitesantennary Site, it is even possible that the Bitesantennary cave was connected by tunnels to Neville's Garden. Diverse kangaroos, bandicoots, diprotodontids, horned turtles, marsupial lions and masses of possums indicated the complexity of the surrounding forest.

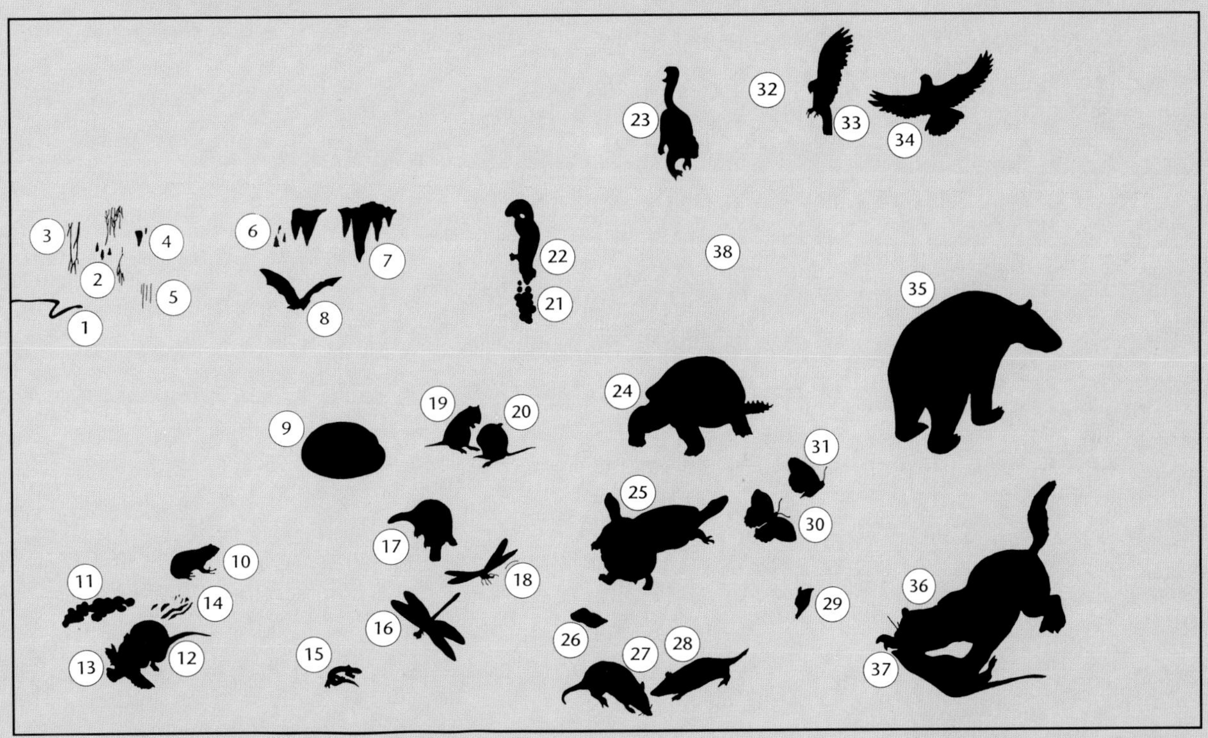

1 Pythonid snake.
2, 4 Leaf-nosed bat (*Brachipposideros* sp. 1)
3 Roots from forest plants that have penetrated limestone.
5 Crystalline cave decorations called 'straws'.
6 Leaf-nosed bat (*Brachipposideros* sp. 2)
7 Stalactites developed under overhung edge of cave entrance.
8 Ghost bat (*Macroderma* sp.)
9 Madtsoiid snake.
10 Tree frog (*Litoria* sp.)
11 'Cave pearls', laminated, rounded lumps of limestone.
12 Small dasyurid marsupial.
13 Swiftlet-like songbird.
14 Travertine ridges developed as lime-enriched water flowed over surface.
15 Small skink (scincid)
16, 18 Dragonfly.
17 Toothed platypus (*Obdurodon* sp.)
19-20 Omnivorous musky rat-kangaroo (*Hypsiprymnodon bartholomaii*)
21 Fig fruits.
22 Brush-tailed possum (*Trichosurus* sp.)
23 Ektopodontid possum (*Ektopodon* sp.)
24 Giant horned turtle (*Meiolania* sp.)
25 Freshwater turtles.
26 Water dragon (*Physignathus* sp.)
27-28 Bandicoot marsupial.
29 'VD'-type bandicoot.
30-31 Butterfly.
32 Ramparts of Cambrian limestone near the Neville's Garden Site (source of the reworked trilobite fossil)
33-34 Species of white cockatoo (*Cacatua* sp.)
35 Herbivorous diprotodontid marsupial (*Neohelos* sp.)
36 Marsupial lion (*Wakaleo* sp.)
37 Balungamayine rat-kangaroo.
38 Dense, species-rich rainforest.

Dunphy

CHAPTER 7

Backdrop to the Riversleigh Story

Drift, Climates and Environmental Change

James Lovelock developed the notion that the whole Earth, its creatures, atmosphere and rocks, was a single living being he called Gaia, the Earth Mother goddess of the ancient Greeks. While his concept is highly controversial, it serves to remind us that the developmental patterns of animals are inextricably intertwined with the histories of the continents, climates and vegetation upon which they depend.

In this chapter we present a brief overview of Australia's rainforests, living and extinct. Rainforests are a crucial part of Riversleigh's ancient environments and were the source for its most diverse prehistoric biotas. Then we present a brief summary of the history of the Australian continent in terms of its wanderlust since the Triassic. The positions of the continent and its changing relationships to other landmasses have been important factors in changing the balance of creatures that shelter in its heart. Finally, we consider how the catastrophic environmental changes of the last 15 million years affected the creatures that once populated Riversleigh's ancient forests.

Rainforests: cradle for Australia's creatures

As the work at Riversleigh unfolded and we came to understand it, it became evident that nearly every Australian group of terrestrial vertebrates, no matter what habitat they occupy today, apparently had their origins in the ancient rainforests of this continent—an understanding we have summarised as the 'Green Cradle' concept. The glimpses we have, peering backwards down the tunnel of time, almost invariably reveal the first-known members of all of this continent's unique lineages either dangling from the boughs of dense rainforest crowns, scurrying or bounding around in the half light of ancient forest floors or burrowing through soils that supported the massive roots of buttressed giants. The first-known koalas, kangaroos, skinks, emus, parrots and even marsupial moles were denizens of green 'mansions'. On balance, it is becoming clear that this continent owes much of its biotic heritage to rainforest.

The rainforest legacy today: fertile wombs under siege

Although it used to be thought that 50% of all the world's species were contained in the tropical forests that occupy a mere 7% of the Earth's dry land surface, research now suggests that a more reasonable estimate is 90%—although the 7% figure is declining at a distressing rate. In these remnant forests occur the most complex ecosystems in the world—webs of life that are so tightly

Rainforests are the richest expression of land-based life on our planet. The diversity of plant types in this canopy commonly reflects comparably high animal diversity.

interwoven that it will take many lifetimes of study before the ways in which they interact and develop become even remotely understood.

To put this incredible biodiversity into context, some facts and figures can be offered. There are, for example, more species of woody plants on the sides of a single volcano in the Philippines than occur in the entire continental United States. While most temperate forests are characterised by 10 to 30 species of trees, tropical forests have hundreds. In a single hectare plot of rainforest in the central Amazon basin of South America, researchers identified 300 species of tree. Recent studies in Malaysia recorded 835 species of trees in 50 hectares. It has been estimated that a typical patch of Malaysian rainforest, just 6 kilometres square, may contain as many as 1500 species of flowering plants, 750 species of trees, 400 species of birds, 150 species of butterflies, 100 species of reptiles and 60 species of amphibians. The number of insect species in small areas of tropical rainforest is staggering. One estimate based on the study of a single hectare concluded that there were at least 42 000 species of insects alone within the area. But study of the biodiversity of the world's rainforests is only just beginning and judging by the rate at which new species are turning up, it now appears likely that as many as 90% of the species which occur in these forests remain to be discovered. Translated into numbers, it has been estimated that only 500 000 of the probable 3 million species that occur in rainforests have been described.

Once rainforest covered all of the tropical lands of the Earth. Most occurred in three regions: South and Central America (three-fifths of total), South-east Asia and West and Central America. There were also smaller patches in Australia, the Far East and the Indian subcontinent, as well as on scattered islands throughout the tropics. Now, however, because of human activities this 'green girdle' is unravelling. On current estimates, half the world's tropical rainforests have been destroyed within the last two centuries. This destruction is occurring at a rate of approximately 40 hectares per minute—every minute of every hour of every day. The biological diversity that is the Earth's heritage and now in our trust is slipping through our fingers along with an incalculably valuable reservoir of knowledge and genetic resources.

A rapidly emptying treasury

Australia's modern rainforests occur as 'islands' within a sea of less diverse habitats. They extend from the north-western to the south-eastern corners of the continent and occupy less than 1% of the continent's land surface and 0.02% of the world's total rainforests. Since Captain James Cook successfully introduced the hand of European economic interests on the continent, 75% of Australia's rainforests have been destroyed. This is even more of a tragedy when we consider that our rainforests have many features that are unique within the world. Besides containing the biotic heritage of possibly as much as 50 million years of isolated evolution, Australia is the only country in the world with rainforests that range from tropical to cool temperate latitudes.

Differences in rainfall, temperature, altitude, seasonality, soils and nutrients, drainage patterns and even the specific types of plant and animal associations have all interacted in distinctive ways to produce a wide variety of rainforest. Much of this variation can be summed up in terms of broad climatic types. Hence, for example, there are 'tropical', 'subtropical', 'monsoonal', 'temperate' (cool and warm) and 'dry' categories of rainforest commonly distinguished in Australia.

Tropical rainforest forms the largest single area of Australian rainforest, representing about one-third of Australia's total. It occurs in a coastal strip

Cool temperate rainforest, such as this is one of many types found in Australia.

extending from Townsville to Cooktown. Most of its bulk lies in the 'wet tropics' between Ingham and Cooktown. The wet tropics contain the most luxuriant of all of Australia's rainforests. They have an average annual rainfall of 2800 millimetres (but sometimes receive as much as 10000 millimetres). Temperatures average around 24°C but never fall below 14°C, even in the coldest months.

The rainforests of the wet tropics contain many primitive kinds of Gondwanan plants. For example, they have the highest concentration of primitive flowering plant families (13 out of the world's total of 19). They also contain what appear to be close relatives of the ancestors of Australia's sclerophyllous plants, including the ancestors of the eucalypts and other plants now more characteristic of the semi-arid regions of the continent. They have the highest diversity of ferns, epiphytic plants and the world's largest cycads (20 metres tall).

The wet tropical rainforests also have the highest diversity of animals of any area in Australia today: more than 230 vertebrate species, including 35 mammals (which is nevertheless barely half of the diversity that characterised the ancient wet tropical forests of Riversleigh!), 140 birds, 25 frogs and 30 reptiles. Of these, 160 species are dependent on these rainforests for their survival. Of the 230, 54 are found only in the rainforests of this region including as many as 9 mammals, 10 birds, 19 frogs and 16 reptiles.

Forms unique to this region include the Musky Rat-kangaroo, four kinds of ringtail possums, tree kangaroos, the Atherton Antechinus, the Thornton Peak Melomys, Southern Cassowary, Golden Bowerbird, relict skinks and even

TROPICS, CAIRNS

The wet tropics of north-eastern Queensland (right and far right) contain the most luxuriant and complex of Australia's rainforests. These represent a living link with the ancient forests that once covered Gondwana.

TROPICS, CAIRNS

representatives of ancient groups of Gondwanan invertebrates, such as the Stag Beetle (*Sphaenognathus queenslandiae*) whose closest living relatives are found today in the Andes of South America. In fact, the rainforests of the wet tropics have the greatest concentration of plants and animal relicts of ancient Gondwana.

The ancient rainforests of Australia: Hosts to dispersing dinosaurs and drifting devils

The existence of many now isolated islands of rainforest in Australia is evidence that these forest islands are remnants of what was once a much wider, more uniformly spread forest. Of the islands that remain, those of the tropical, subtropical and temperate regions are clearly the descendants of the ancient, primordial types of rainforest that Australia shared with the other southern continents at least as recently as 100 million years ago. More than any other forests in the world, the tropical rainforests of north-eastern Queensland are the best-known living link with the once vast Gondwanan forests that spread across the southern continents including Antarctica.

About 100 million years ago at least the northern parts of Gondwana (including Australia) were covered in vast forests similar to the tropical rainforests that characterise the wet tropical regions of modern Australia. Within these forests were the earliest flowering plants (or angiosperms) which may in fact have originated in these Gondwanan forests 120 million years ago. Between 45 and 38 million years ago, Australia broke free from East Antarctica and rafted north into tropical latitudes. By this time, it was dominated by tropical rainforests. Fossils of leaves characteristic of tropical rainforest trees have been found, for example, at Maslin Bay in South Australia, in deposits 45-50 million years ago.

Australia's continuing northwards drift towards lower, warmer and wetter latitudes tended to offset the general cooling that was affecting the rest of the Earth. As the world headed towards the ice ages of the Pleistocene, sea temperatures dropped, precipitation declined and in general many lands became cooler and drier. Australia's northwards drift enabled it to maintain its tropical conditions, at least in its northern areas, for longer than many other southern lands. For example, many of the ancient tropical forests gave way to cool temperate forests in South America, Antarctica, New Zealand and southern Australia. Tasmania today preserves one of the world's last remaining cool temperate rainforests of this kind. As a consequence, new kinds of plants and communities developed in these cooler, sometimes drier habitats, drawing for their source on the ancestral plants and animals that formerly characterised the wet tropical forests.

65 million years of continental walkabout

Australia's changing geographic position and climates over the last 65 million years is a complex history. Not only did Australia separate from Gondwana to become an island affected by the ameliorating influences of a new maritime southern climate, but the continent also moved through various latitudes, oceanic currents and regionally differing climatic belts.

The end of the Cretaceous Period of Earth's history witnessed the *decline* of the dinosaurs (birds are surviving dinosaurs!) and other giant reptiles. After ruling their realm for over 100 million years, most of the great reptiles vanished from the face of the Earth 65 million years ago. But it was not just the dinosaurs and their allies that disappeared. Mass extinctions occurred in many groups — including plankton, belemnites, ammonites, clams, toothed birds, brachiopods, sponges, mammals, freshwater fish, gymnosperms and laurels.

Mass extinctions have occurred reasonably regularly during Earth's history, the most significant being those of the Cambrian, Devonian, Permian, Cretaceous and Eocene. Although the mass extinctions that marked the end of the Permian were far more extensive, the Cretaceous extinctions have, particularly in recent years, attracted most attention because they included the extinction of the large dinosaurs.

Astrophysical phenomena have taken centre stage in recreations of the extraordinary events that occurred at the end of the Cretaceous, although temperature changes, sea level fluctuations and volcanism have also featured strongly. Probably the most popular is the spectacular asteroid impact theory. This proposes that a 10-20 kilometre long asteroid (weighing 1000 billion tonnes)

Since its origin about 3500 million years ago, life on Earth has undergone spectacular changes, from single-celled bacteria to arrogant animals bent on conquering space. Here we note some of the more significant events in the evolutionary history of life on Planet Earth.

ERA		PERIOD	Significant Events in Australia's Biological History
CAINOZOIC	QUATERNARY	HOLOCENE 10,000 years	18. Also called Recent. In Australia climates and biotas were more or less as at present.
		PLEISTOCENE 2 million years	17. Aborigines arrived between 50 000 to 120 000 ybp. Extinction of most megafaunal species (e.g., of kind represented at Naracoorte) around 35 000 ybp — perhaps because of climatic change and human activities. Dingos introduced about 4000 ybp. Cool/dry climates and low sea levels alternated with warm/wet climates and high sea levels as polar ice caps expanded and contracted.
	TERTIARY	PLIOCENE 5.2 million years	16. Australia drifts into lower latitudes, cools and dries out. Rainforests continue to decline and eucalypts and grasslands spread. First appearance of a variety of modern types of animals including specialised grazers. Many lineages becoming gigantic.
		MIOCENE 23.3 million years	15. Early to middle Miocene characterised by lush forests (e.g., as at Riversleigh). Australia crashes into southeastern Asia soon after which rodents enter. By late Miocene, rainforests decline and grasslands begin to spread.
		OLIGOCENE 35.4 million years	14. By end of Oligocene, koalas, kangaroos, possums and other modern families present. Fossil deposits in Tasmania, South Australia and Queensland contain diverse vertebrate faunas, most indicative of forest communities.
		EOCENE 56.5 million years	13. Rainforest covers much of southern Australia. Australia's oldest marsupials, bats, frogs and snakes (Murgon). Non-marine mammals known from Antarctica — where forests still flourished.
		PALAEOCENE 65 million years	12. World climates cool. Mammals begin to diversify on all continents following decline of dinosaurs. First horses, primates, carnivores and other groups appear. Marsupials diversify at least in South America.
MESOZOIC		CRETACEOUS 145.6 million years	11. Early in this period, much of Australia is covered by shallow seas. Giant aquatic reptiles abound. Flowering plants appear and spread rapidly; conifers and cycads decline. Donosaurs and many other groups become extinct by the end of this period. Australia's oldest known mammal is a platypus-like animal. Australia's oldest birds. At least some exchange of plants occurred between Australia and lands to the north.
		JURASSIC 210 million years	10. Australia's climate is warm and wet. Conifers, cycads and ferns abundant. Large plant-eating and aquatic reptiles abundant. Ray-finned fishes present. Earliest birds found in Northern Hemisphere. Mammals diversify but as tiny, mouse-sized creatures.
		TRIASSIC 245 million years	9. Mammal-like reptiles present on all continents. Australia's climate warmer and drier. Insects, primitive amphibians and reptiles well-represented. Gondwana begins to break up. First mammals known from many continents but not Australia. Many groups became extinct or steeply declined at the end of this period.
PALAEOZOIC		PERMIAN 290 million years	8. Glaciers cover parts of Gondwana. Climate later more temperate with swamp forests. Insects, fish and early amphibians plentiful but, in Australia, no reptiles. Trilobites become extinct. Major period of extinction at the end of this period.
		CARBONIFEROUS 362.5 million years	7. Warm conditions at first; later, glaciers cover much of Gondwana as it breaks away from Laurasia and drifts towards the South Pole. Club moss forests die out and are replaced by hardier seed ferns. Many kinds of fishes and amphibians. First reptiles appear in Northern Hemisphere.
		DEVONIAN 408.5 million years	6. Plants spread from water margins into swampy areas forming thick vegetation with tree-like club mosses and ferns. Fish of many kinds, including lungfish, sharks and armoured fish in sea and freshwaters. First amphibians develop and venture onto the land.
		SILURIAN 439 million years	5. Life colonises the land. Land plants evolve from seaweeds. Jawless fish and sea 'scorpions' live in freshwaters. First fish with jaws appear in seas. First coral reefs formed.
		ORDOVICIAN 510 million years	4. Varieties of sea life become extensive including brachiopods, bivalves and crinoids. First jawless fish evolve, Australia's being among the earliest.
		CAMBRIAN 570 million years	3. No life on land but seas teeming with life including jellyfish, sea anemones, sponges, trilobites, brachiopods and molluscs.
PRECAMBRIAN		PROTEROZOIC 2000 million years	2. Oxygen becomes abundant in atmosphere and life diversifies. Multicellular organisms develop from single-celled organisms. Algae, jellyfish, primitive worms and sponges appear in the oceans.
		ARCHAEAN 4500 million years	1. Development of Solar System: formation and development of Earth's crust, atmosphere and oceans. Life evolves from spontaneously formed organic compounds at least as early 3.5 bya. Some bacteria produce oxygen that forms ozone layer that shields Earth from lethal UV radiation.

collided with the Earth, causing a cloud of dust and debris that enveloped the globe, and like a nuclear winter, blocked sunlight for weeks, months or even years. Such a catastrophe would obviously cause major collapses in most of Earth's ecosystems.

The fact that mass extinction events appear to have occurred regularly in Earth's history (some believe as regularly as every 26 million years) has prompted, among other theories, the notion of astronomical regularities. Nemesis, a small, cool, unseen companion star to the Sun, is thought to have been ploughing through the Oort cloud of comets at the edge of solar space, regularly disturbing comets to leave their normal orbits and enter a collision course with Earth.

Although few doubt that mass extinctions did occur in many animal groups in the late Cretaceous, some researchers maintain that plants did not experience dramatic mass extinctions, but merely a higher species turnover rate. Other palaeobotanists, however, believe that the effects of the event were as catastrophic and immediate for plants as they were for animals. Whatever the cause, by 65 million years ago, at the beginning of the Tertiary, major changes had occurred in plant and animal communities.

By Eocene time, about 50 million years ago, the world had taken on a much more modern appearance, although Australia and South America were still joined via Antarctica. Laurasia was basically a single landmass, being connected at least by land bridges across the northern Pacific and northern Atlantic. India had (about 55 million years ago) crashed into southern Asia, creating the Himalayan mountain chain in the process.

In the early Tertiary, during the Paleocene and Eocene, conditions were warm and wet, and broad-leafed rainforests covered most of Australia. Australia's eastern highlands continued to be uplifted and in the east and south-east there was a great deal of associated volcanic activity. Fossil plant communities at Angelsea in Victoria have been compared, on the basis of leaf shape and size and on taxonomic composition, with today's humid tropical rainforests in the vicinity of Noah Creek north of the Daintree River in the wet tropics. Leaf associations from Eocene-aged deposits at Maslin Bay near Adelaide and at Nerriga near Braidwood in southern New South Wales resemble those of modern notophyll vine forest but were of a kind otherwise unique because they also contained *Nothofagus* and conifers (Araucariaceae and podocarps).

During the early Tertiary, a cool temperate rainforest corridor, probably dominated by southern beech trees of the genus *Nothofagus*, extended from Australia to South America via Antarctica. To what extent this 'southern beech route' was interrupted at intervals by archipelagos, volcanoes or ecologically unsuitable habitats is uncertain but the eastern part was closed when Australia finally broke away between 45 and 38 million years ago. The South American-Antarctic leg of the southern beech route was disrupted probably by Oligocene time when the Drake Passage and the Circumpolar Antarctic Current were established. But even after severance, a corridor involving islands may have existed as a filter barrier to dispersal of many groups between the two landmasses, at least until the Pliocene. As for Antarctica itself, forests persisted around its margins until at least the middle Pliocene.

When Australia broke away from Antarctica between 38 and 45 million years ago, it took with it forests containing many different kinds of trees including species of *Nothofagus*, a vegetation type that was also subsequently isolated in southern South America. For this reason, the modern forests in southern Chile are in many ways very similar to those that survive today in Tasmania.

Many vertebrate groups shared by South America and Australia must have populated Antarctica. They would have included marsupials, some frogs and turtles, and, perhaps, flightless birds (called ratites). So far, of these groups, only two marsupials (polydolopids) have been found as fossils along the Antarctic leg of this route. These first-known Antarctic terrestrial mammals were found in the Eocene sediments of Seymour Island, part of the Antarctic Peninsula. More recent discoveries there suggest that placental mammals were also present, although so far they appear to represent groups otherwise only previously known from the fossil record of South America.

About 35 million years ago, during the Oligocene, South America pulled away from western Antarctica while Australia finally separated from eastern Antarctica sometime between about 45 and 38 million years ago. Since that time, Australia has been drifting in a generally northerly direction at about 11 centimetres per year—the rate at which human hair grows. If this rate of northward movement persists, in another 40 million years Australia will straddle the equator and crocodiles will be chasing swimmers in Melbourne!

By 25 million years ago, in the late Oligocene, Australia was about 10° further south than it is today. As it moved north, the northern edge of the Australian plate nose-dived under the southern edge of the plate to its north forming a major subduction zone. Continued subduction in this area today results in the frequent violent earthquakes that rattle many of the areas to the north of Australia, such as New Guinea.

Until 25 million years ago, Australia was covered by rainforest. Riversleigh's fossil deposits suggest that parts of northern Australia were covered by a tropical, species-rich lowland rainforest. Fossil deposits in what is now the Simpson Desert suggest that at this time central Australia was also covered by closed forest but, if the relatively low numbers of possums and other arboreal animals in these central Australian deposits is any indication, these forests were probably biotically less diverse than those carpeting northern Australia. This pattern persists today, albeit on a different scale, and reflects the greater rainfall and solar energy influx into the northern areas of the continent—factors conducive to increasing biological diversity in any community.

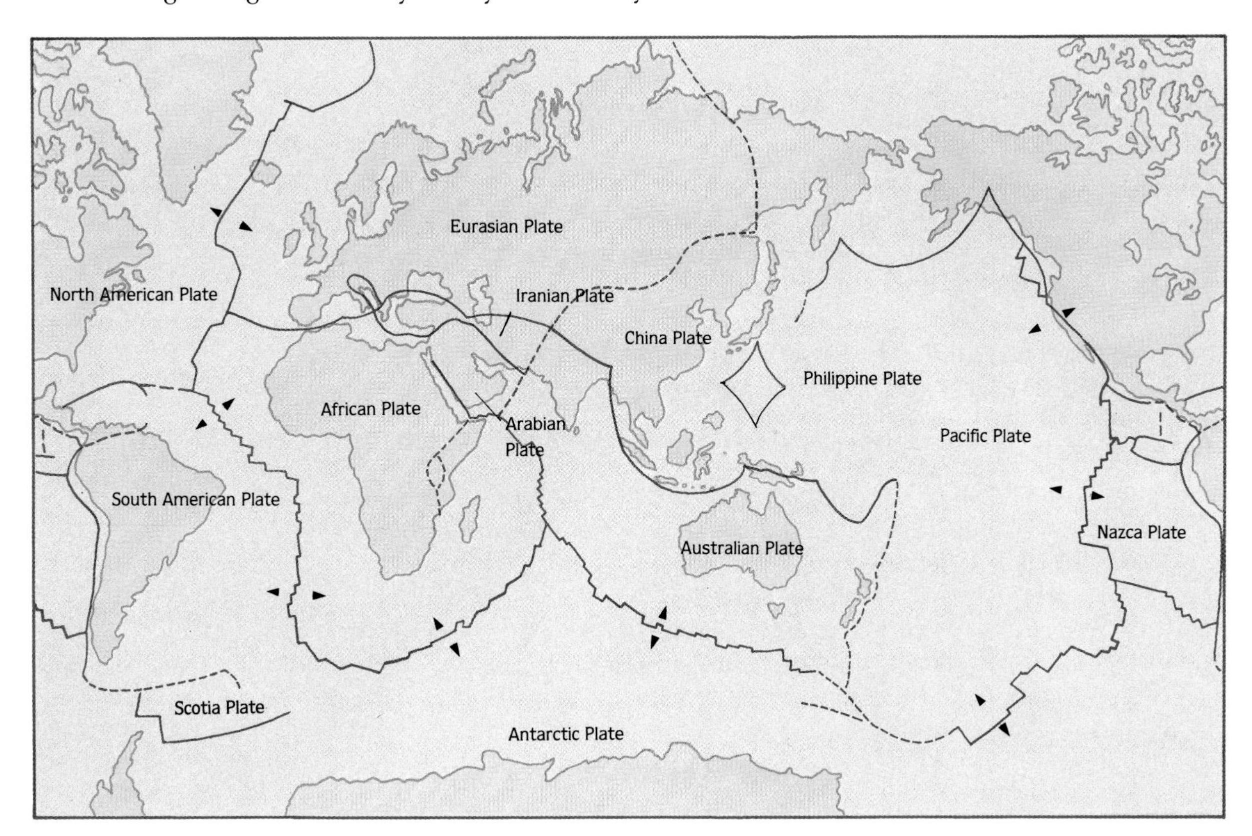

The major tectonic plates of the Earth's crust and their directions of movement (arrows). The Australian Plate which includes India is sometimes called the Indo-Australian or Indian Plate. As part of this Plate, Australia is drifting northwards at a rate of approximately 11 centimetres a year.

Four stages in the history of Australia's global wanderings. When Australia was still green, it separated from Antarctica between 45 and 38 million years ago, the last contact being via Tasmania.

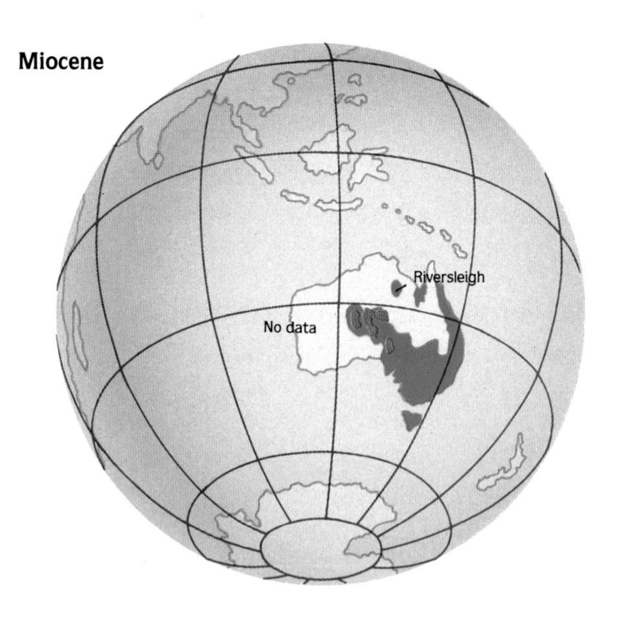

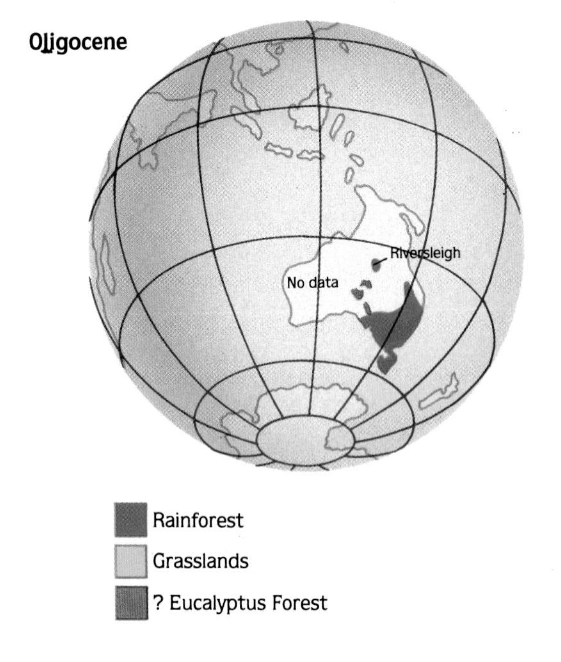

During Miocene time, Australia drifted northwards until it collided with south-eastern Asia about 15 million years ago. As a consequence, what was to become the New Guinea Highlands began their rise into high altitudes. By late Miocene time, the continent was beginning to dry out and the ancient rainforests to retreat.

The modern rainforests of wider Australasia reflect the complex history of the region's rainforests. While a Malesian (islands between Australia and south-eastern Asia) assemblage of rainforest plants characterises the lowland rainforests of modern Papua, an association of predominantly Australian kinds dominates the lower montane and montane rainforests, mostly above 900 metres in elevation. However, while many of the tree genera in New Guinea's modern rainforests above 1500 metres have Gondwanan origins, a number represent northern, often Malesian invaders that have penetrated the remnants of this continent's ancient rainforests.

Clearly, north-eastern Queensland and the mountains of New Guinea have become refuges for what is known as the 'Palaeoaustral' rainforests, those ancient forests that once dominated at least northern Australia in the Oligocene. But it now also appears, from studies of the relationships of the plants in tropical forests of south-eastern Asia and Australasia, that while the rainforests of the other areas may not on balance be able to trace their origins to the ancient forests of the northwardly drifting Australia, they may have nevertheless arisen from parts of Gondwana. It seems likely that most of the Gondwanan rainforests of south-eastern Asia invaded this region after being brought to southern Asia by drifting microcontinental fragments of Gondwanan origin and by India when it crashed into this region perhaps 40 million years ago.

By 15 million years ago, the lighter continental portions of the Australian crustal plate began to crumple up (to become in part the highlands of New Guinea, the highest rocks of which are marine limestones) as they collided with

The Break-up of Pangaea

Successive stages in the break up of Pangaea. During the Triassic, about 240 million years ago, most of the continents were temporarily joined together as Pangaea, a supercontinent consisting of a northern section called Laurasia and a southern section called Gondwana. Over the next 200 million years, Gondwana gradually split up, the Atlantic Ocean formed, the Mediterranean Sea was isolated and India moved north towards Eurasia. Throughout this interval, Australia remained bound to Antarctica, failing to completely separate until 38 million years ago.

The movements of the Earth's 12 major plates are driven by thermal convection currents in the mantle. Molten rock rises, most commonly as mid-oceanic ridges, and spreads out to form new oceanic crust. As a result, the sea floor spreads and the continents are moved away from the ridges. As the crustal plates grow, they push against the margins of plates on the side away from the ridge and the edge of the plate is shoved beneath the other in what is known as a subduction zone. Most of these occur as deep sea trenches where recurring earthquakes and volcanic activity are common. Because the continents are made of lighter materials than the rest of the crustal plates on which they ride, they are rarely dragged under at subduction zones. Rather, the continental crust buckles here to form mountain ranges such as the Himalayas, Andes and New Guinea Highlands.

Pliocene

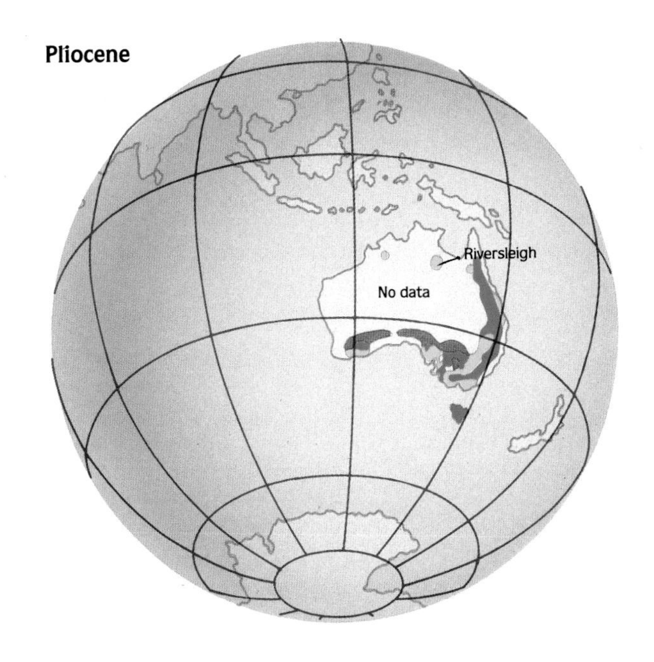

By early Pliocene time, much of Australia was covered by grasslands. The still northward-moving continent was cooling as the world headed for the ice ages.

Holocene

By the Holocene (the last 10,000 years), Australia had reached its present position and was recovering from the climatic disasters of the Pleistocene. By this time, the ancient rainforests were a green rim just clinging to the edge of the continent.

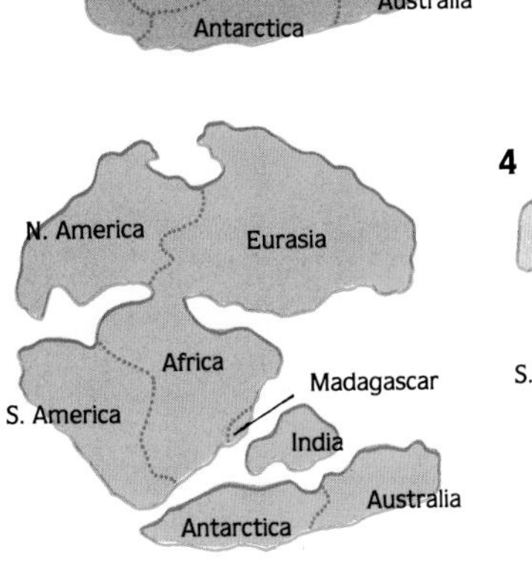

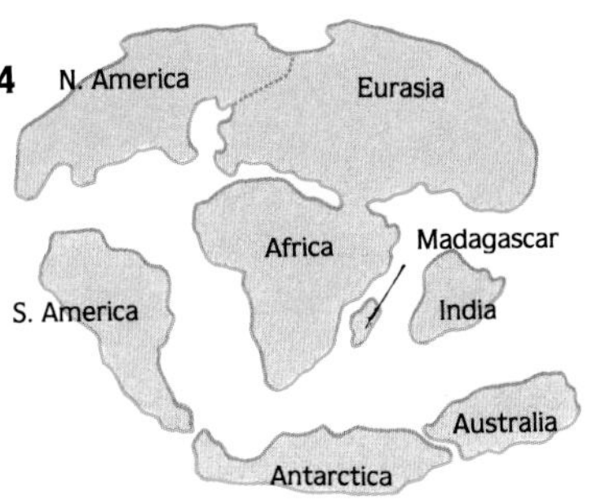

similarly light materials riding on the south-eastern Asian plates. Bits of land that had never been near each other were steadily shoved down each other's throats. Strange hybrids were created, such as the island of Sulawesi which is in fact a marriage of an Asian landmass (the western half) and a piece of ancient Gondwana (the eastern half).

The mountain building in New Guinea may have been one of the most important factors in causing the decline of central and northern Australia's rainforests—in part because it would have produced rainshadows along the northern Australian continental margin. Some of the cyclonic rain clouds that previously would have swept south across the continent would have been forced to dump much of their moisture on the New Guinea Highlands.

To add to the problems of this continent's declining central and northern Gondwanan rainforests, throughout the world the Miocene was a period characterised by rapid cooling. The polar icecaps were now permanent and rapidly growing, tying up more and more of the world's supply of water. Ocean temperatures were steadily dropping, thereby further reducing the amount of evaporation from the seas and hence providing less for precipitation. The net effect was that the whole world began undergoing a severe bout of cooling and drying.

Over millions of years, the contact with Asian lands provided migration routes to and from Australia for many groups of animals as well as plants, including some frogs (e.g. ranids), reptiles (e.g. elapid snakes and varanid lizards) and many birds (e.g. ducks, fowls, rails and pigeons). More significantly, it provided a means of entry to Australia of this continent's bats, rats, mice, the Dingo and humans.

A number of groups were able to take advantage of these potential migration routes even *before* the Australian-Asian plates collided. Accumulating evidence now suggests that long before the Australian ark ran aground on south-eastern Asia, perhaps even before it left Antarctica, it sported a biota that included Eurasian elements, as well as its ancient legacy of Gondwanan groups. Most of the evidence for this comes from the Australian palaeobotanical record and, recently, from the vertebrate-rich fossil deposits of Riversleigh.

The palaeobotanical record demonstrates at least sporadic interchange of plants between Australia and regions to the north during the last 80 to 100 million years. Further, there appears to be no evidence for a great influx of Asian taxa into Australia's plant communities within the last 15 million years—evidently because those groups that would have invaded this southern land had already done so long before Australia's collision.

Similarly, Riversleigh's fossil deposits indicate that a number of vertebrate groups probably first entered Australia from south-eastern Asia more than 30 million years ago. Among the more proficient of these early vertebrate dispersers were birds and bats.

By about 6 million years ago in the latest Miocene, there had been a marked increase in the size of the Antarctic ice-sheet which led to a further reduction of available precipitation. Most of Australia's vast rainforests were disappearing, ultimately leaving only pockets along the east coastal regions and possibly small patches in the south-western and north-western corners of the continent. In combination with the highland rainforest refuges that persisted in New Guinea and north-eastern Queensland, these remnant areas of rainforest were just sufficient to maintain representatives of some of the more distinctive and archaic rainforest groups that were once widespread across the continent, such as the musky rat-kangaroos, cuscuses and giant pythons.

Northern Australia began to develop highly seasonal climates with summer rains. Central Australia lost virtually all of its forests and began to yellow under a spreading cloak of arid grasses and sand. In the eastern areas of the continent, climates were less severe and conditions in some areas fluctuated so that areas of rainforest alternated with dry sclerophyllous vegetation (e.g. in the Lake George area of New South Wales). In contrast, in areas of north-eastern Queensland, such as Butcher's Creek on the Atherton Tableland, relatively cool, complex submontane rainforest with *Nothofagus* persisted despite the changes that swept over the rest of the continent.

Where the rainforests disappeared, they were replaced first by wet then dry sclerophyll forests, then woodlands, savannah woodlands, grasslands and ultimately, in wide areas of central Australia by approximately 2 million years ago, by the vast inland desert systems such as the Simpson, Gibson, Great Sandy, Great Victoria and Tanami. Seventeen thousand years ago, these deserts were even more extensive than they are now. It was in these new, semi-arid and arid habitats, which at present comprise approximately 44% (which equals about 3.3 million km^2) of Australia's surface, that major new adaptive radiations of animals and plants took place.

Bass Strait formed, cutting off Tasmania from the mainland of Australia—an oceanic separation that proved rather fickle because it subsequently vanished and re-developed at least eight times. The ease with which animals could pass back and forth between Australia and New Guinea is less easily determined but the number of identical species of small terrestrial mammals shared between the lowlands of New Guinea and northern Australia suggests that these connections have been substantial and relatively recent. At times of extensive glaciation and cooler, drier periods in the Pleistocene, much of what is now the Gulf of Carpentaria was dry land, its river systems draining to the west towards the Arafura Sea.

The Pliocene Epoch, between 5 and 2 million years ago, was characterised by relatively rapid swings in climate. It appears to have begun with a warm, wet phase which soon gave way to drier, cooler climates. By the beginning of this epoch, most of Australia's vast rainforests had vanished. They survived only along the eastern margins of the continent and possibly in the south-western and north-western corners. New Guinea was the only region to retain vast tracts of this ancient habitat. Where the rainforests had vanished, they were replaced with open forest, woodlands and grasslands in which the grazing kangaroos began to diversify. It is not at all clear whether there were any areas of sandy desert at this time although it is reasonably certain that these very arid habitats had become established by 2 million years ago and were in full development 17 000 years ago.

The Pleistocene Epoch, that period of time between 2 million and 10 000 years ago, is commonly known around the rest of the world as the 'Ice Ages'. While Australia did not endure continental glaciation during the Pleistocene in contrast to the northern continents which were scoured four times by colossal ice sheets, the cycling of cold and warm climates nevertheless had a profound effect on the development of our biotas. The mountainous parts of New Guinea, mainland Australia and Tasmania did, however, suffer glaciation which alternately pushed the montane habitats downwards and then allowed them to expand again up the slopes. These periodic strangleholds of ice may be one of the reasons why faunal diversity in the south-eastern Divide and Tasmania is relatively low compared with that of the non-montane areas west of the Divide.

Unfortunately, there are few dated fossil plant deposits representing early Pleistocene time. The oldest dated plant record is 700 000 years BP from Lake

George while Lynch's Crater on the Atherton Tableland provides a record back to 190 000 years BP. From the thick sequences in these deposits, it is clear that rainforests underwent severe contractions during the Pleistocene although the precise history of this vegetation type is different in different areas. In the more temperate, southern regions the Pleistocene rainforests became considerably less diverse than those in northern regions. On the Atherton Tableland prior to 30 000 years ago there was a now extinct type of rainforest with abundant *Araucaria* and other conifers. When it disappeared, rainforest as a whole nearly vanished from the region, being widely replaced by sclerophyllous vegetation.

In general, around 17 000 years ago, during the peak of the last glacial episode of the Pleistocene, all of Australia's water-dependent habitats underwent severe contraction. The deserts and arid grasslands reached a peak in size while the rainforests suffered extensive reduction. The rainforests on the Atherton Tableland nearly vanished while those in south-western Victoria did vanish. However, about 10 000 years ago, complex rainforests reappeared on the Atherton Tableland, presumably having survived the arid phase in pockets in valleys or along stream margins. Even today in areas of eastern Australia, providing fires are relatively infrequent, rainforests can often outcompete sclerophyll forests for space. In some forests along the Great Divide (e.g. those on Mount Glorious), where fires have not been allowed to run riot, giant remnant gum trees may be found deep within areas of aggressively spreading rainforest.

Our own origins are no less tightly interwoven with these world crises than are those of the grassland kangaroos of Australia. In Africa, the world's cooling and drying climates carved vast holes in that continent's ancient rainforests and, among other groups of opportunistic animals that took advantage of the rapidly spreading areas of grassland, were our own ancestors, the australopithecines. Even after their descendants, *Homo erectus/sapiens*, first escaped the continent of Africa perhaps 1 million years ago, it was not until sometime between 120 000 and 60 000 years ago, probably during dry periods of the Pleistocene, that the first humans gained entry from south-eastern Asia to Australia. This they may have done by coming south from Java (home of 'Java Man'), Timor or perhaps New Guinea, perhaps using canoes or accidental rafts of vegetation to cross remaining water barriers.

The interval of time represented by Riversleigh's fossil record, that is, from approximately 25 million years ago to the present, clearly spans the major periods of change affecting this continent's climates and biotas. During this time, the great forests of ancient Gondwana and their archaic inhabitants struggled with the rigours of a drying land, competed with aggressive colonists from northern lands for declining resources and, ultimately, sought toe-holds in a world progressively depleted by the needs of an invading avaricious primate.

The fickle finger of fate: three ways to deal with a bad hand

As the rainforests declined, the animals that depended on them underwent one of three responses: extinction, retreat or adaptation.

Extinction. The radical loss of most of Australia's rainforests during the last 10 million years has resulted in or at least contributed to the extinction of many ancient inhabitants. These extinct groups, like the wynyardiid marsupials, pilkipildrid possums and Thingodontans, often had pedigrees extending back at least 20 million years. Destruction of their habitats was evidently not a change to which they could adapt.

The extinct forms, however, did not all vanish at the same time as one another. Some, such as the ilariid marsupials and yingabalanarid mammals, vanished even before the rainforests disappeared. Others, such as the ektopodontid possums, hung on until two million years ago (at least in southern Victoria). Still others, like the marsupial lions (thylacoleonids) and most of the diprotodontids (giant wombat-like marsupials), did not become extinct until about 35 000 years ago, well after Aborigines had arrived on the continent. Perhaps the arrival of Aborigines, with their later introduction of the Dingo as a pet about 4 000 years ago, was the final straw that broke the back of the last of the ancient inhabitants already under stress from changes in climate and vegetation.

Retreat. Unlike the animals that went extinct following the changes to Australia's ancient environments, others hung on in the remnant patches of rainforest.

MARSUPIAL LIONS

flesh-eaters of forgotten nightmares

Among the most ferocious of marsupials ever to have lived were the marsupial lions, members of the family Thylacoleonidae. In Riversleigh there were at least two different kinds: the cat-sized species of *Priscileo* and the leopard-sized species of *Wakaleo*. The species of *Wakaleo* may have given rise to those of *Thylacoleo*, lion-sized carnivores that co-existed with Aborigines before stepping, as the last representatives of a bizarre lineage, over the shadowy edge of existence perhaps 30 000 years ago. The reasons for their demise probably included progressive loss of suitable habitat and prey.

THINGODONTANS

weirds from an older world

Yalkaparidontids (informally known as Thingodontans) are perhaps the most extraordinary of all the groups that did not survive the loss of Australia's inland rainforests. Although they are known from whole skulls, which reveal remarkable specialisations that must relate to a very peculiar diet, their body shapes, lifestyles and food preferences remain mysteries. All we know is that they appear in the Riversleigh rainforests and then just seem to vanish. Yet their very distinction indicates that they must have had a long prior history on this continent. Why did they die out? Without understanding the role they played in the ancient rainforest communities of Riversleigh, we cannot begin to guess.

These survivors are today found mainly in the tropical wet rainforests of north-eastern Queensland and New Guinea. When New Guinea rose to high altitudes and the mountains of north-eastern Queensland rose, they preserved, in their cooler, wetter altitudes, rainforest habitats once widespread across the lowlands, including the area around Riversleigh. As a result some animals vanished from the lowlands but persisted in these remnant high altitude forests. Examples include: the cuscuses (very large herbivorous to omnivorous possums distantly related to brushtail possums), which were once present at Riversleigh but today occur only in the rainforests of north-eastern Queensland, New Guinea and surrounding tropical islands; and the 'woolly' ringtail possums (in the genus *Pseudochirops*), specialised leaf-eaters that were once widespread and diverse at Riversleigh but are now confined to the rainforests of north-eastern Queensland and New Guinea.

We know from the fossil record that there were other refuge areas in south-eastern Australia (for example at Hamilton in western Victoria) that at least until 4 million years ago maintained populations of some of these 'retreaters' but these last southern remnants vanished, presumably with their rainforests, before the middle Pliocene.

Adaptation. Unlike the animals that went extinct or those that hung on within the remnants of shrinking rainforest, others gradually adapted to the non-rainforest environments that began to dominate most of Australia. These 'adapters' were the source of the majority of Australia's living animals. Before the rich fossil record of Riversleigh's rainforest animals began to unfold, we had in

THE CUSCUSES
just hanging in there

In the Riversleigh area 20 million years ago there were many kinds of possums including at least one cuscus (***Strigocuscus reidi***), a distant relative of the brushtail possums but a close relative of the ground cuscuses (e.g. ***S. gymnotis***) of New Guinea. The Riversleigh cuscus was a Koala-sized omnivore that may have spent as much time on the rainforest floor as it did in the trees. Although long gone from this area today, its descendants survived until 4 million years ago in the now defunct rainforests of south-western Victoria and they persist today in the rainforests of New Guinea and north-eastern Queensland.

MUSKY RAT-KANGAROOS
cryptic survivors with an ancient pedigree

20 million years ago, the Riversleigh rainforests were pinball machines to little, dark-furred kangaroos that rocketed from place to place in a most un-kangaroo-like manner. These small (pademelon-sized), omnivorous rat-kangaroos did not hop like other kangaroos—instead they galloped. Also unlike other kangaroos, they had five instead of four toes on the hind foot and gave birth to twins rather a single young at a time. These were species of the genus ***Hypsiprymnodon***. With the extinction of the great inland rainforests, their descendants were among those that retreated with their forests. They persisted for a while in western Victoria but by 4 million years ago they were evidently confined to the highland rainforests of north-eastern Queensland where their only living descendant is the Musky Rat-kangaroo, ***Hypsiprymnodon moschatus***.

several cases erroneously concluded that the specialisations of Australia's dry-country creatures meant they had descended from lineages with a long history in arid habitats—despite a lack of evidence for significant antiquity of Australia's arid habitats.

Examples of living non-rainforest animals which have representatives in the fossil rainforest deposits of Riversleigh include: scaly-tailed possums, marsupial moles, bettongine rat-kangaroos, elapid (poisonous) snakes, mountain pygmy-possums, thylacines (Tasmanian 'tigers') and koalas. Of these, some highly specialised forms, like the marsupial moles (notoryctids), appear to have been 'pre-adapted' within their ancestral rainforest habitat for subsequent survival in drier habitats. Others, such as the brushtail possums, appear to have needed only relatively minor adaptations to survive without rainforest.

The balance of extinctions, retreats and adaptations is constantly changing, even today. Yet one basic message is clearly spelled out by what we now understand to have happened: the genetic resources of the animals that inhabited Australia's rainforests were sufficient not only to fill those ancient habitats with a diversity of animals and plants higher than that found in any other area of Australia but resilient and responsive enough to populate, within probably less than 10 million years, the whole of the newly developed drier habitats that came to characterise nearly 99% of the continent.

MARSUPIAL MOLES

do holes make the moles?

Today, one of Australia's 'mystery mammals' is certainly the Marsupial Mole (*Notoryctes typhlops*), a mouse-sized, burrowing animal that is virtually blind and deaf and lives only in the deserts of central and western Australia. Because many of its features seem to have evolved as ideal adaptations to a life spent burrowing through the desert sands, palaeontologists presumed its evolutionary history lay within ancestral deserts—unlikely places for fossils to form. We were therefore stunned to discover ancestral marsupial mole remains in several of the Oligo-Miocene fossil rainforest deposits of Riversleigh. Although undoubtedly part of the ancient rainforest communities, the Riversleigh mole shows many (but not all) of the burrowing specialisations exhibited by the living form. The distinctive shape of its teeth and jaws indicate that it could have been directly ancestral to the living species. This suggests that marsupial moles were 'pre-adapted', presumably as burrowing marsupials in the rainforest floor, for life in the dry lands that superseded the rainforests.

THE KOALA

ancient companion of the gum trees

Today, no Koala (*Phascolarctos cinereus*) lives in rainforest—all exist in the open forests and woodlands of eastern Australia as specialised feeders on (mostly) gum trees. But as we have found at Riversleigh, this was not always so. Discovery in the older Riversleigh fossil deposits of a new species of *Litokoala* demonstrates that this previously poorly known ancestral koala was in fact quite closely related to the living genus *Phascolarctos*. Palaeobotanists (scientists who study the fossil history of plants) have concluded that the ancestors of the modern gum trees (*Eucalyptus*) lived in the same ancient rainforests that were home to the species of *Litokoala*. It seems at least possible that these ancestral koalas had singled out the rainforest gum trees as their special food resource and developed a dependent relationship that persisted when these ancient gum trees colonised the drier habitats.

CHAPTER 8.

Stones, Bones and Catacombs

Understanding the Layers of the Landscape

The bushfire that swept over the spinifex-covered hills of Riversleigh Station in the summer of 1984-5 left more than just ashes. Its legacy was a totally new understanding of the Riversleigh fossil deposits. The fire cleared the limestone bluffs of hip-high, prickly spinifex and for the first time we were able to see the Riversleigh deposits as they really were: a thick vertical sequence of gently inclined layers of limestone, many of which were fossil-rich.

As we now understand it, Riversleigh presents a vast complex of rock strata and fossil faunas. Currently we recognise more than 150 fossil-rich deposits, representing: at least three periods during Oligo-Miocene time (between 12 and 25 million years BP); one Pliocene assemblage (about 3-5 million years BP), i.e. the Rackam's Roost Local Fauna; many Pleistocene assemblages (between 2 million and 10 000 years BP) from cave and ancient riverine deposits such as the Terrace Site; and many Holocene assemblages (less than 10 000 years BP), some of which appear to span the arrival of introduced mammals and Europeans.

Riversleigh's sediments range from brilliant white Oligo-Miocene freshwater limestones to brown Pleistocene river gravels and black, barely consolidated cave floors. Of Riversleigh's sediment types, the limestones stand out in providing hundreds of thousands of remarkably well preserved bones. Some have clearly developed as layers in a cake, younger levels being cemented or plastered on top of older ones. Sometimes caves appear to have been cut by groundwater when the limestones were still young and then filled with younger bone-rich deposits.

Unlike most of Riversleigh's diverse vertebrate fossils, the majority of Australia's Tertiary vertebrates accumulated in muddy pools and abrasive river gravels. When a skull begins to fossilise in clay, it is almost guaranteed a rough future. As more clay accumulates above the entombed object, the increasing weight squeezes the underlying sediment. Water is forced out and the loss of volume causes the clays to compact. The rigid, perhaps slightly mineralised skull then begins to fracture and collapse as the clays that contain it are compressed. As a result, the best skulls retrieved from the late Oligocene clays of central Australia emerge from the laboratory looking like refugees from the floor of a steam-roller factory. Worse still, if the skull was anywhere near the surface of the deposit before it was discovered, it will have undergone perhaps hundreds of years of alternate expansion and shrinking as the clays in which it resides alternately expand when wet and shrink when dry. The result is often discovery of a jigsaw puzzle of hundreds of tiny bits that will, over weeks, drive some conscientious preparator nuts.

In contrast, when skulls accumulate in limey pools, the first thing that happens is that they rapidly accumulate a hard crust of limestone over their whole surface, inside and out. Subsequently, this hard-coated object is cemented with more limestone into the relatively rigid fabric of a limestone matrix, the whole becoming harder than cement. Short of a Himalayan mountain accumulating above the deposit, skulls entombed in limestone often emerge from the acid laboratory looking as if they fell out of their owner's head barely a week before.

The calcium carbonate-charged waters of the modern Gregory River readily precipitate limestone—in this case as tufa dams which emerge from below to isolate pools.

Disturbances along the path of the Gregory River result in limestone cementation of sand and pebbles such as around the roots of trees.

Right: Crusts of limestone develop as evaporites on the surfaces of isolated pools alongside the Gregory River. Potentially treacherous surfaces of this kind covered some of Riversleigh's prehistoric pools.

Far right: Limestone pavements near the present crossing of the Gregory River, containing cemented river cobbles and Quaternary-aged bones and shells.

Delicate bones, such as this skull of an Oligo-Miocene ringtail possum, were often first encased in a protective shell of limestone (the white halo) before being finally buried. (R. ARNETT)

Beds of freshwater limestone containing most of Riversleigh's fossil deposits, such as Site D, entombed the bones of a wide variety of aquatic and forest creatures. (R. ARNETT)

However, what we have been very slow to find at Riversleigh are articulated skeletons of the older vertebrates. Although a few partially articulated skeletons have been found (such as the new zygomaturine mother and pouch young from VIP Site), most skeletons appear to have become disarticulated before they were actually fossilised. Perhaps the carcasses were scavenged by turtles before the separated bones finally slid out of sight into the limy mud at the bottom of the pool.

This disarticulated state is typical for most Tertiary vertebrate deposits throughout Australia. In contrast, there are a few Oligo-Miocene sites in central Australia that retain articulated skeletons. One of these is Lake Pinpa in South Australia. Here, following several very dry and windy years in the early to mid-1970s, Dick Tedford and his crew found partial skeletons punched into a late Oligocene green clay. Although a spectacular find in providing information about the skeletal anatomy of some of the more 'primitive' marsupials of the time (e.g. ilariids), the conditions ever since at Lake Pinpa have changed. Years of 'good' seasons have resulted in tonnes of sand being washed onto the normally dry salt lake's surface, burying beyond reach the skeletons of Pinpa.

In the meantime, although we hope that the winds of central Australia will again expose the Pinpa skeletons, naturally acid rains are etching Riversleigh's limestones to expose and, if not collected, destroy the treasures these northern rocks contain.

A Warm, Leafy Past: Riversleigh's Oligo-Miocene Lake, Pool and Cave Deposits

We define the three main periods of Oligo-Miocene time represented at Riversleigh as Systems A, B and C, with System A being between late Oligocene to early Miocene in age, System B possibly early to middle Miocene and System C possibly middle Miocene to early late Miocene (see site map and page 18). There are four major regions within Riversleigh Station where most of the Oligo-Miocene sediments have been found.

The first of these is the flat-lying D-Site Plateau (which contains Site D as well as many others), west of the track that runs between Riversleigh and the neighbouring Lawn Hill Station. Here, exposures of limestone deposited during late Oligocene to early Miocene time (System A limestones) abound on the southern and eastern margins. Early to middle Miocene units (System B units) are exposed on the northern edge of the plateau (Neville's Garden Site is possibly of this age). On the south-eastern and northern edges of the D-Site Plateau, 'cave' deposits of Oligo-Miocene age occur (i.e. Microsite and Bitesantennary Site respectively).

The second major region of Oligo-Miocene deposits are the flat-lying mesas to the east of the Riversleigh/Lawn Hill track. Here, additional exposures of what appear to be System A and B limestones occur. The third major region is the slightly dipping Godthelp Hill and Hal's Hill series west of Site D, where at least seven metres of System B units lie on what are called D-Site Equivalent units. The fourth region is the slightly dipping Gag Plateau to the north-west of the D-Site Plateau where approximately 12 metres of System C units are superimposed on D-Site equivalents, in subregions such as Ray's Amphitheatre, Archie's Amphitheatre, Megabyte Hill and Jeanette's Amphitheatre.

Each fossil-rich Riversleigh site or isolated faunal assemblage (e.g. Upper Site) has been treated as a distinct local fauna. Different local faunas potentially accumulated at different times and/or in different microenvironments (e.g. a forest floor and adjacent pool) and hence are potentially unique in composition. Some of these assemblages are less than five metres apart in limestone that is otherwise devoid of fossils and seemingly homogeneous. Other local faunas are isolated by hundreds of metres from their nearest neighbour.

In eroded or excavated exposures, many of these Oligo-Miocene bone-rich deposits appear to be roughly lens-shaped in cross-section and are commonly 3-5 metres in diameter and 0.5-1.0 metres thick. Looking down on the deposits, they appear to be oval, suboval or arcuate in shape. Often the lower levels of these lens-shaped assemblages contain large clasts or masses of yellowish-white limestone (which experience tells us indicates relatively large amounts of brown silt) cemented by whiter,

The majority of Riversleigh's fossil-bearing sediments, such as the lungfish-rich Melody's Maze deposit, formed in shallow pools on the rainforest floor.

Sedimentary features, such as these spheroidal 'oncolites' at Bull's Eye Site, help us understand the nature of the environments in which some of the fossils accumulated.

Thick sheets of 'flowstone', a type of travertine limestone, such as this at Ray's Amphitheatre, suggest that at intervals, slow-moving fans of calcium carbonate-charged water may have spread out over wide areas.

Cave sediments, such as this massive, bone-rich bat guano pile in Carrington's Cave, have preserved some of Riversleigh's more distinctive fossil assemblages.

Modern caves that perforate the older Riversleigh limestones continue to accumulate bones although, in the present arid climate, few of these will be fossilised. (R. ARNETT)

Initial discoveries, such as Bitesantennary Site deposit (upper right corner), often lead to discovery of others nearby, such as Low Lion Site (lower slopes to right) and Elaine's Elation (near dead tree in the centre). (R. ARNETT)

Some fossil deposits, such as Judy's Jumping Joint on the crest of Hal's Hill, are distinctly unlike the majority and suggest a very different depositional environment.

more crystalline limestone. These lower levels grade upwards into whiter more homogeneous, although sometimes silty, limestones. The lower levels often contain more complete skulls and bones while the bones in the higher more crystalline levels are often broken with the 'clean' edges of the breaks lacking abrasion (which might otherwise reflect transportation after the bones were buried).

These accumulations appear to represent the remains of isolated, possibly steep-sided pools of lime-rich water and detritus within the Riversleigh rainforest. Around the edges of deeper pools, silt-rich limestone accumulated and perhaps became periodically unstable, sliding *en masse* towards the centre of the pools. Here, the semi-consolidated yellowish clasts may have been secondarily cemented by more crystalline limestone precipitating in the spaces between the clasts. In the final stages of pond filling, the upper units may have become more crystalline as the shallowing water became more concentrated with dissolved limestone.

Small, still pools supersaturated with dissolved limestone and surrounded by warm tropical rainforest appear to have supported fragile, surface crusts of limestone. These crusts may have been augmented or initiated by floating mats of limestone-secreting algae. In a few sites, the more crystalline phases contain thick sequences of what appear to be shattered evaporite surface crusts that have periodically settled to the bottom with the bones of the animals that may have caused the surface to break. While evaporites are normally associated with saline waters, limestone evaporites are known to form on the surfaces of cave pools. We have also found evaporite crusts of limestone on small isolated pools of the modern

Younger (in this case paler) fissure-fill deposits that fill crevices in older limestones commonly contain fossil bones and teeth of animals representing distinctive periods of time.

Gregory River system in open air situations.

Most of the System B and C deposits suggest that the pools were either shallow or short-lived because they rarely include bones of *large* fish, crocodiles or turtles. However, in some exceptional sites, particularly in System C sites such as Ringtail Site, Bob's Boulders Site, Crusty Meat Pie Site, Quentin's Quarry Site, the assemblages are dominated by larger animals. These may reflect episodes of raised watertables, consequently deeper pools and longer periods of growth for fish, turtles and crocodiles. Changes in the level of the watertable may have been due to a tectonic lowering of the region or a worldwide change in sea level.

A few Oligo-Miocene deposits, such as the Bitesantennary Site, are bounded by what appear to be thin, travertine laminations, suggesting that the bone-rich sediments filled travertine-lined cavities developed within the parental limestone. In these situations we visualise that caves may have developed in the older Oligo-Miocene units. Roof collapse may have subsequently provided access to bats, which roosted in the cave, and rainwater may have accumulated as ponds containing ostracods, snails and possibly rafts of algae. Terrestrial reptiles, birds and mammals may have stumbled over the edge of the cave's mouth or been brought in by mammalian or avian predators to become rare elements in these particular deposits.

Apparent exceptions to these generalised models include the faunal assemblages in System A, such as Site D, and some of the D-Site equivalents, including Sticky Beak Site and Bone Reef Site. In these the bones are on average larger (almost exclusively fragments of large crocodiles, flightless birds, turtles and diprotodontids) and rarer. The parent limestone in these sites forms thicker units (at Site D the fossil-rich unit is at least three metres thick) with some evidence of flood-type sediments.

A model that might explain these features would be a large lake that filled the central basin of the area (through which the Riversleigh/Lawn Hill road now runs) fed by springs, perennial streams or periodic rushes of surface water from the surrounding forested terrain. The western margin of this lake's sediments would have extended at least as far as the thick units containing Tedford's Sites B-F (see Chapter 2) and possibly further west as the underlying Tertiary sediments of the D-Site Plateau. Erosion of these lake sediments to the east of the road may have produced the present blacksoil valley with its isolated mesas forming erosional remnants. (However, not all of the System A and D-Site equivalent deposits are of this kind—Burnt Offering Site appears to be more similar to the lens-shaped deposits common in the sediments of Systems B and C.)

Three recently discovered deposits promise to extend our understanding of Riversleigh's complex geological history. One, Low Lion Site, appears to be lower in the sequence, and hence possibly older, than any other bone-bearing deposits known from Riversleigh; another two, Dredge's Ledge Site and Cleft of Ages Site, appear to represent fissure fills in Oligo-Miocene limestone which may be appreciably younger than most Riversleigh deposits.

Low Lion Site outcrops on the western side of the D-Site Plateau, below Burnt Offering and Bitesantennary Sites, and almost certainly below the level of Site D. As yet, little limestone from the Low Lion deposit has been processed but the partial skull of a marsupial lion, which in 1989 gave the Site its name, appears to be more primitive than others collected from Riversleigh. Fossil material collected from the Site in 1990 promises to provide a great deal more information.

Dredge's Ledge, discovered in 1989, is a sheet of bones and teeth cemented into a vertical crevasse on the south-western slope of Godthelp Hill. As yet only a small amount of limestone has been processed from the Site but it seems likely that Dredge's Ledge represents a younger deposit that filled a crevasse eroded into the older limestones of the area. Similarly, Cleft of Ages Site, discovered on the Gag Plateau in 1990, appears to have accumulated within a limestone fissure. Among many specimens exposed at the Cleft of Ages Site were an incisor and the cheekteeth of a type of wombat, maybe Riversleigh's first. Pre-wombats, however, are well represented in what are now interpreted to be Riversleigh's slightly older Oligo-Miocene deposits. Needless to say, the Cleft of Ages limestone has been given priority in the acid baths at the University of New South Wales and we expect soon to be able to determine the age and importance of the deposit.

Crystalline Clues to the Past: Spring Mounds and Stalagmites

Approximately halfway up the System C sequence of the Gag Plateau, a shelf of travertine 'flow' outcrops over a wide area. Several faunal assemblages, such as the Two Trees LF, have been recovered from this layer which in areas may be as thick as 40 cm. It may represent an episode of widespread outpourings of lime-rich spring water. Fractured surfaces of this material reveal a highly porous carbonate with narrow, vertically orientated columns (possibly diagenetically altered calcite crystals).

Near Jim's Jaw Site on the Gag Plateau there is a roughly circular, metre-wide cauliflowered structure contiguous with this flow. We interpret this to represent a spring head and at least one of the sources for the carbonate rich waters that spread away from this point. A similar flow and spring mound system has been recognised in the System B sequence on Godthelp Hill. In this area we have found fossils (e.g. pseudocheirid dentaries) in the eroding surface of the flow. While we cannot exclude the possibility that these 'flows' are intrusive, it seems much more likely that they represent repeated, possibly regional events that occurred within Systems B and C.

It is possible that non-contemporaneous spring heads formed some or all of the isolated hills capped with System B and C sediments (such as the Gag Plateau and Godthelp Hill). This model involves spring water forced up through the underlying Cambrian and older Tertiary limestones to deposit calcite on the surface as travertine sheets and cauliflowers and then subsequently re-erupting through these deposits to add progressively younger carbonates until the accumulating mass became impenetrable. At that point, the spring water might erupt in another area to start another younger sequence. The faunal assemblages we find might then be interpreted as the cemented contents of isolated rim pools that developed lateral to the spring head. This might account for the oncolite-like concretions we sometimes find developed around bone fragments in some of the deposits (e.g. Grime's Site on the Gag Plateau). Against this model, however, is the fact that while the Riversleigh limestones are clearly capable of preserving surface features (such as the rippled travertine rims and fragments of straws found in Neville's Garden Site sediments), these features are unknown from any of the other sites we have so far excavated.

It is clear to us that many different depositional mechanisms were responsible for producing the variety of Oligo-Miocene sediments at Riversleigh. The task now will be to test the competing models and to contrast them with the palaeoecological interpretations independently arising from study of the equally diverse faunal assemblages.

A fortuitous collection of food scraps: the Rackham's Roost cave deposit

Rackham's Roost, the only Pliocene site recognised, appears to be the floor of a long, narrow cave that at one end opened onto a vertical face along the ancestral Gregory River (see page 51). Another larger, lower opening may have been developed more than 50 metres away in the opposite direction. Most of the millions of bones accumulated in this cave appear to have been brought in by carnivorous megadermatid bats which dropped uneaten bits and defecated the indigestible teeth and bones to the floor below their roosts. There are rarer bones of snakes which may have preyed on the bats and very rare, large kangaroos which may have used the cave for shelter or fallen through holes in the roof.

River gravels and fissure fills: Riversleigh's Ice Age deposits

The Pleistocene deposits are of two main kinds: isolated cave deposits and perched fluviatile terraces. The cave deposits represent fissure fills, megadermatid roosts and/or owl-pellet accumulations of the kind that now accumulate in the general region, for example in the caves around Camooweal.

In some instances (e.g. at Two Trees Site), what may be Pleistocene (or Pliocene) fissure fills have been found as buff- or rust-coloured heterogeneous carbonates intruded into eroded spaces within the Oligo-Miocene sediments. Some of these fissure materials contain rodents.

The fluviatile terraces are of undetermined extent but each is almost certainly defined by subsequent channelling. Terrace Site, for example, presents an approximately 3-metre thick cross-section through stream bed deposits comprising a poorly sorted basal conglomerate (containing bone, freshwater mussel shells and charcoal) which grades upwards into finer sands and silts. The bones in the fluviatile deposits may represent the discards of crocodile meals, accidental drownings or seasonal flushing of the surrounding countryside by rain-fed creeks. Some (but not most) of the bone fragments are very worn, which invites an interpretation of long-distance transport and/or re-working from older Cainozoic deposits.

Throughout the region there are many other presumably Pleistocene (or late Tertiary) bone-bearing sediments. One of the most widespread is a rust-coloured, possibly indurated soil horizon commonly found on subsoil surfaces or flanks of deeply fissured blocks of the Oligo-Miocene limestones. When it occurs as a cement on the surfaces of bone-rich limestone (e.g. as it does at Helicopter Site), it sometimes contains broken pieces of teeth and bone of Oligo-Miocene taxa. We presume that these had previously been lying free in the soil, having been naturally etched from the receding surface of limestone before being re-cemented as an overlying, probably Pleistocene, soil horizon.

In addition to these deposits presently being examined, we have encountered many others in the region that may prove to be significant when adequately assessed. Most commonly these are fissure deposits of unclear age, some of which contain bones. Other presumably Tertiary rock types in the region include zones of secondary iron enrichment in isolated patches of Cambrian (and possibly Tertiary) limestones. Wang Site itself appears blackish-grey because of primary deposition or subsequent concentration of ferromagnesium minerals. We have not seen undoubted exposures of laterite beneath System A units, although they were reported by Tedford in 1967, but we have seen a limited exposure of what appears to be laterite immediately north of Megabyte Hill (the location of Wang Site) on the north-western edge of the eastern portion of the Gag Plateau. Its stratigraphic relationships to the fossiliferous deposits have not yet been determined.

Work is presently under way to test and refine these preliminary concepts. Dirk Megirian from the Museums and Art Galleries of the Northern Territory in Darwin is concentrating on the geology while about 30 students, colleagues and the rest of us are focusing on the faunas. All of the many kinds of evidence that might clarify understanding about Riversleigh's stratigraphy and palaeoecology will be examined.

Preliminary results from Dirk's studies suggest that the older Riversleigh sediments may have formed as alluvial fans of carbonate mud spread perhaps by seasonally torrential rains. He has also determined that the region has had a complex history of minor fracturing (faulting) that may have produced small basins within which some of the younger, fossil-rich sediments accumulated.

GAG SITE

By 15 million years ago, the hint of a cool, perhaps slightly drier change was in the air. While rainforest still dominated the area, some of the stranger creatures that inhabited the older rainforests, such as ilariids and wynyardiids, had vanished and several groups common today, including dasyurids, had begun their rise to dominance. This was also a spectacular time for possums. In the green canopies that cascaded over the little pool at Gag Site there were at least 18 different kinds of possums including more than 9 kinds of leaf-eating ringtails. Plant species diversity would have been correspondingly high, making these forests richer than any in Australia today.

The floor, too, was alive with weird kangaroos, including flesh-eaters and an abundance of browsers. Ancestral thylacines foraged for prey along the water's edge while cow-sized browsers pushed their way through the undergrowth. Bats may not have been as varied at this time but they were still abundant, possibly using tree leaves as well as caves for shelter. All up, the Dwornamor forest at Gag Site was beautiful and busy. But this was the twilight of paradise. Rain, the life blood of diversity, was about to be put on ice—at the poles of a cooling world.

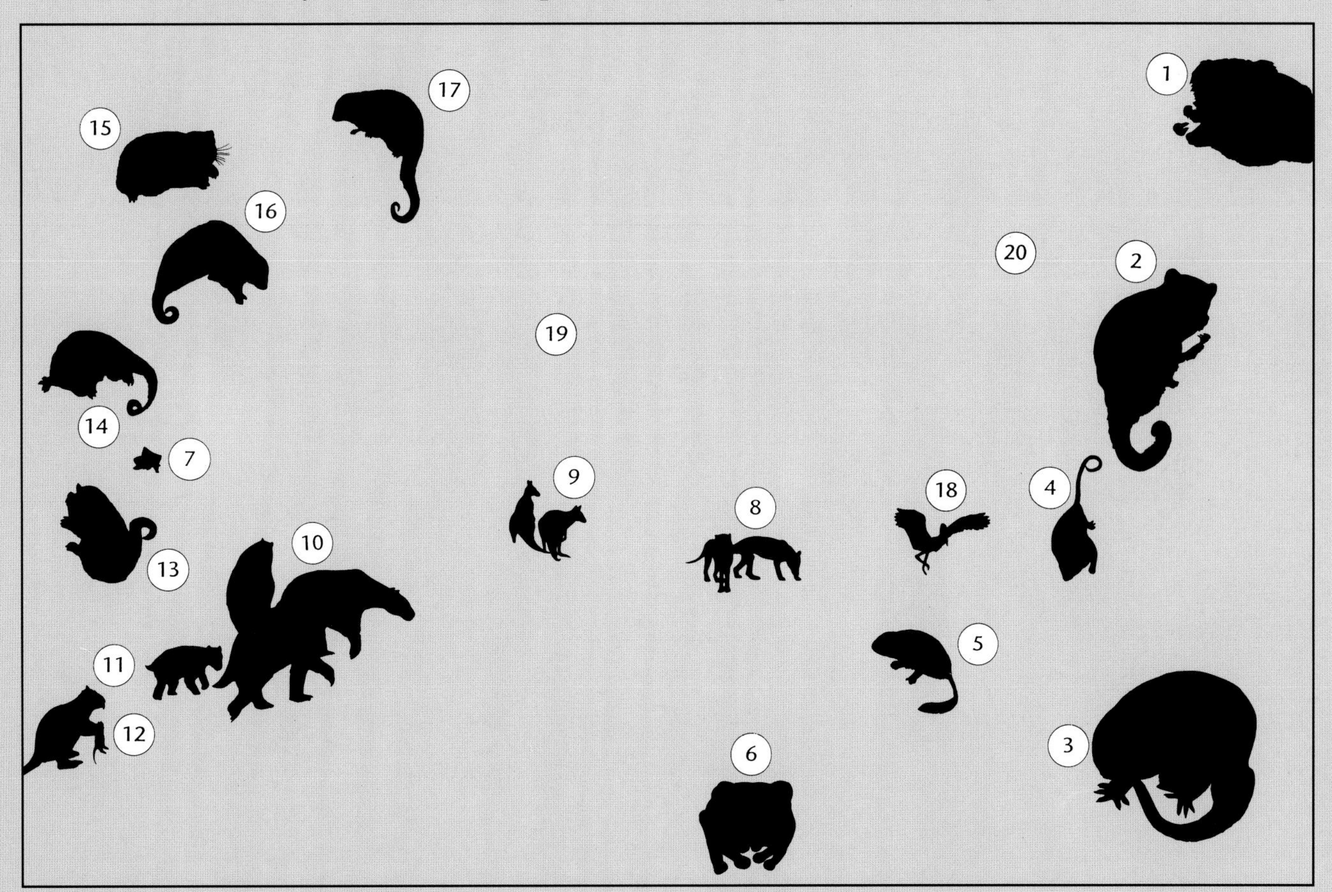

1 Rainforest koala (*Litokoala* sp.)
2 Woolly ringtail possum (*Pseudochirops* sp.)
3 'Ground' cuscus (*Strigocuscus reidi*)
4 Pygmy possum (*Cercartetus* sp.)
5 Feather-tailed possum (acrobatid)
6 Giant tree frog (*Litoria* sp.)
7 Leaf-nosed bat (*Brachipposideros* sp.)
8 Small carnivorous thylacine (*Thylacinus* sp.)
9 Balbarine kangaroos.
10 Large, browsing diprotodontid marsupials (*Neohelos* sp.)
11 Carnivorous kangaroo (*Ekaltadeta ima*)
12 A weird roo's dinner.
13 Brushtail possum (*Trichosurus* sp.)
14 Ringtail possum 2 (*Paljara* sp.)
15 Ringtail possum 3 (*Pildra* sp.)
16 Ringtail possum 4 (unnamed genus)
17 Ringtail possum 5 (*Marlu* sp.)
18 Stork (ciconid)
19-20 Tall, species-rich rainforest.

CHAPTER 9

Riversleigh's Prehistoric Environments

In this chapter, we examine in detail six different-aged assemblages: Site D, Upper Site, Bitesantennary, Rackham's Roost, Terrace Site, and the modern biota of the Riversleigh region. Using this information, we can then consider how the faunal balance of these communities and their environments has changed over the last 25 million years. A list of these changing rainforest mammal communities is provided in Appendix 2 and should be referred to for details.

Five come alive

Although Riversleigh's Tertiary fossils were originally thought to represent one environment and one distinct assemblage of medial Miocene animals (the Riversleigh Local Fauna of Tedford 1967, from his Sites A-F), we now recognise about 150 local faunas from this area representing many different environments. The five examples chosen here are broadly representative of the range of communities that occurred in the Riversleigh region during the last 25 million years. As a basis for comparison, we have also examined the modern mammal fauna of the Riversleigh region. The examples are:

1-2. The suite of Oligo-Miocene sediments (about 25 to 15 million years BP) includes some lake deposits that represent deep-water to near-shore environments (e.g. the Riversleigh Local Fauna in part from Site D); and many pool deposits that represent rainforest environments (e.g. the Upper Site Local Fauna.)
3. A few Oligo-Miocene cave deposits represent rainforest environments (e.g. the Bitesantennary Local Fauna).
4. At least one Pliocene cave deposit represents an open forest/grasslands environment (e.g. the Rackham's Roost Local Fauna).
5. Several Pleistocene fluviatile deposits represent riverine and near shore environments (e.g. the Terrace Site Local Fauna).
6. The modern mammal fauna of the Riversleigh region, drawn from woodland savannah and spinifex grasslands, is also similar to pre-modern Holocene faunal assemblages found in many of Riversleigh's caves and younger River deposits.

1. The Riversleigh Local Fauna from Site D: an overlap of lake and land

Site D was discovered at least as early as 1963, by Dick Tedford and Alan Lloyd and its local fauna was the first to be examined. Estimated to be late Oligocene to

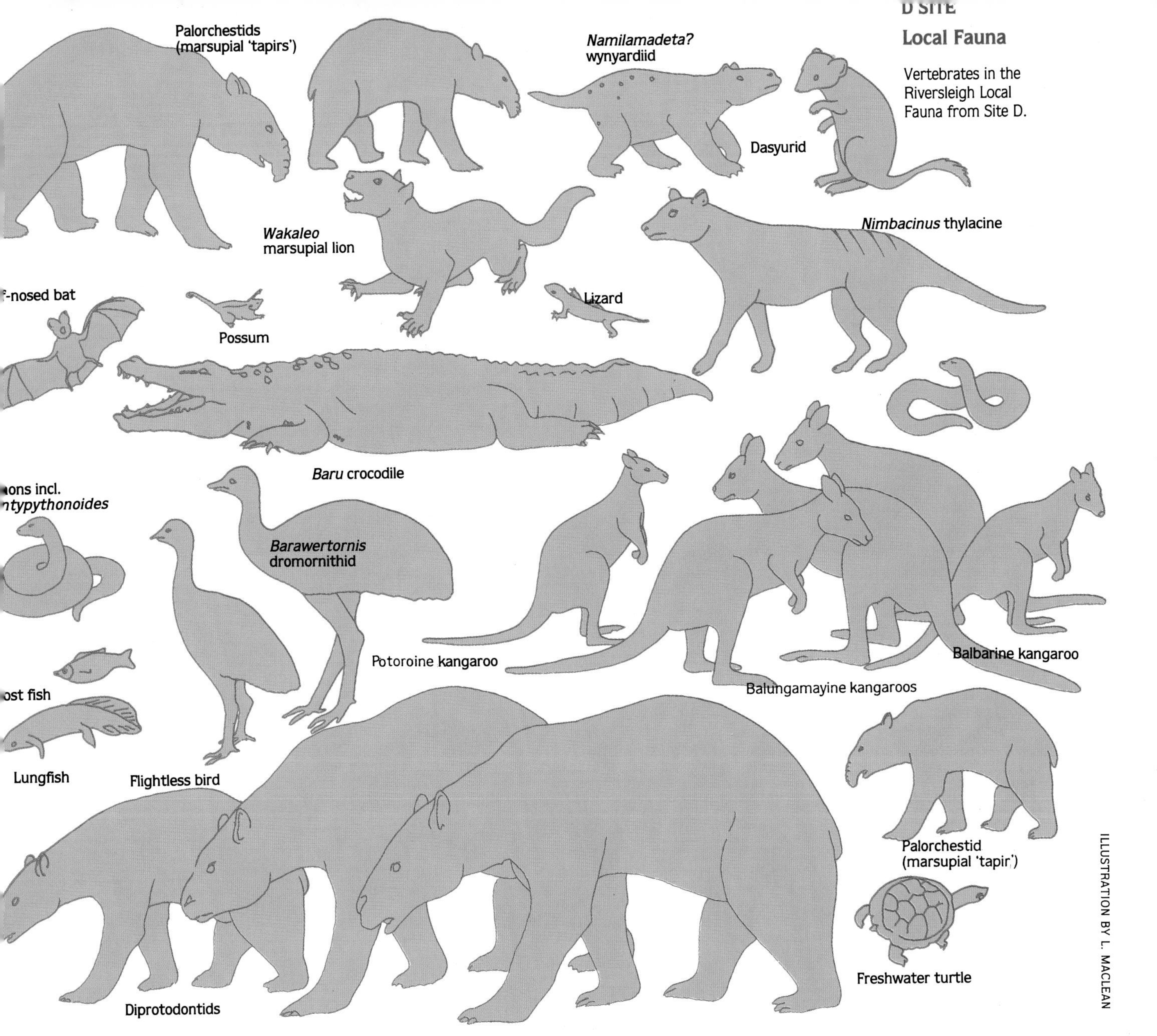

early Miocene in age, it is among the oldest of Riversleigh's local faunas (although perhaps not as old as the Low Lion and Dirk's Towers LFs, both of which have yet to be assessed). Because it was the first local fauna found and was at the time thought to be the only one in the area, it was called *the* Riversleigh Local Fauna. Although we know now there are many others in the Riversleigh region, it must nevertheless retain the name.

Despite its early discovery, the Riversleigh LF is certainly *not* a diverse fauna, perhaps because it appears to have accumulated in a relatively vast sedimentary basin at some considerable distance from the forests that sheltered the majority of the animals of the day. The vertebrate fossil remains here are not concentrated as they are at most of Riversleigh's other Oligo-Miocene sites. The bulk of the vertebrate material visible at Site D represents aquatic vertebrates, such as crocodiles and turtles. The few mammals known are mostly large in size. The bones of smaller mammals, which are such a significant part of most of Riversleigh's other Oligo-Miocene assemblages, failed to accumulate as far from the forest edge as many of the larger bones. The abundance of large bird (dromornithid) bones from this site suggests that at least some individuals actually foraged in the lake or in marshes along its edge.

WHY ARE RIVERSLEIGH'S OLIGO-MIOCENE BIOTAS INTERPRETED AS RAINFORESTS?

We interpret the older Riversleigh local faunas (i.e. those from Systems A-C, including Riversleigh, Upper Site and Bitesantennary) to represent rainforest communities for five main reasons:

1. High species diversity in restricted local faunas (e.g. 63 mammal species in Upper Site LF) indicates a rich, finely-divided and stable (i.e. constant) resource base.
2. As a correlate, there appear to have been complex feeding 'guilds' involving up to six species of small sympatric mammals (e.g. the peroryctid-like bandicoots under study by Jeanette Muirhead), a situation most likely to reflect high resource diversity and environmental stability.
3. The high number of sympatric obligate leaf-eaters in single local faunas (e.g. nine species of pseudocheirids and three other selenodont marsupials in the Dwornamor LF) indicates that many different species of trees occurred within relatively small areas, a common feature of rainforest but not sclerophyll forests.
4. The presence of many groups whose modern representatives are largely restricted to rainforest communities (e.g. species of *Strigocuscus, Hypsiprymnodon, Pseudochirops, Cercartetus caudatus*-like pygmy possums, diverse peroryctid bandicoots and logrunner birds [*Orthonyx*]).
5. While there are vast numbers of browsing marsupials in the early and middle Miocene assemblages, there is a complete absence of grazers (i.e., no macropodine kangaroos) suggesting that the forest canopy was closed. The only possible grazer known, a rootless wombat from Encore Site, may be part of an early late Miocene assemblage (Encore Site) that accumulated after the rainforests of the middle Miocene began to open up.

The last point deserves elaboration. High-crowned toothed kangaroos (macropodines) do not make their first appearance until the Bullock Creek LF (middle Miocene of the central Northern Territory) and even there they are extremely rare (one relatively high-crowned kangaroo tooth!). Undoubted grazers do not become a significant feature of Australia's mammal communities until the early Pliocene. Previous reports of grass pollens in the Oligo-Miocene Etadunna Formation in South Australia may represent aquatic rather than emergent grasses. Certainly, if there were grasslands in Australia when Riversleigh's or central Australia's Oligo-Miocene faunas were accumulating, we see no sign of them reflected in the teeth of the animals that might otherwise have used them for food.

It is possible that Riversleigh's high diversity of fossil mammals resulted from species from distant drier habitats wandering in to be fossilised alongside forest species from the immediate region. However, we have found no evidence for this in the sediments or the biotas and there is no palaeobotanical evidence for contemporary drier habitats. Further, some of the highest levels of diversity are among leaf-eating possums which seem unlikely to wander between habitats. Finally, mammal diversity in modern rainforests (e.g., mid-montane New Guinea and the lowlands of Sarawak) is *higher* than that found in the Riversleigh deposits which suggests that there is probably nothing unusual about the levels of diversity found in the assemblages of Riversleigh. The contrast between the diversity in Riversleigh's assemblages and those of modern Australia, which *is* striking, may be accounted for by the severe contractions these rainforests endured during dry periods of the Pleistocene.'

We are unclear about the precise nature of the Oligo-Miocene rainforests of Riversleigh. Fossil plant material collected from Riversleigh's new Dunsinane Site during the 1990 field season is yet to be analysed. However, the high mammal diversity in single locations, particularly for sympatric obligate tree-leaf eaters, and the presence of a wide size range of herbivores suggest a comparably high diversity of plant species and multi-level distribution of these resources within the forest. This argues for a complex lowland tropical rainforest with a partially open understorey.

2. The Upper Site Local Fauna: in the heart of the green cathedral

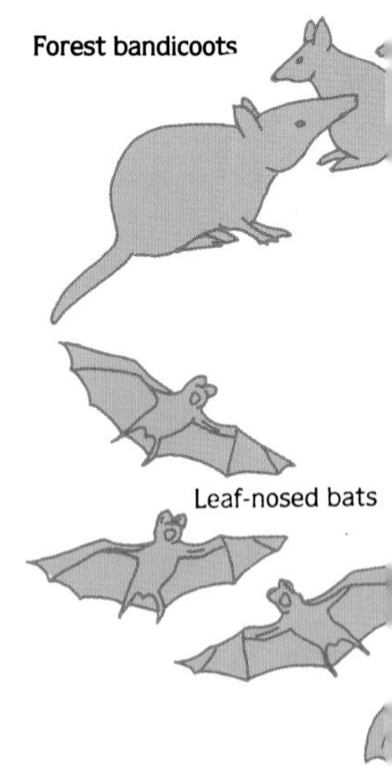

Upper Site was discovered on the inclined slope of Godthelp Hill by Henk in 1985. It stands proud of the surrounding slope as an erosional remnant of a slightly higher level, thus earning its name. We quarried approximately 1 tonne of fossiliferous limestone from this site in 1985 and another 0.5 tonne in 1986, almost all of which is now processed.

By modern standards, this early Miocene assemblage is diverse with almost twice the number of mammal families and twice the number of marsupial species of any surviving Australian or New Guinean ecosystem.

The animals recovered included a diverse assemblage of vertebrates as well as, perhaps most surprisingly, uncrushed arthropods. Wrinkled sheets of what we interpret to be algal mats were also recovered. Many of the invertebrate remains, in particular, dozens of coleopteran larvae occurred in vertical tube-like extensions of these mats which may have grown down into the bottom mud of the pond or hung below the algal mats if they floated. Although similarly preserved invertebrates are now known from other sites, the Upper Site invertebrate fauna is so far the most diverse.

The palaeoenvironmental conditions under which the Upper Site Local Fauna accumulated appear to be reasonably clear. This was a small pool on the floor of a complex, species-rich, lowland tropical rainforest.

Thingodonta

Upper Site Local Fauna

Vertebrates in the Upper Site Local Fauna.

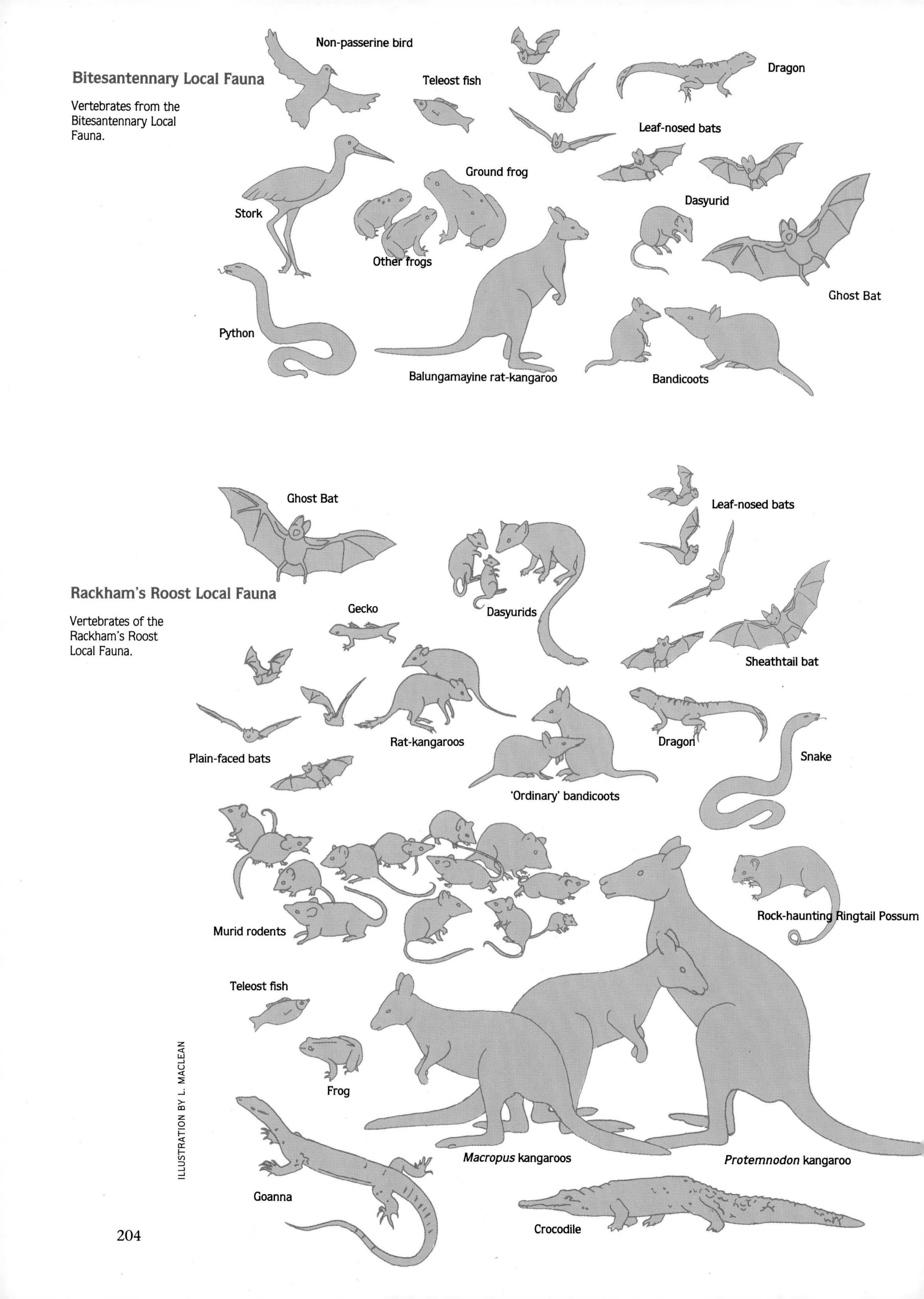
Bitesantennary Local Fauna
Vertebrates from the Bitesantennary Local Fauna.
Non-passerine bird
Teleost fish
Dragon
Leaf-nosed bats
Ground frog
Stork
Dasyurid
Other frogs
Ghost Bat
Python
Balungamayine rat-kangaroo
Bandicoots
Ghost Bat
Leaf-nosed bats
Rackham's Roost Local Fauna
Vertebrates of the Rackham's Roost Local Fauna.
Gecko
Dasyurids
Sheathtail bat
Plain-faced bats
Rat-kangaroos
Dragon
Snake
'Ordinary' bandicoots
Murid rodents
Rock-haunting Ringtail Possum
Teleost fish
Frog
Macropus kangaroos
Protemnodon kangaroo
Goanna
Crocodile
ILLUSTRATION BY L. MACLEAN

RACKHAM'S ROOST SITE

By 5 million years ago, Riversleigh's rainforests had almost certainly vanished, taking with them nearly all of the forest animals. Open forest with sparse trees and yellowing grasslands filled the void. Caves were now essential for the survival of many of the diverse kinds of bats that still dominated the night skies. Ghost Bats were lords of the ancient Rackham's Roost cave, for they were ferocious carnivores. As moonlight flickered from the leaves of the gum tree in which rested the last of the region's ringtails, and large browsing kangaroos trampled the dew-wet grass underfoot, these flying predators hunted small marsupials and the recently arrived rodents that scurried among the rocks and grasses beyond the cave entrance. Captured prey was carried to the deep recesses of the cave. Discarded bits steadily accumulated on the floor as guano. Near the larger entrance facing away from the river, pythons waited patiently for careless bats, as they had done for millions of years. Here blocks of limestone that had fallen from the roof and walls provided hiding places for wary mammals. Like the ancient forests that had gone, the cave's life, too, was winding down. Soon it would be little more than the bone-rich floor that an excited Alan Rackham would discover.

1, 21 Plain-faced bat (vespertilionid sp. 1).
3-9 Leaf-nosed bat (*Brachipposideros* sp. 1).
10-12 Leaf-nosed bat (*Brachipposideros* sp. 2).
13, 22 Ghost Bat (*Macroderma gigas*).
14 False mouse dinner (*Pseudomys* sp.).
15-17, 20 Sheath-tailed bat (*Taphozous* sp.).
18-19 Orange Horseshoe Bat (*Rhinonicteris aurantius*).
23-25 Plain-faced bat (vespertilionid sp. 2).
26 Fig tree clinging to entrance face of Rackham's Roost cave.
27 Gum tree (*Eucalyptus* sp.).
28 Goanna (*Varanus* sp.).
29 Wallaby (*Prionotemnus* sp.).
30 Pythonid snake.
31 Bone-rich guano pile accumulating beneath roosts of ghost bats—future fossil deposit of Rackham's Roost Site.
32 Scavenging insects of various kinds.
33 Rock rat (*Zyzomys* sp.).
34 False mouse (*Pseudomys* sp.).
35 Carnivorous phascogale (*Phascogale* sp.).
36 Grasses—abundant in surrounding countryside.
37 Rat-kangaroo (*Bettongia* sp.).
38 Insectivorous dunnart (*Sminthopsis* sp.).
39-40 Short-footed giant kangaroo (*Protemnodon* sp.).
41 Open savannah woodlands.
42 Gecko (gekkonid lizard).
43 False mouse (*Pseudomys* sp.)

Dunphy

Terrace Site Local Fauna

Vertebrates of the Terrace Site Local Fauna.

3. Bitesantennary Local Fauna: black hole in the forest floor

Bitesantennary Site, although in fact discovered in the Australian Bicentennary year of 1988, was actually named after a bizarre incident involving the consumption of a raw 'prawn'. It may once have been part of a chain of small caves or tunnels that led to what we now know as Neville's Garden Site. Its structure suggests that it was a pot-like chamber, open to the surface via a collapsed roof and half filled with water. If so, it may have been an ideal pit-trap—a 'hole' in the forest floor. Bats which dominated the local fauna almost certainly roosted in colonies under the edges of the hole and watched as forest creatures wandered off the normal paths at night to suddenly find themselves hurtling down into the dark pool below.

Although this scenario is plausible, at this stage we know very little about the local fauna. To date we have only processed about 0.5 tonnes (i.e. about 6%) of the available material. It may take several years before the fauna is sufficiently well known to test this hypothesis. Similarly, many more non-bats may have to be discovered before we can more confidently determine the age of this assemblage.

4. Rackham's Roost Local Fauna: bat penthouse with excellent views

Rackham's Roost represents a Pliocene cave deposit. The evidence for this is overwhelming, with remnants of the ancient walls and floor still in place. On one side the cave evidently opened onto a cliff facing the ancestral Gregory River. Remnants of the floor material, rich with fossil bones, fill cracks and crevasses all the way out onto the present cliff face. At the other end, it appears that the cave opened out into an amphitheatre with a flat, grassy area of woodland surrounded by steeply inclined slopes leading up to limestone rubble and cliffs. The fossilised bone-rich cave floor now laps up against the remnants of the Cambrian limestone walls. The roof has long since gone, eaten away by millions of years of naturally acid rain.

The vast majority of bones fossilised in the floor were either brought in by carnivorous ghost bats or were those of other bats that lived, presumably warily, in the same cave. Although a few creatures appear to have entered under their own steam, such as pythons and a couple of kangaroos that may have used the entrance as a shelter, only creatures small enough to have been the prey of ghost bats are represented. This means of course that the Rackham's Roost Local Fauna is not a fair representation of the creatures that lived in the habitats surrounding the cave. While we cannot be certain, it seems likely that on the plains outside marsupial lions, large diprotodontids and trunked palorchestids roamed. In the trees there may well have been koalas and brushtail possums but of these, as yet, we have no trace.

Although dominated by the small animals of the surrounding environments, Rackham's Roost does reveal a great deal about the way in which the older Oligo-Miocene assemblages were transforming into those more characteristic of the modern world.

The nature of habitats outside the Rackham's Roost cave can be deduced by the nature of the small, non-flying mammals present. None of the lineages suggest rainforest; all animals in the fauna have as their closest living relatives forms that live today in more open habitats. For example, while *Sminthopsis leucopus* and to a lesser extent *S. murina* live in seasonally moist habitats, the other more than 13 living species prefer the drier, even desert, environments of modern Australia. Similarly, no living species of *Planigale* lives in closed forest. *Planigale maculatus* appears to prefer sclerophyll forests and coastal grasslands but all the other species prefer the semi-arid areas of cracking blacksoil in inland Australia.

Of the rodents present, *Zyzomys* species today are inhabitants of dry rock piles in northern Australia. Further, the teeth of all of the rodent species recovered indicate that they were granivores, which are found more typically in grasslands than rainforests.

Seemingly contradictory indications are given by the persistence here of several species of *Brachipposideros* bats, the group that dominated the Oligo-Miocene rainforests of this region. But considering that the only living member of this ancient group, the Orange Horseshoe Bat, survives in this arid region today by sheltering during the day in hot and very humid caves, perhaps a microenvironment of this kind characterised the Rackham's Roost cave in the Pliocene.

Two apparent anomalies in this list warrant comment: the crustacean and the crocodile. The crustacean is not really unexpected. Beneath the roosts of living Ghost Bats we have collected among food debris crab carapaces and limbs, the remains of meals evidently regularly enjoyed by these bats. The crustacean gastrolith may have belonged to a terrestrial crab grabbed from seasonally wet black soils outside the cave's entrance. But what is a crocodile doing in a cave perched almost 30 metres above the level of the Gregory River? Three possible explanations come to mind. Perhaps the ancestral Gregory River and its banks at this time were in fact much higher than they are now. With a river like this, charged with so much calcium carbonate that it deposits at the flick of a ripple, it readily builds tufa dams that are then self-generating with the consequence that the river level can rise behind the dams. Alternatively, the croc may have been a quinkanine that was partially terrestrial and able to climb the hills to the level of the cave entrance. Two Pleistocene cave occurrences of quinkanine crocodiles are already known from other areas of Australia. Yet again, and perhaps most likely, the croc and the crustacean were brought into the cave which may have been used as a den by terrestrial scavengers like large dasyurids and thylacinids.

On balance, we have concluded that the Rackham's Roost LF was culled from a dry sclerophyll forest or woodland with a grassy understorey, probably not too unlike the environment that dominates Riversleigh today.

5. Terrace Site Local Fauna: from the cellar of an ancient river

Although Pleistocene river deposits are very common over the whole continent of Australia, they are relatively rare at Riversleigh. Perhaps this is because, as far as we can determine, the present river system is not very old. The diverse Oligo-Miocene deposits suggest that Riversleigh's drainage system was mainly subterranean. This is why Riversleigh's Terrace Site is so distinctive; it clearly is a river deposit, albeit a relatively young one.

The fauna is dominated by riverine animals, including freshwater mussels, fish, large turtles and a variety of crocodiles. It also contains the bones of animals that were washed in by floods, fell in along the banks or were dragged in by crocodiles. Consequently, there is again a size bias, as there is at Rackham's Roost, but this time for the larger animals, creatures more likely to be the target of aquatic predators and less able to extricate themselves once they fell off the bank. This 'imbalance' was further exacerbated by the carrying capacity of the ancestral river itself. The fossil bones are bound into a matrix of coarse cobbles and sand. Presumably bones of smaller animals would either have been washed further downstream or were pulverised by the cobbles as they churned along in the river's bed load. However, there are enough small mammal bones to suggest that with continued processing of this material, a more balanced fauna will be recovered.

There is very little in the Terrace Site Local Fauna (Appendix 2) to suggest an environment significantly different from the one that dominates the same area today. Agile Wallabies (*Macropus agilis*) today dominate the grasslands surrounding many of northern Australia's rivers. The Water Rat (*Hydromys chrysogaster*) is present in the rivers of coastal Australia. Freshwater Crocodiles (*Crocodylus johnstoni*) are similarly common in the freshwater rivers of northern and parts of eastern Australia. While the habitat requirements of *Diprotodon optatum* are not clearly understood, it is known as a late Pleistocene fossil from areas as widespread as the Warburton River near Lake Eyre to the Kimberley in northern Western Australia to Tasmania. It may have been a browser/grazer that lived a hippopotomus-like existence as a member of herds that wandered along the margins of lakes and rivers.

6. The modern Riversleigh mammal fauna: cats & rats & lots of bats

It is difficult to compare fossil faunas with the living animals of the same area because the fossil deposits will almost always contain a limited representation.

Information about the living fauna of an area can be gathered from a much wider area and in a much more thorough manner. For example, invertebrate animal groups that do not fossilise as readily as vertebrates will be seen to form a far greater part of the living fauna.

We have restricted our tally of modern animals to the mammals of the Riversleigh region, the group so far best surveyed. This group also has one of the best records in the region over the last 25 million years.

Although the modern habitats represented at Riversleigh are quite variable, we have pooled all of the mammals known from the region into a single list. Despite the fossil faunas being accumulated from perhaps no more than a hectare of relatively uniform prehistoric habitat, we doubt that it will cloud the overall pattern of change through time.

Five species – Water Rats, Orange Horseshoe Bats, Ghost Bats, Yellow-bellied Sheathtail Bats, Rock-haunting Ringtail Possums – known from the region of Lawn Hill Gorge National Park immediately north of Riversleigh Station are included because it is highly likely that they are also on Riversleigh Station. Droppings of the Rock-haunting Ringtail have been found along the cliffs of the Gregory River and Water Rats have almost certainly been sighted in the Gregory River itself. We have also listed all introduced mammal species that have established feral populations. The modern bat fauna was compiled with the help of Ray Williams and Susin Churchill.

Naturally, there is relatively little difficulty in interpreting the environmental conditions under which the contemporary biota lives! Riversleigh Station occurs in the southern watershed of the Gulf of Carpentaria where on average it experiences hot wet summers and warm dry winters. The major vegetation types of the region vary from remnant vine scrub along the margins of the Gregory River, dominated by paperbarks, figs, pandanus and eucalypts, to savannah woodlands on the flats to spinifex grasslands with some eucalypts and bauhinias on the hills. The mammal fauna includes a few species shared with central Australia, such as Red Kangaroos, but more shared with other areas of northern savannah woodland and black soil plains. This is now, on balance, a hot, dry land. Unpredictable but frequent droughts are the primary natural factor that limits the mammal diversity of the area.

Riversleigh's last 25 million years: fading shadows of green mansions

From the albeit patchy summaries presented above, there appears to have been a marked reduction in diversity within the Riversleigh region over the last 25 million years. But, three of the six local faunas considered here—Bitesantennary, Rackham's Roost and Terrace Site—almost certainly significantly underestimate the real diversity of their original habitats.

In order to more fairly compare changes in diversity over the period of time represented by Riversleigh's Cainozoic deposits, we have tallied, in Fig. 1, mammal diversity for seven different-aged local faunas spanning the last 25 million years. To the six discussed already (including the modern fauna), we have added the Dwornamor LF from Gag Site, one of the richest known in the System C sequence.

Ignoring the three obviously under-represented LFs (Bitesantennary, Rackham's Roost and Terrace Site), species diversity in the region appears to have dropped from a high of at least 68 at any one time in the late Oligocene/early Miocene faunas to a low of 31 in the Pliocene/Quaternary—a drop of at least

FIG. 1

Changes in mammal species diversity at Riversleigh

Family of mammals	Riversleigh LF: 25 myo	Bitesanten. LF: ?20 myo	Upper Site LF: 20 myo	Dwornamor LF: 15 myo	Rackham's Roost LF: 3-5 myo	Terraces Site LF: ~ 20 000 myo	Modern
Carnivorous mars.							
Thylacinidae	1	-	1	2	-	-	-
Dasyuridae	1	1	5	6	3	-	2
Bandicoots							
cf. Peroryctidae	-	2	6	7	-	-	-
Peramelidae	-	-	-	-	2	-	-
Perameloidea	-	-	2	3	-	-	-
Marsupial moles							
Notoryctidae	-	-	1	-	-	-	-
Koalas							
Phascolarctidae	-	-	2	2	-	-	-
?Phascolarctoidea	-	-	1	-	-	-	-
Diprotodontoids							
Diprotodontidae	3	-	3	2	-	1	-
Palorchestidae	3	-	-	?1	-	1	-
Wynyardiids							
?Wynyardiidae	1	-	1	-	-	-	-
Wombat-like							
Vombatomorphia	-	-	1	-	-	-	-
Marsupial lions							
Thylacoleonidae	1	-	2	-	-	-	-
Possums							
Burramyidae	-	-	2	2	-	-	-
Phalangeridae	-	-	2	3	-	-	-
Pilkipildridae	-	-	1	1	-	-	-
Pseudocheiridae	-	-	5	9	1	-	1
Petauridae	-	-	1	-	-	-	-
Petauroidea	-	-	1	1	-	-	-
?Petauroidea 2	1	-	2	1	-	-	-
Acrobatidae	-	-	?2	1	-	-	-
Kangaroos							
Potoroidae	3	1	8	8	1	2	-
Macropodidae	?2	-	2	2	3	-	6
Thingodontans							
Yalkaparidontidae	-	-	1	1	-	-	-
'Weirdodontans'							
Yingabalanaridae	-	-	1	-	-	-	-
Bats							
Hipposideridae	?1	5	7	4	4	-	2
Megadermatidae	-	-	1	2	1	-	1
Molossidae	-	-	1	-	-	-	1
Emballonuridae	-	-	-	-	2	-	2
Vespertilionidae	-	-	?1	-	4	-	8
Pteropodidae	-	-	-	-	-	-	1
Rodents							
Muridae	-	-	-	-	13	2	4
Dingoes							
Canidae	-	-	-	-	-	-	1
Humans							
Hominidae	-	-	-	-	-	?1	1
Echidnas							
Tachyglossidae	-	-	-	-	-	-	1
Families represented	10	4	27	19	10	4-5	13
Species (min. numbers)[1]:	**17**	**8**	**63**	**58**	**33**	**6-7**	**31**

[1]From Riversleigh local faunas in addition to those noted here, we have representatives of: a new diprotodontian family (e.g. Hiatus Site 1, ?System A); Ektopodontidae (*Chunia* sp. from Dirk's Towers, ?System A; and *Ektopodon* sp. from Waynes Wok, System B); ?Miralinidae (e.g. Wayne's Wok, ?System B); Ilariidae (new genus & species; e.g. White Hunter, ?System B); a new family of vombatiform diprotodontians (e.g. Boid Site East, System B); a new phalangeridan family near to but distinct from phalangeroids (e.g. Wayne's Wok, System B); and Ornithorhynchidae (*Obdurodon* sp., e.g. Ringtail Site, System C). We can also presume that palorchestids were present in System B LFs because they bracket this period of time in the region. Introduced mammals are not included in the totals for the modern Riversleigh fauna.

58%. The drop in family-level diversity over the same period is even more striking: from at least 34 in the Oligo-Miocene to no more than 13 in the modern community—a drop of at least 62%. These marked drops in species-level and family-level diversity appear to mirror the decline of the region's rainforest and its replacement with open forest/savannah habitats.

As mentioned previously, diversity of the Rackham's Roost LF is almost certainly under-represented because of the reduced likelihood of finding middle-sized to large animals in a deposit primarily accumulated by ghost bats. Similarly, the low number of species in the Riversleigh Site LF probably reflects peculiarities of the way in which the Site D fossils were accumulated.

Looking at Fig. 1 more closely, it is evident that most of the losses in species-level diversity have been among the marsupials and most of the gains among the placentals such as rodents and bats. Many marsupial groups, such as thylacoleonids, thylacinids, wynyardiids, diprotodontids and pilkipildrids, suffered total extinction over the last 25 million years. Many others vanished from the Riversleigh region but persisted elsewhere, such as some kinds of bandicoots, marsupial moles and feathertail possums. In fact most mammal groups show reductions in diversity in most Australian habitats over this period of time. Only *one* group of marsupials, the grazing kangaroos, shows a striking increase in diversity, presumably in response to the reduction of rainforest and concomitant spread of grasslands.

The low diversity of marsupials known from the Pliocene Rackham's Roost LF may be anomalous because of limited prey selection by megadermatid bats. However, if we compare species- and family-level diversity at this site with that from Gotham City Site, another megadermatid-accumulated assemblage at Riversleigh but one from the Oligo-Miocene, the marked drop in marsupial diversity is still evident. Hence we suggest that these groups were absent from the habitats around Rackham's Roost.

Within the placental groups over the last 25 million years, a change in bat diversity is evident. High diversity of leaf-nosed bats (hipposiderids) in the older deposits gives way to higher diversity of 'ordinary bats' (vespertilionids) in the younger deposits. This inversion reflects a similar change during the same period in Europe and probably reflects a world-wide late Cainozoic change from more tropical conditions in the early to middle Tertiary to drier habitats in the late Tertiary to Quaternary.

The *proportional* increase in the number of bat species is also clear although the absolute number stays more or less the same. However, the nature of these similar-sized bat assemblages changes quite significantly as we have pointed out in Chapter 4 with vespertilionids increasing and hipposiderids decreasing.

In summary, Riversleigh's environments and their biotas appear to have undergone a major change from complex, species-rich lowland tropical rainforest in the Oligo-Miocene, to open forest in the Pliocene and monsoonal savannah woodlands in the Quaternary. Today the area is seasonally as dry as central Australia, although in the summer it normally blossoms under the influence of the 'Wet', when monsoonal rains sweep across the bottom of the Gulf of Carpentaria. All that remains today of the vast rainforests that once carpeted the region are remnants of vine scrub, such as the Cluster Figs that precariously cling to the edges of the Gregory River and, of course, a few metamorphosed descendants from those ancient forests, such as the living mammals that now peer at us from beneath dust-covered lashes.

CHAPTER 10

Excavation and Collection in the Field

The rate of discovery of new sites, weird creatures and biostratigraphic understanding depends on the efficiency of field and laboratory techniques. The palaeontologist must consider a range of often critical questions about the individual site.

Sampling the Fossil Deposits

How much should be collected from a site to obtain understanding about its significance? How many sites need to be sampled to provide an adequate understanding about palaeodiversity and palaeoecology? Important factors to consider when deciding how much should be sampled from a particular site include quality of preservation. We must also watch for 'mixed' samples from different aged sediments. Questions of this kind are important because the answers may set limits on the adequacy of information arising from the work.

In practice, we make an initial effort to sample, even if only superficially, all isolated bone-rich sites to determine whether they are either faunistically and geologically similar to an already known site—at which time sampling normally ceases, or distinctive in either respect. Initial sampling normally involves about 50-150 kilograms. If this sample suggests that a larger one would reveal a distinctive biota or palaeoenvironment, the following year we may take a further 0.5-2 tonnes. In most cases, experience teaches us that this is enough to indicate *most* of the taxa present, depending of course on how densely bone-rich the deposit is. For some sites, like Camel Sputum, probably 4 tonnes would be required before a fauna as diverse as that from Upper Site was obtained, although the material from Camel Sputum is generally better preserved.

Inevitably, no matter how large a sample is taken, some taxa will be poorly represented. This could either reflect general rarity in the prehistoric community or 'intrusion' of a foreigner from a distant environment—something that could happen if, for example, a predatory bird dismantled an animal at one site that it had caught at another. By way of example, despite almost 2 tonnes of material processed from the bone-rich Upper Site, we have only recovered a single tooth of the extraordinarily distinctive *Yingabalanara richardsoni* and barely three teeth of a yet to be named new family of pre-wombat-like marsupials.

There is a wrinkle here. Because we normally cannot tell how 'deep' into the cliff face a particular bone deposit goes, it is sometimes hard to gauge what percentage of the total material at that site we have collected. Normally, however, it is clear that at least 50% of the bone-bearing sediment remains at the site, and usually far more. An exception to this self-limiting stricture was Upper Site where more than 50% was inadvertently sampled.

This peculiar deposit was sandwiched between an overlying bone-rich but less diverse level and an underlying bone-poor limestone. After we had removed about a tonne from blocks jumbled up at the front of the cliff, we could see the wide, distinctively black-banded bone-rich level continuing straight into the cliff. When we returned the next year all fired up to sample a bit more, we were mortified to find that although the black-banded layers did indeed continue into the cliffs, they became relatively bone-poor the further in we went. So, although Upper Site is still capable of being worked, the rewards for effort have fallen off steeply. That said, it is of course entirely possible that the level will pick up again further into the cliff. If it does, others after us may yet recover a second tooth of *Yingabalanara richardsoni*!

Also considered is the presumed palaeoenvironment of a particular deposit. For example, some sites are dominated by aquatic animals. It has become customary to wait until we have on the expedition an authority on aquatic vertebrates before these sites are touched. For example, although the Melody's Maze Site was discovered in 1983, we resisted the temptation to sample it until Anne Kemp arrived in 1988 because it was dominated by fossil lungfish—Anne's specialty. Similarly, the spectacularly turtle-rich Crusty Meat Pie Site was discovered in 1987 but not sampled until Arthur White could join the expedition of 1989. These authorities have a better idea about what would be an appropriate sample size to collect.

The Importance of Collection Data

The cardinal rule of palaeontological collecting is to take nothing away unless you can guarantee to be able to put it back where it came from at any time in the future. To be able to do this, it is essential to accurately label material at the time it is collected and to ensure that the relationships between a fossil, its site and geological setting are recorded. Museums have far too many specimens with labels that say 'Locality unknown'. Without site data, fossils degenerate to mere curiosities without age, a home or palaeoenvironmental context. In essence, they are critically devalued.

To maintain site data, we routinely do a number of things. First, we make sure that samples taken from a particular location are painted with a symbol unique to that site. The sites themselves have this same symbol painted on the quarry face. We are at the moment investigating the possibility of installing brass bars with key information at every site from which we have collected.

All sites are also mapped with respect to each other and, where possible, with respect to an absolute datum point. In the first few years, we used a Brunton compass to create a base map for the Gag Plateau. In 1986, surveyor Murray Bannister helped map all accurate vertical and horizontal information about each site.

When we started, we had inadequate base maps on which to record this sort of data. Available black and white and colour air photos had been taken from too great a height to be useful in discriminating the kinds of fine detail we needed to accurately pinpoint individual sites in a fossiliferous region. Happily, this has recently changed thanks to a low altitude photographic survey done of the area—courtesy of the Australian Heritage Commission.

Collection Techniques—from Limestones

When a site is discovered, every effort is made to understand the shape and nature of the deposit because this may help us to anticipate the shape of the bone-rich part of the deposit. Is it an irregular cave fill, a vertical or horizontal fissure deposit or a primary horizontal lacustrine unit? Exposed bones are examined to identify any that may need special attention. These are circled with a felt marking pen and, if they appear to be fragile, protected with a burlap patch soaked in Aquadhere. Then any loose bone-filled lumps are loaded into labelled hessian sacks.

Bone-rich boulders always provoke debate. Should they be taken as they are or broken up into smaller pieces that will fit into the bags? Or should they be left because they do not appear to contain material any different than that showing on the sides of other, more easily retrieved rocks?

If a large block needs a corner broken off or if a single specimen needs retrieval from within a mass of otherwise poorly fossiliferous rock, we often use the 'plug and feather' technique. This involves using an electric drill to drive many holes in a ring around the object of interest, placing two metal wedges and a central chisel in each hole, and then pounding each chisel in turn. As the chisels are driven in, they spread the wedges against the sides of each hole. With each round of hammering, the ringing of the chisels changes pitch until finally there is a sound like a snapping cracker as the piece sought pops free.

If large boulders or rock faces need to be more thoroughly broken up, it is normal to use light explosives. Holes are drilled but this time as deep as the long drill bit will allow. There are all sorts of hazards here that must be avoided. Sometimes there are 'vugs' or spaces within the rock caused by solution of the limestone through ground water. If the drill bit enters these, it may 'bind', causing the drill itself to spin around whacking the unwary operator and putting the drill bit itself at risk of breaking. Rock drilling is an art that needs strength, planning and, ideally, a lot of experience.

When the drilling is done, loops of ICI Red Cord are pushed down into the holes and joined across the top so as to make certain that they blow simultaneously. Then the workers abandon the pit for distant cover. An electric detonator is taped (by a licensed shot-firer) to one end of the sequence and, when all else is ready, the wire leads are attached to the electric charge generator. When everyone is at a safe distance, 'Firing!' is shouted. The key is turned and seemingly simultaneously, there is a loud 'Blam!' and an explosion of dust.

In the quarry, the best site to greet us is one of almost no apparent change—except for cracks linking all the drill holes. Crowbars are then wedged into the cracks and teams wiggle and heave the bars until the large slabs of rock begin to separate. Even then we may not see the goodies because, if the drill holes were well placed, the cracks through the rocks will surround rather than transect the bone layer. Soon, slabs of bone-rich material are removed from the quarry and, if appropriate, the area is cleared for another operation.

Alan Rackham, the most experienced driller and blaster working with our group and one of Mount Isa Mines' explosives supervisors, comes to the task like a jeweller or sculptor. As a diamond cutter would study a rough gem before deciding where best to strike, Alan decides how best to 'carve' the unwanted stone from above and below the targeted level. After he has spent a morning at a rock face, on one side of him there will be a pile of square-sided blocks of relatively bone-less limestone whose faces are lined with grooves from the drill holes. On the other side will be similarly neatly cut blocks of the bone-bearing level, ready for labelling and transport. Working with him will be volunteers cleaning the rock dust away from the drill holes as he goes crow-barring up the freed slabs, sledging off unwanted, relatively bone-less sections, flagging and labelling items for special attention and relating separated slabs for potential reunion of divided bones in the laboratory.

On balance, the use of light explosives to 'carve' large masses of limestone is by far the least damaging and most predictable of any technique available. This is because with very few drill holes, the rock can be cleanly cracked along predetermined lines, with minimal loss of material. Although sledge-hammering is routinely used to break large rocks into smaller manageable lumps, it can result in damage. Fractures are much less easy to control and small pieces of potentially useful material are sometimes lost.

For us, it is sometimes necessary to see cross-sections of bones before we can determine whether a newly discovered site is worth the investment of energy, time and resources to quarry. In the same way, while it is certainly possible to collect all lumps of rocks at a site, whether or not they reveal cross-sections of bones on their surfaces, we would end up wasting a phenomenal amount of time, money and acetic acid if we did. Hence we are normally selective about what is quarried and retrieved. The basis for this selection is most commonly the nature of bones exhibited as cross-sections on the blocks.

While every effort is made to keep the vast majority of bones as intact as possible, as long as all of the pieces of a fractured jaw or bone are recovered, their reconstruction is easy and results in a final product that may, because it involves access to information about the cross-section of items, be more revealing than the same specimen recovered intact.

As our capacity in the laboratory progressively increases, we try to retrieve larger blocks from the field, some weighing as much as 150 kilograms. To do this many factors must be just right. The rock must be manageable, using whatever equipment can be gotten into the site. Although a light helicopter can reasonably easily lift blocks up to 50 kilograms, larger blocks must be transported on the ground. If we cannot get the helicopter or truck directly to the site, which is frequently the case, large blocks must be 'waddled' to the nearest transport by groups of sweaty enthusiasts.

Some samples require still different means of transportation. For example, because Rackham's Roost Site is at the top of a very high hill, a 'flying fox' seemed an appropriate way to start the 2 tonnes of sacks on their journey to Mount Isa. Accordingly, Alan and his group of volunteers hooked each sack onto a metal hook and then sent it wildly careening down an inclined length of fencing wire to a stack of mattresses waiting below. The sacks were then transported back to camp by a very slow boat.

Collecting from Riversleigh's river deposits

Quite different are the tasks that face workers at, for example, Terrace Site (these folks being known as Riversleigh's 'Terrorists'). Here they face a perched river terrace of sands and gravels about a half kilometre from what are now the banks of the Gregory River. They spend most of their time kneeling or slumped over a gravel bench, using fine picks and brushes to systematically quarry loose sediments in search of normally much rarer bones. Fossils here include turtle shells, limb bones and jaws of *Diprotodon*, a maxilla of *Palorchestes*, kangaroos, crocodiles and freshwater mussels.

Because the sediment that contains these fossils is unconsolidated and because the bones are frequently heavily fractured within the loose sediment, nearly every large bone requires special treatment. Normally this involves at least a coating of dilute Aquadhere with a reinforcement of torn pieces of fibrous paper. If the bone is large or particularly fragile, it is pedestalled by careful quarrying away of sediment from around its widest points. To do this most efficiently, awareness of the nature of the bone is helpful so unseen extensions can be anticipated. Damp paper is tenderly tamped over any exposed surfaces of the bone. Then several layers of wet, plaster-soaked bandages, made of hessian strips, are wrapped around and over the object and allowed to dry hard. A few hours later, the plaster-hardened pedestal is undercut using picks and hand trowels, care being taken to ensure that no more of the bone extends further into the sediment than was expected. The block is then turned over and non-fossiliferous sediment removed from the base until the underside of the bone just shows. Then the process of plastering is repeated until the whole object becomes in effect a plaster egg safely containing the fragile bone. Thus secured, the bone can be safely transported to the laboratory for careful preparation.

But we do not presume that all we see is all that is there. As we quarry, the sands and gravels are also sieved to recover tiny bones that might otherwise have been missed. These include the teeth of rodents and fish and the vertebrae of snakes.

Mussel shells and pieces of charcoal are also collected and placed in sterile, sealed plastic containers for potential radiocarbon dating. In addition, analysis of the pebbles carried as part of the river's bed load may give us information about their source. It is also possible that statistical analysis of the stone fractions will reveal artifacts indicative of the presence of humans. As quarrying proceeds, maps will be constructed of the channel deposit and attempts made to reconstruct the direction and possible strength of the ancient river's flow, using pebble and bone orientations.

Contemplating Collecting in Carrington's Cave

As we write this, we are contemplating a series of major excavations in two of Carrington's Cave three main chambers—the entrance chamber and the rear 'Great' chamber. Carrying out this work requires different procedures from those previously employed.

The floors of both chambers will be surveyed and care taken to pick the best places to begin 'test excavations'. Then, in the areas where the main excavations are intended to take place, the floors will be criss-crossed with a string grid to serve as a reference system for recording position details for all materials recovered.

In this work, fine brushes, awls and trowels will be used to carefully excavate layer by layer. All of the material will be sieved after being taken well beyond the cave entrance. All stone and bone will be retained for laboratory analysis. All charcoal and speleothems will be kept for dating or determination of palaeotemperatures. Intact samples of the sediment will also be collected for later recovery of fossil pollens.

All in all, this will be a very different kind of collecting from that which has so far occurred on Riversleigh but a kind that has been successfully practised in many other areas of the continent. The team of research workers currently gearing up for this project will be very diverse and includes geologists, anthropologists and palynologists as well as the usual mob of palaeontologists.

It's a long haul from field to final reconstruction but the results can be enormously rewarding. The skull of this new herbivorous diptrotodontid from VIP Site is one of the most perfect fossil marsupials known.

R.ARNETT.

Limestone block containing part of the snout of the VIP skull, in side view with the tooth row exposed.

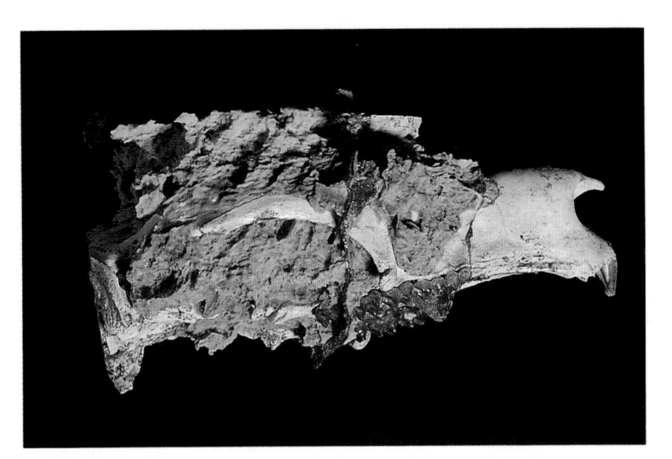

Limestone still partially encases the VIP skull, all parts of which were joined by epoxy resin prior to acid preparation.

Underside of the fully prepared VIP skull now ready for detailed study and comparison. For the full story on how this magnificent specimen was restored see Pampered Preparation for Special Fossils, page 223.

Right Anne Kemp from the Queensland Museum uses water-soluble Aquadere glue to temporarily protect tooth plates and partial skulls of fossil lungfish found at Melody's Maze Site.

Below Crowbars are used by palaeontologists Paul Willis and Stephan Williams to pry up slabs of fossil-rich limestone at Neville's Garden Site.

A light helicopter is vital to the Riversleigh field work. Not only does it help carry bags and blocks of limestone from otherwise inaccessible sites, but it is used extensively each field season to find and survey new areas of fossil-rich limestone.

R. ARNETT

Below Cathy Mobb uses an electric drill, driven by a petrol generator, to push holes into a limestone boulder, an almost constant activity on Riversleigh fossil digs. The holes will either be filled with coils of Red Cord or used for plug and feathering.

R.ARNETT

R. ARNETT

Experience enables Henk to accurately glue back small pieces of bone onto what turned out to be an almost perfect diprotodontid skull found at VIP Site during sledge-hammering.

R.ARNETT.

Elaine Clark inspects blocks of limestone from Neville's Garden Site before placing them in clearly labelled bags.

R.ARNETT

A solid bone-rich boulder is muscled, with the help of an aluminium ladder, to a point on the edge of the D-Site Plateau at which the helicopter can land and load.

R. ARNETT.

Coils of ICI Red Cord, a light linking explosive, are pushed into drill holes. The explosive is used to crack massive blocks of Riversleigh limestone. Crowbars can then be used to pry the blocks apart.

A winch is used by Alan Rackham Snr and Syp Praeseuthsouk to raise a huge slab of limestone onto the back of the *Australian Geographic Bonesmobile* at Grimes's Site on the Gag Plateau.

'aught in the act, Vice-Chancellor of the University of New South Wales, Michael Birt, helps clear rocky verburden from the fossil-rich layers of limestone at Neville's Garden Site, while Pro-Vice-Chancellor Tony Vicken (left in red hat) chooses his next missile.

R.ARNETT

Right A delicate fossil emerging from a partially acid-dissolved block of Riversleigh limestone is reinforced with sticks and painted with acid-insoluble glue by preparator Cathy Nock.

Left A cloud of dust accompanies an explosion on Godthelp Hill. On balance, light explosives cause less damage to the Riversleigh fossils than any other technique available.

R.ARNETT.

From field to lab reconstruction: here we see how scientific knowledge builds upon itself. Peter Murray's three-stage reconstructions are of the freshwater crocodile *Baru wickeni* (right) and a marsupial lion – an unnamed species of *Wakaleo* – both from Riversleigh's Site D.

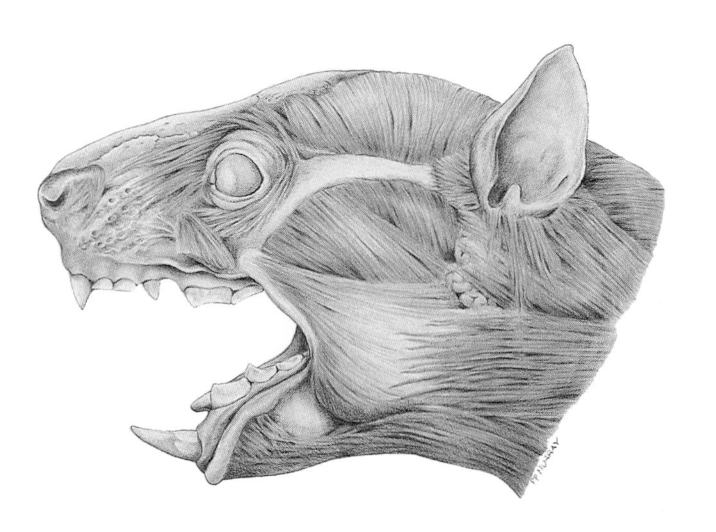

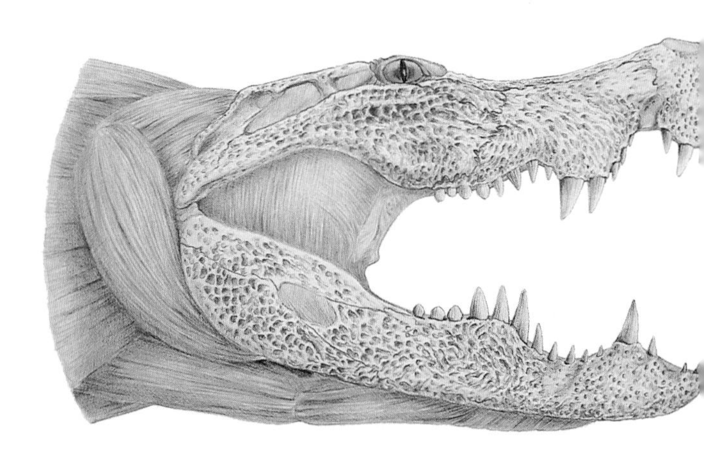

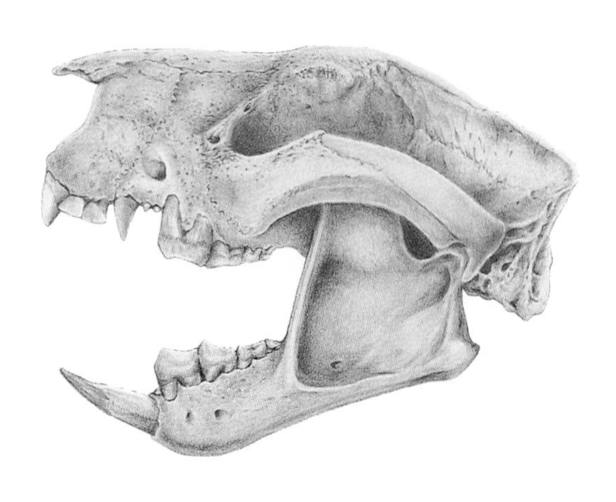

CHAPTER 11

Techniques of Laboratory Preparation

What we see in the field at Riversleigh is a bare lick and a promise of what eventually develops as palaeontological treasures. Certainly the occasional spectacular find is first noted in the field but normally the vast bulk of mind-boggling stuff emerges in the acid laboratories at Sydney and Mount Isa. A dreadful amount of patience is required in the field when a good site is found because it takes at least two months before the real significance of what was found can be assessed.

The primary factor in deciding how much of the initial sample gets prepared is how distinctive the emerging fauna appears to be.

The exact nature of the laboratory work depends mainly on the type of rock or sediment that contains the bones. If it is limestone, we can usually acid-process it. If it is sand or clay, we can usually dry- or wet-sieve it.

When we recovered our first Riversleigh jaw in 1976, we were told that Riversleigh limestones were silicified. This meant that they could not be dissolved in any of the relatively mild acids, such as acetic, hydrochloric or nitric, to obtain what we presumed would be chemically more resistant fossil bones (modern bones being of course readily soluble in most acids). As a result, we mechanically prepared our first jaw using a vibrotool—a noisy, hand-held mini jack-hammer that uses old gramophone needles for chisel points. It took about eight days of non-stop rat-tat-tatting in a dimly-lit back room in the old Queensland Museum and as many days for the buzzing in our hands to stop. Worse, we damaged a piece of enamel on one tooth because of a 'rogue' needle and finally abandoned efforts to remove a lump of limestone that still clings tenaciously to the specimen.

Somewhere further down the line, we doubted community wisdom and dropped a piece of Riversleigh limestone, containing a fragment of 'junk' bone, into hydrochloric acid—with spectacular effect: the solution went mad with reaction froth. Unfortunately, after we examined the assaulted lump, the bone had dissolved just as quickly as the limestone. Clearly, neither was silicified.

So, we shifted gears and tried milder acids, soon discovering that acetic acid, at any strength, dissolved all of the limestone but none of the bone. The chemical reaction is: $CaCO_3$ (limestone) and $2CH_3COOH$ (acetic acid) produces $Ca(CH_3CO_2)_2$ (calcium acetate), CO_2 (carbon dioxide) and H_2O (water). The by-products of the reaction are non-toxic, so the way seemed clear. Eventually, we settled on a 5-10% solution of acetic acid in water as the standard.

That welcome discovery was a major turning point in our work. Now, in the Sydney labs alone, we use approximately $15 000 worth of acetic acid every year—about 10 000 litres—to process about 10 tonnes of limestone. Thankfully, ICI Australia, one of the Riversleigh Project's biggest corporate sponsors and the major producer of acetic acid, provides that amount to our laboratory on a yearly basis. Without that support, progress on the whole Project could have been crippled.

When the sample bags arrive from the field, representative sacks from each site are taken to the laboratory. The rest are placed in temporary storage (commonly in freight containers kindly made available by Ansett/Wridgway). Plastic butchers' vats are then labelled to identify the material they will hold during the acid-processing. Metal and plastic tags are made up with the names of each of the sites. These tags will travel subsequently with all materials removed from a vat, so that at all times the identity of the material being processed is known.

In practice, we commonly put about 20 kilograms of limestone in the butchers' vat which is then half filled with water to which is added enough acetic acid to make a solution of about 5-10%. This occurs in a well-ventilated room because acetic acid fumes may be irritating to eyes and lungs. A plastic lid is then fitted over the vat. Depending on the temperature (we sometimes use aquarium heaters in winter to accelerate the reaction), the acid will have spent most of its strength within four days.

As this process goes on, the bones in the blocks begin to emerge from receding surfaces, sometimes in the thousands. Some project starkly like fingers clawing their way out of a tomb in which they have been sealed for 20 million years. Brightly coloured teeth coalesce out of a haze of limey mud, emerging in rows along amber-yellow jaws. Chains of python bones wind around the neck vertebrae of a long extinct dark-eyed kangaroo and sheets of fine bones intertwine over what was once an ancient surface of white mud. A small femur emerges with one end pointing suggestively into the heart of the rock where, soon, a tibia with which it worked in life is found, propelling a bizarre bandicoot on its way to the treacherous mud and an appointment with eternity. Each bone has far more to tell about its archaic world than we will ever know but it is our delightful task to extract as much information as we can, as the bones emerge into the light of day for the first time in many millions of years.

SOME REVELATIONS ABOUT THINGODONTA

Revelation 1. Soon after we began preparation of the first of the Gag Site material in 1983, a very odd tooth appeared in the concentrate accumulating beneath one of the blocks being acid processed. Its peculiar V-shaped crown was like nothing we had ever seen before. But, while desperately anxious to know more, there really wasn't much that we could do apart from hope that more teeth like it would turn up. And that is exactly what did happen. Soon there were five isolated teeth, all roughly the same size and more or less similar in shape but, because they differed slightly in shape, some of the V-shapes being more open and asymmetrical than the others, it seemed likely that they represented isolated upper and lower molars. At this point, someone began to refer to them as 'Thingodonta' and the name stuck.

Revelation 2. The second clue about the nature of Thingodonta came a few months later when a lower jaw began to appear on the side of a dissolving Gag block. It was short, oddly shaped, with an enormous, procumbent lower incisor and a single Thingodonta molar firmly inserted by its roots halfway along the otherwise toothless tooth row. It was very exciting—but it was also infuriating. While here was undoubtedly a jaw of Thingodonta, it left us as much in the dark as ever. The procumbent incisor was reminiscent of those of diprotodontians but nothing else about the beast was. The molar was as strange as ever and the jaw, well, it too resembled nothing familiar. Only when the jaw was finally free of the block and could be examined from other views, did an exciting possibility present itself. Clearly, this creature had just three molars. Why might this be important? Primitively, all adult marsupials retained four molars (although some specialised marsupial groups such as the *Petaurus* gliders lost their last molar); no placental is known with more than three. Was Thingodonta possibly a *placental* mammal? Hearts started thumping. If it was, it was the first and only known pre-Pliocene terrestrial placental mammal (apart from bats) known from Australia. Supporting the possibility, we noted that the 'angular process' of the dentary was very small, this being a very unusual condition in marsupials but normal in placentals. And then, of course, there were those weird teeth—unlike any marsupial known but, now that placental affinities were a possibility, not that unlike the molars of some 'primitive' insectivores, like the tenrecids of Madagascar. Hearts started racing!

Revelation 3. Fortunately, before we could leap into print with an announcement of Australia's first pre-Pliocene (non-flying) placental mammals, we burst our own bubble—in the nicest possible way: a skull turned up. By this time, many other Riversleigh sites were being processed. *It*, bearing the newest bundle of clues, appeared unexpectedly on a Camel Sputum block. An almost complete rostrum with both tooth rows emerged beside a lower jaw and, slightly adrift from the rostrum, the rear half of a skull. When these were finally freed from the block we were stunned. Although the delicate cheek arch (the jugal bone) of the rear half of the skull could not be precisely matched against the jugal process from the front half (with its undoubted Thingodonta molars and incisors in place), the rear half was the right size, all adjacent processes matched in position and size and it was, as noted, fossilised adjacent to the front half and the lower jaw.

What did this skull tell us? Most certainly not what we expected. Whatever else Thingodonta was, from the structure of the rear half of the skull, it was *not* a placental mammal—it was a perfectly good, albeit very 'primitive', marsupial with a marsupial-like process from one of the bones of the floor of the middle ear (the alisphenoid bone). In fact it was so primitive that we felt obliged to make comparisons with other Gondwanan marsupial groups such as those from South America. On balance, the rear half of the skull resembled those of some of the most primitive bandicoot-like marsupials known, while the front half still looked like nothing we had ever seen before.

Revelation 4. Taken together, they appeared to represent a distinct order of mammals and it was that conclusion that we enshrined in a scientific paper we submitted to *Nature*. The journal, however, rejected the paper, saying it was of only parochial interest to Australia! The fact that news of its discovery had filled newspapers and news journals all over the world suggested otherwise. We licked our wounded pride and submitted a slightly revised version to *Science*. It was promptly accepted for publication, with one primary recommendation: that we desist from formally naming the new order Thingodonta because they were concerned it would be misconstrued as a joke! We agreed and it is now Yalkaparidontia with two species, *Yalkaparidon jonesi* (from Gag Site) and *Yalkaparidon coheni* (from Camel Sputum Site).

What will the next revelations be? For starters, we should soon be able to distinguish the limb bones of Thingodonta by comparing ratios of limb and tooth elements within sites. An animal rarely represented by teeth will probably have its limb elements equally rarely represented. This in turn will give us a better idea about lifestyle (as well as relationships)—which we have already guessed at based on the structure of the dentition.

Further, with increasing understanding about the structure of the Oligo-Miocene communities in which Thingodonta lived, we should be able to interpret at least some of its ecological relationships within those communities and perhaps to comprehend why such a distinctive group of mammals died out. In answering that question, we may better understand the whole process of extinction in Australian ecosystems.

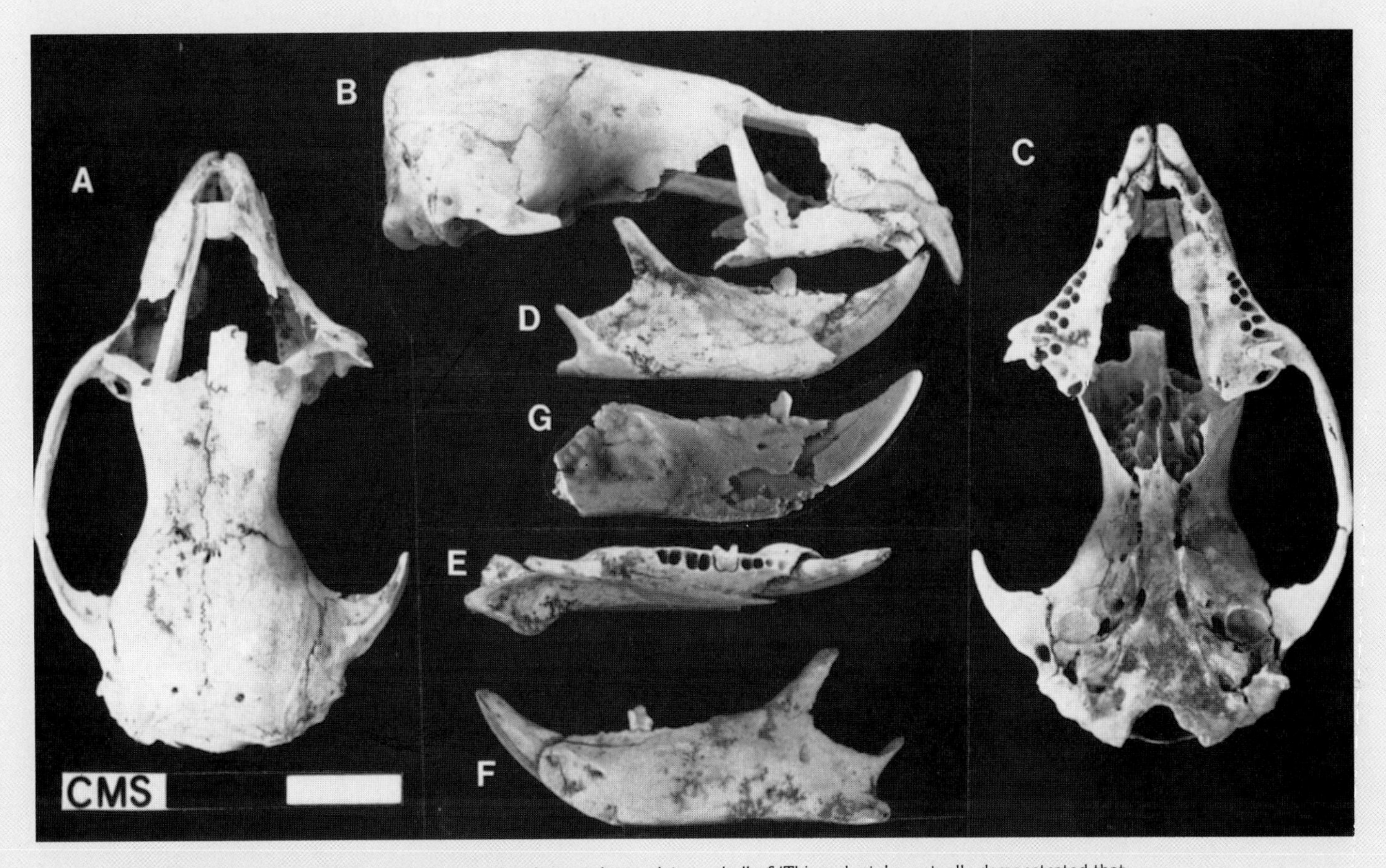

Progressive discoveries of first isolated teeth, then jaws and years later a skull of 'Thingodonta' eventually demonstrated that these Oligo-Miocene Riversleigh creatures represented an extraordinary different type of marsupial. This composite illustration was published in *Science* when the discovery was first formally announced.

Periodically, the surfaces of dissolving blocks are inspected for fragile bones or bones that, because of their proximity to each other, might have been parts of single individuals. When these are detected, the block is taken out of the vat, immersed for at least a day in slowly running water to remove the calcium acetate. Then, after drying, the bones needing attention are gently painted with a soluble plastic, such as Bedacryl diluted with acetone, to prevent any small pieces from being lost. If large pieces of the object are at risk of separating during subsequent acid processing, they are linked together using matchsticks, strips of rice paper and more soluble plastic. Then, when completely dry again, the block is replaced in its vat, a new solution of dilute acetic acid put in and the whole process repeated.

As single bones and teeth are finally freed from the surfaces, they are removed, soaked in water for several days, dried and preserved with soluble plastic if fragile. An initial determination of the fossil's relationships is made and it is given a unique number which corresponds to an entry in the specimen catalogue recording the nature and source of the fossil. It is then added to the accumulating collection of similar fossils or passed directly over to the person studying that particular group.

Not all materials needing preparation are handled in this way. Some, such as skulls accidentally split in the field, require special treatment.

PAMPERED PREPARATION FOR SPECIAL FOSSILS

When a special item returns from the field, it is rarely in one piece. The fact that it was noted at all almost always means it suffered at least one break because otherwise the bone wouldn't have been exposed. Such was certainly the case with the new zygomaturine skull recovered in 1989 from VIP Site.

At first no-one had any idea it was there. We had noted what looked like megadermatid (ghost bat) limb bones on the surface of the slab near the edge of the hill. Alan Rackham was asked to see if he could isolate the bat-bone-bearing slab of rock using a sledge. Michael Birt (Vice Chancellor of the University of New South Wales), his wife Jenny, Tony Wicken (Pro-Vice-Chancellor of the UNSW) and other folk were around to help. After sledging the ghost bat block, Alan turned his attentions to the block next to

it. When Alan's sledge came down, the slab popped into eight pieces. Hearts stopped when the first of these was lifted for inspection, for in the middle was a cross-section of a perfect skull. One tooth row could easily be seen emerging from the side of the vertical snout. Portions of the skull were visible in the other blocks as neat cross-sections.

Care was taken to collect every piece visible. After inspecting these, we also spotted a cervical (neck) vertebra of a size appropriate for the skull. Perhaps further along in this slab was the rest of the skeleton?

Back in Sydney, the eight pieces were taken to the School of Applied Geology to cut away excess limestone with a diamond blade rock saw. To do this most efficiently, it helps to be able to anticipate the dimensions of the embedded bone in order to be able to trim off as much excess limestone as possible without damaging the coveted object contained within. *Unfortunately*, the skull turned out to be marginally longer than anticipated with the result that we neatly trimmed off a 4-millimetre slice from the rear of the skull. Then, after kicking ourselves in the collective leg for the misjudgement, we proceeded to neatly trim the tips off the front incisors! Fortunately, the saw blade was thin enough for very little information to be lost.

The trimmed pieces were then re-associated, matching broken surface with broken surface. Once the pattern of the whole thing was clear, epoxy casting resin was mixed and spread over all facing surfaces. The whole uneasily bound lump was then pulled tightly together with straps and left for 24 hours. By this time, the resin had set and the whole object could be immersed for the first time in dilute acetic acid.

For the next few weeks, the ungainly lump was mothered with extreme care. As soon as the first surfaces of bone began to appear beneath the rapidly dissolving limestone, its condition was checked to see if it needed hardening with soluble plastic—but it didn't; it was in excellent condition.

As the limestone receded, it left thin, intersecting sheets of epoxy resin where the pieces had been joined together. These were trimmed off as they appeared with a hot scalpel blade, finally leaving just dark lines on the skull at the points where it had been joined together.

Finally, most of the skull had made its appearance—and it was nothing short of incredible. It was uncrushed, undistorted, unfractured, unbelievable. We found ourselves asking why we were so sure it was approximately 20 million years old rather than 20 years old. Yet, of course, it was not even bone—which would have dissolved in acetic acid. It was stone but, down to the tiniest detail, an incredibly faithful, molecular replacement of the original skull.

There was also an eerie side to its rebirthing. In its left orbital region, until the last few days of preparation, there was a sphere of epoxy resin that had apparently filled a space in the original rock. It rested near the front edge of the orbit but did not actually touch any of the surrounding bones. It was in turn connected to the rear of the orbit by a millimetre-thick cylinder of resin. It became impossible to think of this bizarre thing as anything other than a spooky reincarnation of the left eye!

And, maybe it could have been. If muddy, carbonate-rich water had infiltrated the orbital region before it had completely macerated, it is just possible that a surface of micro-crystals of calcite might have formed on the inner surface of the eyeball before its tissue frame vanished. Unfortunately, although this suggestive epoxy cast was not trimmed off, when the whole skull was finally dipped in acetone-soluble plastic to harden it in preparation for the next few centuries of palaeontological probings, the epoxy 'optic nerve' withered and the 'eyeball' collapsed.

Once the skull was finally ready, we began the task of trying to comprehend its relationships to all other creatures previously known. Immediately it was obvious that this animal matched none that had so far been named but interestingly it did, for the first time, make sense of two quite different finds from Riversleigh. In 1967 Tedford had described, but not named, a very poorly preserved maxilla fragment from Site D which he regarded to be a palorchestid marsupial. And in 1977 we had collected two associated lower jaws of what were unmistakably a small, previously unnamed zygomaturine diprotodontid marsupial. Now it was clear that Tedford's maxilla is very similar to, if not the same as, the VIP skull and our Site D lower jaws fit perfectly against the VIP skull. As might then be expected, the skull shows a unique mixture of palorchestid and diprotodontid features which may challenge contemporary understanding about the relationships of the two groups. (For photos see page 217)

Others are large blocks weighing upwards of 150 kilograms, often with skulls visible at the surface. These are much too large to fit into the normal butchers' vats so we designed movable, metal gantries that suspend the blocks over a very large vat of acid. As the block dissolves from below, it can be raised in order to preserve emerging items and then lowered again for continued preparation.

But more comes to light using this method than just the spectacular items noted on the receding surfaces of the blocks. In the bottom of each vat, after a few days, is a sludge containing whatever was insoluble in the acid. Commonly this includes many bones and bone fragments, isolated teeth, tiny crusts of algal-like mats and rare bits of chert re-worked into the Tertiary deposits but originally from the much older Cambrian rocks of the Riversleigh area. This sludge is gently recovered from the bottom of the vat after the vat is emptied between changes of acid, washed in a fine sieve with gently flowing water and dried. It is now called a 'concentrate' and is sorted meticulously under a binocular microscope. It is at this stage that most of the tiniest mammals, lizards, frogs and birds are recovered. Again, every distinctive item is identified as far as possible, given a catalogue number and placed with other items representing the same group of animals.

Although most of this preparation goes on in the palaeontological laboratories of the University of New South Wales, the Mount Isa Mines Palaeontological Laboratory run by Alan Rackham also processes as much as it can, being primarily limited by the availability of acetic acid and technical support. In Sydney, some members of the Riversleigh Society, such as Jim and Sue Lavarack and Reg Angus, also help to prepare materials in vats, which they monitor in their own backyards.

In all, from large laboratories to backyards, about 12 tonnes of Riversleigh limestone per year are processed.

CHAPTER 12

Returning the Breath to the Beast

Reconstructing Ancient Creatures and their Environments

Wild imagination?

It has been said that palaeontologists seem too willing to reconstruct in detail—down to the last whisker—an entire creature from nothing other than a single tooth. Sometimes this is justifiable criticism.

Yet it is important to realise that most conclusions drawn about the species-level diversity, relationships, intraspecific variation, soft-tissue structure and the ecological roles of extinct creatures are of necessity probability statements. That in itself is no condemnation of their value because most observations made about living creatures are also probability statements.

In any case, probability statements are falsifiable, just as are most aspects of a palaeontological reconstruction. Falsifiability is one of the most important aspects of science for it is the means by which we can select the best of two different hypotheses that compete to explain the same data. New discoveries can prove limb proportions wrong and reassessments of relationships can lead to a revised idea of probable eye colour. These reconstructions are thus models, in the scientific sense, comprised of many hypotheses each of which can be tested and rejected or modified as a result of new discoveries or understanding.

We consider that most of the better reconstructions are the creations of committees. This was the way we coordinated our scientific understanding with the skills of the principal artist for this book, Dorothy Dunphy. After spending years coming to terms with the nature of Riversleigh's animals, we provided rough 'sketches' and poses for the animals required. We also identified the closest living analogues to the extinct animal as a basis for the initial drawing but then itemised the ways in which the extinct animal would have differed. After Dorothy completed her first sketches, she brought them back for critiques, at which time we discussed any changes needed before she finalised the reconstruction. All in all, we are delighted with Dorothy's final products as presented here but will not lose sight of the fact that they *are* hypotheses and not, as we would like to think, divine revelations!

Putting Flesh on the Bones

Peter Murray, Deputy Director of the Museum and Art Gallery of the Northern Territory, is first a professional palaeontologist, second an anatomist and third, an artist—a devastatingly efficient combination. This means that behind each of his drawings is a great deal of research and anatomical understanding.

On page 220 is his study of Riversleigh's crocodile, *Baru wickeni*, from Site D. While we were aware that it was a specimen of a large crocodile, we did not know how much of the skull was in the two blocks we found loose on the slope. In this case, a curious hiccup interfered with the initial preparation. One of the blocks ended up in the Queensland Museum and the other in the University of New South Wales. When the latter was prepared, it turned out to be a large portion of a maxilla with a battery of huge, recurved teeth. In the meantime, mandibles (lower jaws) of other individuals had been found at Riversleigh. A partial skull recovered at Bullock Creek in the Northern Territory *apperared* to represent the same taxon as the maxilla from which Peter attempted a reconstruction based on all of this material in anticipation of the animal from Bullock Creek being named *Baru darrowi* by Paul Willis, Dirk Megirian and Peter.

After the other Site D block was relocated and prepared several years later, it turned out that the restored specimen was the best part of a skull. Further, it was then clear that it was *not* the same species as the Bullock Creek material which has been named *Baru darrowi*. This revised understanding led to the need for a revision of the reconstruction of the Riversleigh crocodile, which is what Peter has done here, and to the naming of the Riversleigh crocodile as *Baru wickeni*.

In this series of drawings, he used three mandible fragments and five fragments of the skull to reconstruct the shape of the original skull and lower jaws of the Riversleigh species. Missing portions were restored on the basis of general crocodilian morphology as well as that of crocodiles most closely related to *Baru wickeni*, such as *Baru darrowi* from Bullock Creek. Large cracks and other damage evident on the specimens have been 'artistically' repaired so that the final drawing of the skull depicts the skull as it probably would have appeared when it was still intact.

Then the muscles, ligaments and tendons were artistically restored using information gained from study of those structures in the living Saltwater Crocodile (*Crocodylus porosus*). Differences from the Saltwater Crocodile condition were interpreted on the basis of differences in bone shapes and skull proportions. Many muscles leave scars on the bones to which they attached in life. The direction of action and size of these muscles is indicated by the size and orientation of these scars and by the size of the spaces between bones left for their attachment and contraction.

Because the proportions of the profile view of the skull of *Baru wickeni* differ considerably from that of a Saltwater Crocodile, Peter used a computer to distort a digitised picture of the Saltwater Crocodile's skull into the profile shape of the Riversleigh animal. This then also enabled him to trace the scale pattern of the living form onto the reconstructed profile. The ornamentation and colour is based on a three-metre Saltwater Crocodile.

In the end, the reconstruction, although as plausible as informed anatomical deduction can make it, is always a hypothesis, never a certainty. As such, it is always subject to challenge and the need for continued revision. This eternal vulnerability is why palaeontological reconstruction is a science rather than a faith.

The other reconstruction on page 220 is of Riversleigh's Oligo-Miocene marsupial lions, an unnamed species of *Wakaleo* from Site D. Coincidentally, the largest skull portion, upon which this reconstruction is based, was found wedged into the articular region of the lower jaw of a specimen of *Baru wickeni*. Perhaps it was that giant crocodile's last supper?

Again, Peter has used species previously known to help with the reconstruction. Specifically, he compared the largest Riversleigh specimens, a portion of the facial region of the skull and a lower jaw, with the corresponding portions of a skull and lower jaws of *Wakaleo vanderleuri* collected several years ago from Bullock Creek.

The musculature is restored on the basis of that which occurs in phalangerid possums. Because there is debate about whether marsupial lions were more closely related to wombats than possums, Peter's choice represents another hypothesis that will eventually be tested. If it turns out that the wombats are closer relatives, this reconstruction will probably require revision. The muscles shown are only the deeper, facial muscles, these being of greater interest than the more superficial ones.

This 'lion's' external appearance was restored on the basis of comparisons with phalangerid possums, some of which have very similar skull proportions. Peter's portrayal of the pointed ears, fur texture and colour are, however, a concession to possible wombat relationships. Although the actual nature of these superficial features of a prehistoric animal will almost certainly never be tested, plausible reconstructions are best constrained by comparisons with close living or ecologically similar creatures.

Reconstructing Behaviour

Sometimes the fossil record provides remarkably good information about the behaviour of prehistoric creatures, although it is often necessary for a palaeontological Sherlock Holmes to detect clues that might otherwise be missed.

For many years dinosaur trackways were regarded as useful tools, mainly for interpreting the variety of dinosaurs and sometimes the stride length of individuals. More recently, careful studies of complex trackways, such as those preserved at Winton in central Queensland, have enabled palaeontologists to interpret the social behaviour of herds of herbivorous dinosaurs under severe stress—being chased by a large meat-eating dinosaur! Age structure of the herd can be worked out from the sizes of footprints of animals moving at the same speed and in the same direction. Social behaviours can be determined, for example, by studying the extent to which movements of one animal affected those of adjacent individuals. Even the extent to which the mob of dinosaurs behaved as a coherent herd rather than as an assemblage of independent individuals provides important information about prehistoric behaviour. Every detail of these tracks, their depth, elongation, angle of plunge into what was once soft mud, all bare witness to the mental reactions of animals long since dead.

More closely related to the subject matter of this book is the debate about the food habits, as well as the lifestyles, of some of Australia's most remarkable mammals, the marsupial lions. Study of the teeth of the Pleistocene leopard-sized marsupial lion (*Thylacoleo carnifex*) suggested to palaeontologists of the 1800s that it was either a ferocious carnivore, hence Richard Owen's name which means 'meat-eating marsupial lion', or a relatively benign herbivore as suggested by Gerard Krefft. The two views naturally led to significantly different reconstructions of the animal going about its business—either stripping the meat off kangaroos or mindlessly peeling melons. Recent studies of microstriations on the teeth and strontium concentration in its bones favour Owen's view.

Similarly, following a study in the 1970s of the hind foot of *Thylacoleo carnifex* and the observations that its overall structure was not that dissimilar to the foot of a possum, it was surmised that it was, like its placental counterpart the leopard, at least partially arboreal. As a result, it has since been commonly reconstructed prowling along the limbs of gum trees, eye-balling potential feathered dinners. Now, however, following determination that some of the environments in which it occurred lacked trees altogether, it is evident that this carnivore must have been terrestrial in at least part and perhaps all of its range.

Other kinds of behaviour can be construed from palaeontological associations. For example, on a quarried surface at Inabeyance Site, Riversleigh, we found many bones of *Neohelos* diprotodontids. Scattered among them was an extraordinarily high number of bandicoot skulls and partial skeletons. Apart from a kangaroo or two, little else was found on this surface. While heavily into the realm of speculation, we hypothesised that this 'death assemblage' probably meant that the bandicoots had opportunistically fed either directly on the carcasses

Artist Peter Schouten's reconstruction of the marsupial lion *Thylacoleo carnifex.* Debate about the food preference of the Pleistocene creature raged last century.

of the diprotodontids or on meat-eating invertebrates. Perhaps the diprotodontids had become mired and died in limy mud, after which bandicoots foraged among their decaying carcasses and in turn occasionally got caught in the same mud.

Sometimes even vocal and auditory capabilities can be inferred from fossils. For example, because the structure of the middle ear and echolocation calls used by modern bats are reasonably well understood, it is theoretically possible for the kind of echolocation used by fossil bats to be determined. With the thousands of limb bones and jaws of the *Brachipposideros* bats from Microsite were masses of isolated periotics, tiny orange marble-like bones. Within this bone are housed the structures of the inner ear, the semi-circular canals and the cochlea. Because these Riversleigh periotics are remarkably well preserved and uncrushed, it has been determined that these bats used a type of echolocation called Doppler-shifting. Armed with this awareness, we can postulate that this bat, dead for 20 million years, detected flying insects by sending out pulses of sound at a constant frequency to establish, by means of frequency changes in the returning echo (Doppler effect), the insect's movement and speed. During the search phase of the bat's hunt, it probably used a relatively low signal rate (some 5 to 10 pulses a second) to survey its air space but once an insect had been detected this would have increased to perhaps 15 to 50 pulses a second in the approach stage, and ultimately sped up to as many as 100 pulses in the capture phase of the hunt, at that time providing the bat with maximum information about its intended victim.

As another example of behaviour interpretable from fossils, when Paul Willis found the spectacular fossil now known as 'Big Bird' at Site D, he inadvertently provided an important clue about the behaviour of this species of giant flightless bird. In the massive grey limestone, next to the well preserved leg and chest bones of the bird, was a discrete pile of many uniform-sized polished pebbles about 1 centimetre in diameter. From knowledge of modern birds and dinosaurs, we realised that these were 'gizzard stones'. In life, food is passed from the mouth to the gizzard where it is churned with these stones (which are sand-sized in chickens) until broken into pieces small enough to digest. In effect, they do the job that teeth do in herbivorous mammals. To get these aids to digestion, the birds must select them from the ground. Hence, finding Big Bird's gizzard stones enables us to positively identify one of its normal, albeit less distinctive behaviours, that of searching the area for suitable pebbles to swallow.

Similarly, when Walter Boles began the study of

Riversleigh's fossil birds, he identified one bone that had more than an average story to tell. It was the leg bone of a new predatory bird but its structure said something singular about the behaviour of this meat-eater. It resembled, in its distinctive articular surfaces, the same bone as in three other predatory birds from South America, Madagascar and southern Africa which shared a peculiar behavioural trait—the capacity to rotate their hind leg to reach deep into tree hollows to gaff and retrieve living prey. Hence, although it is not clear that this extinct Riversleigh predator was closely related to any of the 'rotary-legged' predators of the other continents, this aspect of its behaviour has in all probability been confidently determined. If it turns out not to be closely related to any of the other birds that possess this feature, it will represent an example of convergence like marsupial moles in Australia and golden moles in Africa.

Reconstructing the World of the Beast

The vegetation. As explained in Chapter 9, all lines of evidence suggest that Riversleigh's older vegetation was rainforest.

The dentitions of the mammals in a species-rich fauna provide a reliable indication of the structure of the vegetation in their environments. In particular, the teeth of herbivorous mammals must have a shape appropriate for pulverising the type of plant materials in which their owners specialise. As a result, it is relatively easy to distinguish between grazers whose foods are abrasive grasses and browsers whose foods are softer vegetation. In Riversleigh's Oligo-Miocene faunas, we have not found a single animal we could interpret as an undoubted grazer. Hence we have concluded that there is no convincing evidence that grasses were present in these environments. Grasses are a very minor part of even modern rainforest communities mainly because of the limited light that reaches the forest floor.

Considering modern marsupials in Australia and New Guinea, *all* the herbivores that have selenodont (double W-shaped) molars are browsers that eat the leaves of trees. This includes all ringtail possums, the Greater Glider and the Koala. Consequently, we interpret tree leaves to be the diet of the selenodont marsupials in Riversleigh's fossil record.

From Gag Site we recovered 12 different selenodont marsupials, 9 of which are ringtail possums. This is a remarkably high diversity of leaf-eaters. The closest approximation to this number are the mid-montane rainforest possum communities of New Guinea but even there no more than four selenodont species are found together. We interpret this to mean that plant diversity per unit area must have been extremely high in the Oligo-Miocene rainforests of Riversleigh in order that so many leaf-eaters could specialise on particular types of leaves.

Using the same kind of reasoning to reconstruct vegetation types for the Pliocene and Pleistocene communities of Riversleigh, we can safely say that by Pliocene time, the rainforests had given way to open forests or perhaps woodlands. By Pleistocene time, the mammal faunas were essentially modern in composition, although there was still one very large, possibly semi-aquatic browser, *Diprotodon optatum*. On balance, this suggests open savannah eucalypt woodlands.

Although we can reconstruct changes in the *structure* of the vegetation of the area with time, without actual fossil plants or a pollen record, we cannot determine actual taxonomic representation of the Oligo-Miocene rainforests. Hence, in the reconstructions of these forests by Dorothy, the 'resurrection committee' had to make shrewd guesses based on palaeogeography, the floral composition of remnant biotas from north Queensland and the advice of palaeobotanists, such as Mary White.

Palaeoclimates. The older Oligo-Miocene local faunas suggest species-rich tropical, lowland rainforests at a time when Australia was still drifting towards the Equator. Combined with the dominance of hipposiderids in the bat faunas, which today are most diverse in very warm tropical areas, we have concluded that the region was warm to hot all year round, perhaps with a high average annual rainfall, something like that of the modern lowland rainforests of Borneo.

It seems unlikely that the area was subject to massive seasonal floods of the kind that annually inundate the Riversleigh region today. Although there is some evidence from Gag Site that deposition was episodic, the frequency and regularity of these episodes is not well understood.

The relatively low diversity and the kinds of animals present in the Pliocene Rackham's Roost deposit suggest that Riversleigh's rainforests had been replaced by more open, less species-rich forests or savannah woodlands. Probably the main cause of this vegetation change was the development of a more seasonal climate with a wet, possibly monsoonal, summer and dry winter, and a decline in total annual rainfall.

Although as yet our knowledge of Riversleigh's Pleistocene vertebrate communities is slim, what there is would not contradict the continent-wide increase in aridity during this epoch. It is even possible that sometime between 20 and 35 thousand years ago, deserts extended into parts of the region.

Original biodiversity. Understanding the biodiversity of a local fauna is not necessarily a simple matter of tallying the species recovered from that site. It must first be demonstrated that the death assemblage, or thanatocoenosis as it is more precisely called, fairly represents the community from which it was derived. The tally could be misleading because of at least three main factors.

Firstly, the trapping mechanism, whatever it was, may not have sampled more than a portion of the animals in the immediate community. This could happen, for example, if the trap were a pool of water with sides only high enough to ensnare *small* animals that fell in. Conversely, a limy mud flat might give way beneath the weight of a *large* animal but be safely crossed by lighter creatures. Or large bones might accumulate as a lag on the bottom of a turbulent pool, perhaps through the movements of up-welling spring water, whereas bones of smaller animals might be flushed to more distant areas. Or the fossilisation site could be a pool in a cave where the only bones routinely accumulated are those of bats. Or yet again, if the site is in a cave, the accumulating agent, carnivorous megadermatid bats, might transport prey only small enough to be air-freighted into the cave. All of these biasing factors would result in an under-representation of actual biodiversity in the area.

A second reason why the evidence might be misleading is that the agent responsible for having brought the bones to the place where they are fossilised might have sampled more widely than the immediate area of the fossil site. As a result, species diversity in the deposit will be artificially high. This kind of problem can arise when, for example, a large river collects animal remains from a variety of different habitats at varying distances from the site of eventual accumulation. Normally a hybrid fossil deposit of this kind can be detected because the bones of a species gathered from a distant habitat will be more worn than those of species gathered from habitats closer to the site of deposition.

Thirdly, the trapping site may have accumulated bones over such a long period of time, perhaps thousands of years, that the number of species in the deposit is actually higher than actually occurred at any one time. This could happen if the community surrounding the site changed over this period of time, resulting in a mix of normally non-associated species, or if the lineages present were undergoing speciation processes so that 'parental' and 'daughter' species which did not actually overlap in time ended up being preserved in the same spot. We are unaware of any Riversleigh deposit that presents this kind of confusion but are ever alert.

Palaeotopography. Studies of the older Tertiary rocks at Riversleigh are gradually revealing the basic topographic features of the region present at the time the fossil deposits formed. These evidently included a large freshwater lake (as evidenced today by thick sequences of lacustrine sediments containing the remains of large crocodiles and turtles), with islands of ancient quartzite jutting up as forest-covered ridges (which persist in the region today) and flat-lying, forest-covered hills of Cambrian limestone to the west. Gradually, after the whole region underwent uplift, the lake gave way in the west to a plethora of smaller freshwater pools, evidenced by the bulk of Riversleigh's Oligo-Miocene deposits. All land in the area would have been covered by dense, tropical rainforest. The surrounding hills of Cambrian limestone, as well as the relatively young Tertiary limestones recently formed by the receding lake, would have been riddled with caves whose entrances sometimes faced onto the forest pools.

The former shape of specific Riversleigh sites is relatively easy to determine because of clues preserved in the sediments themselves. For example, while working at Neville's Garden Site in 1989, we discovered several small boulders and sheets of limestone that, when split with a sledge-hammer, revealed inclined surfaces decorated with horizontal travertine ridges, broken calcite straws and small stalagmites. These structures are characteristically found on the floors of limestone caves. Here, however, they turned up at the edge of the main bone-producing part of the deposit which contained, among other things, diverse marsupials, bats, a platypus, birds, reptiles, frogs and lungfish, which suggested that the site of deposition was a pool adjacent to the mouth of a cave.

A particularly interesting item recovered from Neville's Garden was a portion of a tiny but well-preserved trilobite. Trilobites were marine arthropods that died out everywhere in the world approximately 250 million years ago. They are sometimes found as fossils, more or less identical in shape and colour to the one found in the Neville's Garden deposit, in the 530-million-year-old Cambrian limestone exposed on Riversleigh Station. This was an important discovery because it enabled us to say with confidence that although today there is no Cambrian limestone visible in the immediate area of Neville's Garden Site, there *was* when it formed. It is possible that the cave we interpret to have been present on the edge of Neville's Garden Site was developed in a massif of Cambrian limestone that soon after completely eroded away. The little trilobite dissolved out of the Cambrian limestone and simply dropped or was washed into the pool of fresh water below—its first bath after 500 million years in stone and its last for another 20 million.

In this way, using even the smallest clues, we are gradually building up a picture of the ancient topography of Riversleigh. In the same way, examination of Riversleigh's younger sediments provides clues to reconstructing the key features of a progressively more modern terrain.

Where to From Here: the Supreme Reconstruction?

It seems wise to anticipate that a great deal more information will be recovered about Riversleigh's creatures and ancient environments than we have so far managed to do. With new techniques on the horizon, the task of reconstructing ancient worlds will become less of an opportunity for 'good guesses' and more one for shrewd deduction.

Consider the possible future of one extraordinary technique: that of recovering from fossil bones intact proteins and even, in some cases DNA. In *Jurassic Park* by Michael Crichton (and the derivative, block-buster movie by Steven Spielberg), DNA recovered from the bodies of mosquitoes preserved in amber is used to reconstruct living dinosaurs. While the techniques for actually doing this still lie a bit in the realm of science fiction, 'live' or at least not decomposed collagen and DNA have been recovered from the bones of many extinct animals, including dinosaurs. With the rapid growth of genetic engineering, *is* it really too far fetched to suggest that an embryo of a modern descendant could be reprogrammed with fossil genes to reproduce the prehistoric animal? The mind boggles but it would certainly light a fire under the citadels of traditional palaeontological reconstruction!

CHAPTER 13

How We Date Riversleigh's Creatures

Fossils are a mix of two quite distinct kinds of things: tangible, biological shapes and sizes as well as geological substances; and objects from a specific time and place. Understanding the first kind of things, which are called 'intrinsic features', is relatively easy because, providing the physical information is not misleading, the clues required are what the whole thing is made of. Understanding the second kind, which are called 'extrinsic features', is not so easy because these aspects of a fossil are not tangible parts of the beast itself.

In this chapter we consider the kinds of evidence, sometimes direct but more often indirect, that help us determine the ages of the Riversleigh fossils.

Techniques for Dating

To comprehend the full geological and evolutionary significance of a fossil it is often useful to know its age—in relative, if not absolute, terms. Although age is an extrinsic attribute of a fossil (in contrast to its intrinsic shape) and as such not *essential* in considerations of the evolutionary relationships of an organism, it is necessary to determine the rates of evolutionary change and the history of changes in our environments.

Radiometric methods. To obtain absolute dates, the most reliable technique is the dating of radiogenic substances. These may occur within the object needing to be dated or in the sediments surrounding it. Unfortunately, determining absolute radiometric dates for fossils is rarely straightforward because it depends on four prerequisites:

—an atomically unstable element must be trapped in the object or deposit (e.g. radioactive potassium or uranium);

—the element must have a 'half-life' of sufficient duration to enable enough to remain for measurement;

—its decay product must persist in the deposit so that the ratio of parent and daughter substances can be accurately determined as a measure of time elapsed since the deposit formed; and

—it must have been preserved for the *first* time in the deposit that requires the date, rather than have been reworked from an older, pre-existing deposit, otherwise the date obtained will be older than the age of the deposit.

For Australian vertebrate fossil deposits, these conditions are rarely met, although there are the exceptions. Unfortunately, as yet we have been unable to identify a suitable radiogenic element from Riversleigh's limestones to use in this way. Although there are several possibilities being explored, such as ^{10}Beryllium, so far absolute dates have not been obtained. However, we have been able to use radiocarbon to date the late Pleistocene Terrace Site.

Biocorrelation via the sea. There are several commonly used alternatives to radiometric dating for estimating the age of a deposit. Most of these are collectively called 'biocorrelative' methods but they vary considerably in their reliability.

Sometimes fossil vertebrates are found in marine deposits for which there are reliable dates. This is because marine invertebrates commonly have wide geographic ranges and somewhere within that range normally overlap with a radiometrically datable material. In Australia, for example, the late Miocene Beaumaris Local Fauna was recovered from a marine sand exposed along the Victorian coast. The age of the molluscs from the Beaumaris deposit is well known, based on studies of comparable faunas from many other areas of the world.

International biocorrelation. In the Riversleigh deposits, we have no marine fossils—they are all freshwater or terrestrial. But the principle of borrowing an age determination for a group of fossils from one place and extending it to another is not confined to the oceans. This was a vitally important realisation that came initially from study of the Microsite bats.

When Bernard Sigé first set eyes on the Microsite *Brachipposideros* bats, his mind spanned the Indian Ocean. Here was a highly distinctive group of extinct bats with a distribution that now extended from western Europe to Australia. This realisation enabled two conclusions to be drawn.

First, it was likely that these bats once had a range that extended at least across Europe, into Asia and Australia. Some living bats have comparable ranges so this wide range is not really that surprising. It also suggests that sometime in the future, when a lot more is known about the fossil bats of Asia, the same group will be found in the fossil deposits of at least south-eastern Asia.

Second, the very close relationship of the Riversleigh *Brachipposideros nooraleebus* to very similar species in known early Miocene deposits of France suggests, as a first order hypothesis, that they are comparable in age. If they were not, the two isolated groups should have diverged from each other far more than they had. Accordingly, we concluded that the Microsite deposit accumulated not long after this group of bats first arrived in Australia, presumably from southern Europe, Asia or even eastern Africa.

This was the beginning of our efforts to establish an international biocorrelative dating framework. We are now busy making similar comparisons, using many other Riversleigh and European Tertiary bats in an effort to more precisely tie down the age of the Riversleigh bats.

We should add here that the airways of the past have carried more of value to biocorrelation than just bats. Of potentially similar importance are birds and pollen. Walter Boles of the Australian Museum is currently comparing the Riversleigh birds with those from other continents and is already discovering previously unsuspected correlations (see Chapter 4). Similarly, although no pollens have so far been recovered from the Oligo-Miocene limestones, we expect them to roll out in abundance once we start work on Carrington's Cave, at which time they will be studied for biocorrelative information by Helene Martin.

Biocorrelation within continental Australia. In the same way that shared fossils can be used to extend information about age from one continent to another, they are more frequently used to correlate otherwise undated deposits within the same continent. This application of biocorrelation within Australia is a venerable one with revisions being published at fairly frequent intervals, the most recent being by Woodburne and colleagues in 1985.

In order to correlate Riversleigh sediments with those from other areas of the continent, we need to find at Riversleigh animals previously known and dated from other areas. For example, species of the diprotodontoid genus *Ngapakaldia* were previously only known from the late Oligocene/early Miocene Etadunna and Namba Formations of central Australia—until we found one in Riversleigh's Site D. Similarly, discovery of *Wakiewakie lawsoni* from Riversleigh's Upper Site was the second known occurrence of this small potoroid kangaroo, the first being from the early Miocene Leaf Locality in central Australia. Many other kinds of animals have been similarly matched and collectively serve as the basis for correlating the central Australian and Riversleigh deposits. In the same way, once we better understand the ages of Riversleigh's local faunas, this understanding can be used to date smaller non-Riversleigh faunas which share taxa with the Riversleigh but not central Australian faunas.

Biocorrelation is also being used to establish a time framework *within* the Riversleigh region. With more than 150 local faunas of clearly different ages, it is a major task just to understand the time relationships of these assemblages, one to another. To do this, we consider any and all groups that have representatives in more than one local fauna. Bats will probably be very crucial here because they are present, usually in abundance, in almost every one of Riversleigh's local faunas but, as yet, it is early days for this aspect of the study. Looking at details provided in Appendix 2, potoroids, macropodids, wynyardiids, notoryctids and many other groups are showing patterns of distribution that collectively help to correlate the faunas to each other. For example, most of the larger faunas comprising what we have come to recognise as late Oligocene-early Miocene (Systems A and B) contain wynyardiids belonging to the genus *Namilamadeta* (which is also known from the Tarkarooloo Local Fauna of central Australia). In contrast, no wynyardiids are known from the middle Miocene (System C) local faunas, assemblages which we believe to be younger in age than those of Systems A and B. Presumably, therefore, wynyardiids died out in the region sometime between the time of accumulation of Systems B and C.

While biocorrelation does not provide absolute dates for these deposits, it does enable determination of the relative ages of the compared faunas. If any of these can be more precisely dated, it will help to fix the whole biocorrelation matrix in time.

Equally, although interpretations of the absolute dates for particular faunal assemblages may from time to time be revised, it doesn't cause the correlation matrix to collapse—rather it just shifts part or even the whole of the framework up or down in time. This is what recently happened when the age of the oldest central Australian Etadunna faunas were revised, on the basis of illite and foraminiferan dates, from the previous estimate of middle Miocene to late Oligocene. This revision downwards affected the interpreted ages of many of the central and northern Australian Tertiary local faunas, including those from Riversleigh.

'Stage of evolution' hypothesis as a basis for correlation. A less reliable method for determining the relative age of a fossil fauna involves estimating the relative 'stage of evolution' of the animals it contains. For example, if a particular lineage of animals is concluded to increase in size with time, then a newly encountered, previously unknown animal could be slotted into the time/size cline. This is essentially what Reuben Stirton, Michael Woodburne and Michael Plane attempted in 1967 using diprotodontoid marsupials. They hypothesised that most diprotodontid and palorchestid lineages enlarged with time. The smallest known was then *Ngapakaldia* in the older Etadunna Formation and the largest known, *Diprotodon optatum*, was from the late Pleistocene. In this way, most of us came to assume that finding a small diprotodontid meant we were working in a relatively old deposit and, in general, despite more than two decades of further work, that basic pattern still holds albeit with a few wrinkles (e.g. we now know that very small diprotodontids survived into the Pleistocene in New Guinea and at least one relatively large diprotodontid was present in the late Oligocene Ditjimanka Local Fauna).

When this kind of analysis is done for just one group, the resulting biocorrelation matrix must be treated with extreme caution. But if similar 'stage of evolution' analyses of other groups of animals suggest the same relative ages for the faunas, confidence in the method grows. For example, although they are less abundant, differences in size and shape among the extinct marsupial lions appear to broadly corroborate the biocorrelation patterns suggested by diprotodontoids.

The position of the rocks themselves as a guide to relative age. Finally, the rocks themselves can be used to give information about the relative ages of the faunas they contain. This capacity depends on the seemingly simplistic 'principle of superposition' that says that older sedimentary rocks will be found beneath younger rocks because they were deposited first.

In the Lake Eyre Basin of central Australia, for example, this principle has long served to establish the relative ages of the three commonly encountered sediment types: the older, lower green clays of the Etadunna Formation; the superimposed reddish clays and sands of the Tirari sediments; and the yellow and white sands and clays of the Katipiri Sands.

In Riversleigh, superposition of sediments is evident within regions but biocorrelation is required to relate the sequences of sediments *between* the regions. Thus, we can actually see that the rocks containing Main Site lie above those that contain Ringtail Site which in turn occurs above the rocks that contain Last Minute Site, all three being part of System C. But none of the System C rocks can be *seen* to physically overlie those of System B, although this interpreted temporal relationship has been deduced by comparing the faunas of the two Systems.

Future Technologies for Dating?

At present, we are limited by technology. But this is a rapidly changing field. Considering the incredibly short interval of time between the discovery of the possibility of radiometric dating and the modern ability to dismantle and date each amino acid of a protein molecule recovered from a fossil bone, it would be foolish to think there weren't exciting new tools around the corner. At the moment, dating Riversleigh's fossils is a constant challenge. But perhaps, right under our noses, there are new techniques begging for discovery that will make the whole business of dating strangers a breeze.

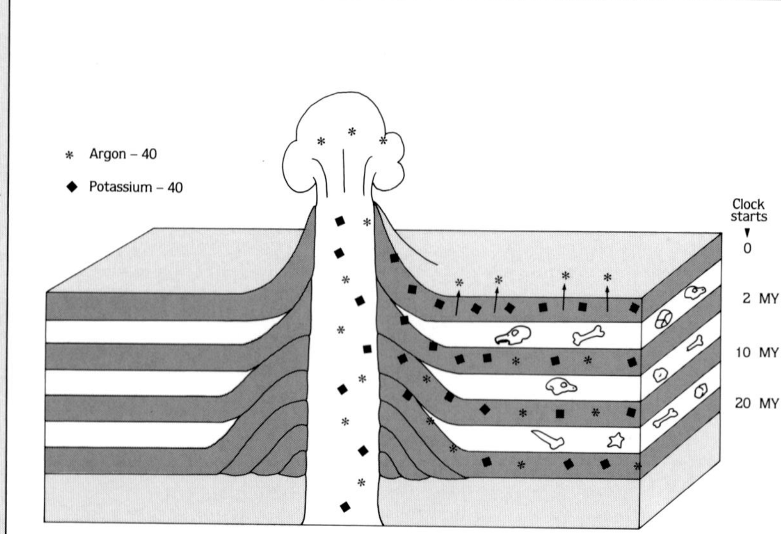

Radiometric Dating

Radioactive elements decay at a constant rate producing, in the process, radiation and daughter products. This constant clock-like rate of decay provides the basis for radiometric dating. In the case of radioactive Potassium, one of the decay products is the gas Argon. When molten volcanic rock reaches the surface as lava, the Argon gas accumulated up to that time can escape into the atmosphere, thereby 'resetting' to zero the radiometric Potassium clock in the molten rock. From the moment the rock cools, however, it is no longer possible for the decay product to leave the rock and from then on it accumulates beside the constantly diminishing amounts of radiometric Potassium. As a result, the ratio of the amount of accumulated decay product to the amount of parental radioactive element gives the measure of time elapsed since the rock first cooled. Other kinds of radioactive clocks work in different ways and tick over at different rates but the basic principle of using the constantly changing ratios of parental and daughter decay products to determine time elapsed since the parental product began to decay is the same.

CHAPTER 14

The Power of Palaeontology

A new tool for conservation

Until recently, palaeontology served rather traditional goals: production of objects of fascination to titillate the imagination and fill the halls and ceilings of museums; documentation of biodiversity through time; correlation of isolated sediments; tests of evolutionary theories; and a challenge to any scientist foolish enough to determine the limits of life's form solely on the basis of what's known at any one time.

In this chapter, we consider what ought to be palaeontology's most vital calling: use of the vast record of life's prehistory to help understand and conserve its threatened descendants. James Garfield, the twenty-second President of the United States, perceptively commented that 'History is but the unrolled scroll of prophecy'. The realisation that prehistoric trends can be used to understand the nature of the present and, with suitable care, predict the future, is a powerful new tool for conservation.

Conservation: Hindsight Makes Better Foresight

It has become almost a truism to say that conservation of Earth's biodiversity is one of the greatest challenges that faces humankind. It is something we all know is important for a range of reasons.

In the rocks at Riversleigh, for those with eyes to see and understand, are the signs of remarkable things that have come and gone: the rocks 'speak' of biological empires once mighty but now in ruins. But the bones of Riversleigh are more than just bits of intriguing lost worlds. Hidden among this wreckage of ancient rainforests is the stuff of prophesy, the means for glimpsing the future. For they are points on a life line—a line stretching towards us from the past through countless great-great grandcreatures to their descendants within whose living loins resides the ever-changing stuff of life's future.

Because palaeontologists focus on the shards of ancient life, dimly shining bits in a black vastness of space and time, they have a distinctive view of life. They view it as a vibrant stream of protoplasm, tying together all that has gone before to all that is now and all that will be in the future. This is a far more realistic and perhaps vital view of life than is commonly perceived by biologists, who often see little more than the three-dimensional creatures that stare back at them, bug-eyed and mindless, from the dark recesses of a hollow log. What we are suggesting is that palaeontological investigations can refine and sometimes redirect understanding about conservation and can produce a database that is an essential resource for this purpose.

Documenting and understanding biological change through time is the business of palaeontology. Preserving the capacity for this biological change must be the prime business of conservation. To be successful, conservationists must consider the significance of long-term trends in biotic lineages and ecosystems or risk misunderstanding and perhaps mismanaging these trends. To make informed judgements about the status of vulnerable species and habitats it is essential to determine much more than the ecology of the living populations. Prehistoric trends must be determined in order to distnguish these from the effects of human activity. Without being able to make this distinction, natural processes may be misconstrued as human-induced and lead to potentially disastrous interference.

Biologists struggling to develop effective conservation strategies often fail to wonder about the type of information they require to accomplish the task successfully. When asked to assess the conservation status of a relatively rare animal, for example, most would attempt to determine: food requirements; home range and/or territory size; reproductive biology; aspects of social behaviour that might affect population size; predation factors (including diseases); and, if possible, some estimate of the size of surviving populations. To add confidence to the initial determinations, they might attempt to assess population size over a two-year period, just in case there are annual fluctuations. If they are particularly conscientious, they might try to determine the historical distribution of the species as determined from old museum records in an effort to see how the species has fared over the last 200 years (and will inevitably conclude that it has suffered loss of some percentage of once suitable habitat).

All of this information about a living species is, without a doubt, required. But it forms only *part* of the basis for establishing an effective conservation program. Unfortunately, biologists rarely, if ever, seek information about changes in the population *through time*.

For every species examined, additional questions that should be asked are:

What have been the *rates of change in population size throughout its known history*? That is, have the numbers and relative abundance of individuals in each ecosystem declined, risen or stayed the same through time? How *rapidly* have changes in population sizes been occurring? Has the rate altered through time?

What has been the *nature and rate of morphological change* leading up to the present? For example, have individuals become larger or smaller through time and, if so, how rapidly has this change taken place?

How has the *nature of relationships* to other species in the same ecosystem changed? For example, does a rise in the number of individuals of one group invariably correlate with declines in another? Understanding these relationships provides a tool for predicting long-term stability in inconstant modern environments.

What were the *quantitative and qualitative responses to former changes in the environment*? How, for example, have changes in vegetation structure or climate affected bird, lizard, crocodile or possum lineages? Answers to these questions may enable us to anticipate changes in modern ecosystems subject to environmental change—such as the greenhouse conditions.

Knowledge of *both* the history and modern conditions of lineages are required to best determine a species' current conservation status and to make reasoned predictions about its future. It is essential to learn as much as we can about the causes and consequences of pre-human ecosystem changes before we contemplate interfering with modern ecosystems. In ecology, no less so than in history, those who do not learn the lessons of the past will in all probability be condemned to repeat them—from loss of a species to wholesale extinction and ecosystem collapse. If we are blind to the long-term dynamic patterns of cause and effect in ecosystems, it would be folly to interfere with natural processes upon which, ultimately, our own survival may well depend.

Case Histories: The Value of Fourth-Dimensional Information for Conservation

The palaeontological record sometimes supports and sometimes challenges current views about the conservation status of particular lineages. We will consider a few examples below. First, let's consider two non-Riversleigh examples.

The Crown-of-Thorns Starfish 'Plague'. The Crown-of-thorns Starfish has been identified as a rapidly spreading predator of tropical coral reefs. There are now millions of these spiny animals munching their way across many of Australia's tropical coral reefs. A common assumption has been that the starfish blooms followed human interference, i.e. collection of the shells of the Triton Snails, an important predator of the Crown-of-thorns Starfish.

As a result, active control programs were initiated here and in Japan. These included injecting starfish with formalin, as well as toxic heavy metal solutions. If there were no other information or if marine scientists were positive that human activity was responsible for the starfish blooms, interference of this kind might be justifiable, even though it involves potentially dangerous consequences. In fact this interference could even be detrimental to the corals. For example, killing the starfish might lead to disruption of normal cyclic processes necessary for the healthy renewal of reefs and the injections of heavy metal and formalin may do more damage to the reefs than the starfish do through eating.

Until recently no effort had been made to determine the prehistoric abundance of the starfish and hence to determine whether the plagues were or were not natural. Then Peter Walbran and his colleagues from James Cook University in Townsville, supported by the Great Barrier Reef Marine Park Authority, cored the reef for the kind of information that should have been sought in the first place—and found a record of more than 4000 years of natural, possibly episodic 'blooms' of this starfish, long before Europeans set foot in these waters.

As a result, there is now concern about the propriety of interfering with the starfish's spread, even though Walbran's initial examination failed to clearly resolve whether the abundance of Crown-of-thorns Starfish spines in the core represents constant high population levels through time or many prehistoric episodic blooms. More work is required to clarify the potentially crucial message of the fossil record but doubts about whether the plagues are human-induced *are* now clear.

Judging the heavy hand of humanity. As with the Crown-of-Thorns Starfish, there have been doubts about the extent to which European invaders have affected Australian terrestrial ecosystems.

Alex Baynes of the Western Australian Museum has spent many years carefully analysing the extensive sequences of late Quaternary vertebrate remains accumulated in the caves of south-western Western Australia. He has found evidence for an amazing stability over time in the assemblages of small to medium-sized mammals from modern times to as far back into the Pleistocene as the cave records go. In fact, changes in distribution patterns *since* the arrival of Europeans are many times greater than any that occurred within the last 10 000 years. Alex's consequent ability to quantify the nature, extent and rate of change in ancestral populations of living species provides a powerful tool for determining the conservation status of remnant populations. As well, he is now in a position to quantify long-term inter-relationships of species within communities and to add this understanding to that of the many ecologists who study the structure of modern communities. Understanding the effects changes in population size of one species have on the others in the same community should be essential information for developing conservation strategies.

Similarly, from study of the Pleistocene biotas of eastern Australia and New Guinea, Tim Flannery of the Australian Museum suggests that the historic extinction of many of Australia's medium-sized arid and semi-arid mammals (e.g. the Banded Hare Wallaby, Pig-footed Bandicoot and Greater Stick-nest Rat) was part of a 'trophic cascade' or secondary chain of extinctions that followed an earlier, more widespread, extinction event about 40-20 000 years BP when most of Australia's larger 'megafaunal' species (e.g. *Diprotodon optatum, Megalania prisca* and *Procoptodon goliah*) vanished. That the trophic cascade did not immediately follow the major extinction event was evidently due to Aboriginal firestick farming. This practice maintained widespread, complex vegetation systems which in turn staved off what would have been many more late Pleistocene extinctions, thereby delaying the trophic cascade for some 40-20 000 years. With the arrival of Europeans and the drastic reduction of Aboriginal land management practices, the trophic cascade picked up where it had left off 20 000 years ago. Tim concludes that an understanding of late Pleistocene environments and events is crucial to effective conservation management. To avoid further extinctions in environments suffering trophic cascades, the fossil and historic records suggest the importance of maintaining Aboriginal land management practices.

Premonitions from prehistoric platypuses. The nature of the modern Platypus, *Ornithorhynchus anatinus*, has long been a puzzle. When first seen in England, scientists of the day tried to remove the creature's bill with a pair of pliers. They were convinced that nothing so weird could be anything other than a hoax. Although modern zoologists no longer doubt its nature, the Platypus's origins and status are still very much in doubt.

The oldest known Platypus-like creature—and Australia's oldest mammal—is represented by an opalised fossil jaw from Lightning Ridge, New South Wales. There, 110 million years ago, *Steropodon galmani* was the consort of dinosaurs, crocodiles and other denizens of a world that looked nothing like the modern powder-dry, semi-desert country that characterises the area today. The robust, toothed monotreme may also have been the world's largest mammal of its day. Clearly, long before Australia broke away from the rest of Gondwana, there were aboard the presumptive ark toothed, egg-laying mammals.

Until recently, the second oldest monotreme, the ancestral Platypus-like *Obdurodon insignis*, was known only from a few isolated teeth and skeletal fragments from central Australia about which little could be told except that this was, like *Steropodon galmani*, a toothed, relatively robust platypus. Then in 1985, Riversleigh's 15 million-year-old limestones produced one of its finest treasures to date: an entire skull of a new species of *Obdurodon* otherwise very similar to the central Australian taxon.

Although in broad features the Riversleigh *Obdurodon* is similar to the living Platypus, the fossil monotreme is much larger, has a more generalised snout and a good set of cheekteeth. The modern Platypus has vestigial teeth as a juvenile but loses them before it is adult. What is more, this fossil skull reveals many of the sutures between bones that are all but unreadable in modern monotremes. This information provides us with further understanding about the relationships and evolutionary history of this enigmatic group.

Overall, the fossil record, slender as it is, gives us cause for concern. What we can see is relatively little change in the basic structure of platypus morphology from 110 to about 15 million years ago—a period of 90 million years during which the lineage remained relatively conservative, effectively toothed and evolutionarily labile.

In contrast, almost all of the changes in the last 15 million years have been in the direction of size reduction and over-specialisation through simplification. If two centuries of palaeontological analysis have given us any useful insight, it has been that when lineages begin to specialise, when they begin to edge far out onto their evolutionary limb, that limb is in increasing danger of breaking off.

Add to this rapid morphological 'decline' the loss of major areas of former distribution (at least central South Australia, north-western Queensland and, within just the last 50 000 years, far western New South Wales) so that it is now confined to the river systems of eastern Australia, and there seems to be even more reason for concern.

Palaeontologists have a somewhat different sense of time perspective than modern ecologists. When we suggest we are worried about the survival potential of the Platypus, our concern may be measured in terms of millions rather than decades of years. Some ecologists, aware that Platypuses appear to be able to survive in mildly polluted rivers near Brisbane and Sydney, suggest that this animal is in no imminent danger of extinction.

But comparing the living Platypus with say, brushtail possums (species of *Trichosurus*), whose ancestors were common in the lowland rainforests of Riversleigh and which today are widely distributed throughout Australia's rainforests, sclerophyll forests and savannah woodlands, the Platypus looks more fragile in terms of durability and evolutionary potential because its morphology and probably its range have significantly declined since the middle Miocene. As far as we can tell, apart from relatively recent loss of the Common Brushtail Possum from areas of central Australia (although they are being reintroduced there), those of brushtails in general have not. It is this relative difference, evident only from the fossil record, that concerns us about the Platypus's future.

The Koala: fuzzy worries about a worried fuzzy. As we noted in Chapter 4, koalas as a group (family Phascolarctidae) have persisted for at least the last 25 million years in more or less the same state of diversity and morphology as they do at present. Most of central Australia's Oligo-Miocene deposits include one or two species. Their teeth and what little is known of their skulls, although indicating three different genera (*Perikoala, Madakoala* and *Litokoala*), appear to have been reasonably similar to the living animal (*Phascolarctos cinereus*). They were all arboreal, selenodont tree leaf-eaters like their living counterpart.

Koalas and koala-like animals were very rare in Riversleigh's Oligo-Miocene environments compared with the representatives of other marsupials families found in the same deposits. Because of this, we have speculated that some of the early koalas specialised on a relatively uncommon resource in Australia's Oligo-Miocene rainforests—which may well have been eucalypts. Today there are still eucalypts in rainforests but of course they now dominate the open forests and woodlands of the rest of the continent. This may explain why modern koala densities seem very high compared with those indicated for the Oligo-Miocene closed forests of Riversleigh and perhaps central Australia. Because modern rainforest eucalypts appear to specialise on poor soils, when the Australian climates began to deteriorate in the late Tertiary, these once rare gums, with their equally rare koalas, may have been pre-adapted to persist and spread in the rapidly expanding areas of nutrient impoverished soils.

The Tertiary fossil record of koalas adds information and, ideally, perspective to questions about the conservation status of the living Koala. While Quaternary changes in Koala distribution and density are important to consider when the status of the living Koala is assessed, our perspective of long-term changes throughout the Cainozoic suggests that in terms of *density*, modern Koalas are more abundant now than they were in the Tertiary. Similarly, in terms of dental morphology, the living Koala is little more specialised than its ancestors which suggests that, in contrast, for example, to the Platypus it has not edged as far out along its evolutionary limb. Further, because most Cainozoic local faunas have between one and two species at most, the existence of one relatively abundant species today does not in itself suggest that the group is in decline; in fact the fossil record suggests the opposite conclusion.

These views must be tempered, however, by consideration of pre-1788 changes in the range of the modern species *Phascolarctos cinereus*. Individuals, albeit rare, are known from late Pleistocene cave deposits of south-western Australia (e.g. Koala Cave near Yanchep, north of Perth) and even from a cave on the southern edge of the Nullarbor ('Madura Six Miles South Cave'). The loss of Koalas from the central southern regions was almost certainly caused by Pleistocene environmental 'declines'—arid periods that correlated with glacial advances in the northern continents—that isolated the south-western from the south-eastern forests. The late Pleistocene loss of Koalas from the south-western forests is not as easy to explain and may instead have been related to the arrival of Aborigines into this region. Reintroduction of Koalas to the native bushland around Yanchep has been reasonably successful suggesting that they were not lost from this area because of extinction of an essential resource.

By 1925, Europeans had exterminated koalas from the whole of South Australia, thereby demonstrating one among many of the less humane 'skills' of humanity. Habitat destruction in other areas of eastern Australia, such as the logging of the forests of south-eastern New South Wales, must be controlled to avoid more tragedies of this kind.

Clearly, concerns about historic declines in the distribution of the Koala should influence modern ecologists in their task of defining appropriate conservation programs for this gum muncher. That said, however, the longer perspective of the whole of the Cainozoic fossil record suggests that the koala lineage is a relatively robust, phylogenetically healthy one compared with, for example, that of platypuses. In fact, providing that human philistines do *not* continue to chew away at the Koala's preferred habitat, the living Koala ought to have no problem pacing if not outlasting humans.

Thylacines: Tiger, tiger lost from sight. It is provocative to wonder what would have happened if a palaeontologist had jumped ship when Captain Cook's team sailed up the mouth of the Endeavour River and struck out overland to discover the messages in Riversleigh's limestones. After weeks in the limestone wilderness reading the record of the rocks, this palaeontologist would have returned to the *Endeavour* a wiser but worried naturalist. For, among other revelations sad and glad, the rocks would clearly have warned that the light of life was rapidly dimming for the Tasmanian 'Tiger', the last of the once great and noble house of Thylacinids.

Jeanette Muirhead, a palaeontologist with a 'vision' of life's history, has been reading Riversleigh's 25-15 million-year-old record of carnivorous mammals, discovering and understanding the plethora of meat-eaters that ranged from mouse to dog in size. Among this assemblage is a variety of marsupial 'tigers' (thylacinids) one of which provided the genetic resources for the Thylacine that first appeared approximately 4 million years ago. In itself, this Riversleigh assemblage is interesting enough, particularly in terms of what it suggests to Jeanette about the complex interplay through time that took place between thylacines of various sizes and the struggling dasyures that eventually came to dominate Australia's carnivorous mammal niches. But for our present purposes, the observation of most importance to our AWOL palaeontologist from the *Endeavour* would have been the simple fact of the high diversity of thylacines between 25 and 15 million years ago. As Europeans were soon to find out, in 1788 there was only a single species of thylacinid in Australia (*T. cynocephalus*) and that was hanging on by its whiskers to the vulnerable island of Tasmania.

We know from the Alcoota LF that by late Miocene time, perhaps 5 million years later than the youngest of the middle Miocene Riversleigh assemblages, just one large thylacinid (*Thylacinus potens*) remained in central Australia. By early Pliocene time, this species had given rise and way to the modern Thylacine (*T. cynocephalus*) which, by at least late Pleistocene time, was widespread throughout Australia and New Guinea. Then, sometime in the middle Holocene, the Dingo (*Canis familiaris*) was brought in as a pet by Aborigines and the Thylacine vanished from the Australian mainland, perhaps because of competition. By at least 4000 years BP, Thylacines could only be found on Tasmania, the only large land mass that Dingoes failed to reach.

All up, by the time Cook sailed into Australian waters, a dreadful disaster was poised to unfold. Although warned by his phantom palaeontologist that Riversleigh's rocks showed the Thylacine's lineage to be in steep decline and dreadfully vulnerable to extinction, having dropped from as many as six species in the middle Miocene to one geographically restricted taxon in 1788, there was

TERRACE SITE

By 25 thousand years ago, most of Riversleigh was dry—except for the lush margins of the spring-fed Gregory River. Along the river's edge were pandanus palms, tall paperbarks, cluster figs and gum trees. Away from the river, the country was a sea of prickly spinifex and sparse, stunted trees. The change in climate that brought ice sheets to the Northern Hemisphere brought powder dryness to Australia. The limestone cliffs with bones of rainforest creatures were slowly being dissolved by the irregular rains of the 'wet season'. In the caves and shelters that developed like cavities in teeth, humans wove present and past into a dreamtime history of their new land. Huge broad-snouted crocodiles made the lives of turtles hazardous. The largest marsupial known, the rhinoceros-sized *Diprotodon optatum*, occasionally misjudged the sandy banks along the water's edge. Its bones were then added to those of Agile Wallabies, marsupial 'tapirs' and Water Rats on a gravel bar not far downstream. Sands and clays soon buried the gravels, preserving the deposit from the next season's floods. Thousands of years later, long after the Gregory had cut more deeply into its valley, local stream erosion once again exposed these gravels with their now precious load of fossilised bones and teeth.

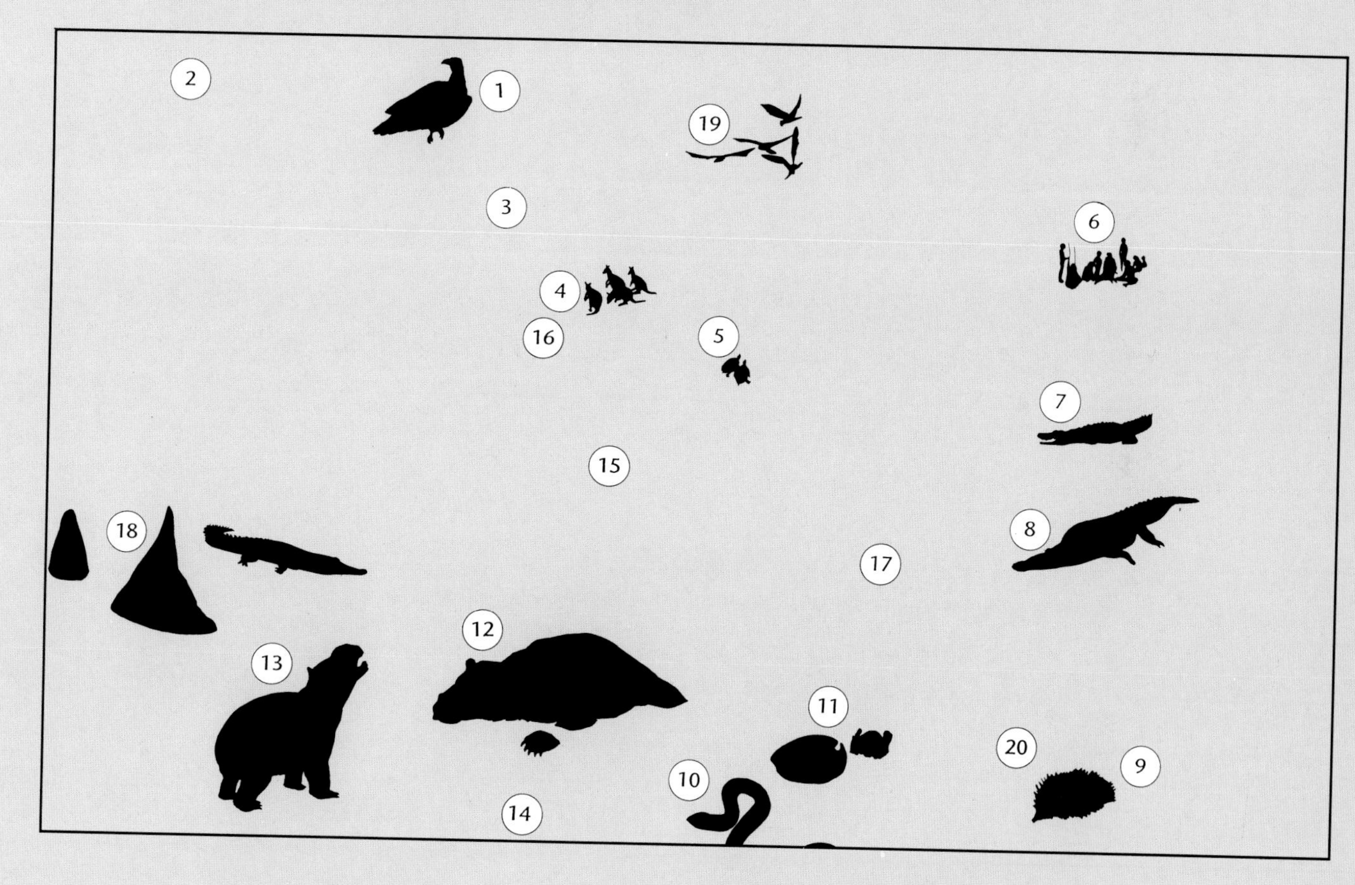

1 Eagle (presence inferred but yet to be confirmed).
2 Gum trees (*Eucalyptus* sp.).
3 Cliffs of Tertiary age confining basin for ancestral Gregory River.
4 Agile wallabies (*Macropus agilis*).
5 Giant Elseya turtle (*Elseya* sp.).
6 Humans (based on presence of possible stone tools).
7-8 Giant, broad-snouted crocodile (*Palimnarchus* sp.).
9 Echidna (presence inferred but yet to be confirmed).
10 Poisonous snake (elapid; presence inferred but yet to be confirmed).
11 Carapace of the giant Elseya turtle (*Elseya* sp.).
12 Giant herbivorous diprotodontid (*Diprotodon optatum*).
13 Juvenile *Diprotodon optatum*.
14 Treacherous sands along the edge of the ancestral Gregory River.
15 Ancestral Gregory River draining to the Gulf of Carpentaria.
16 Older terrace deposits that the River has transected.
17 Pandanus palms.
18 Termite colony.
19 Birds.
20 Midden deposit of freshwater mussel shells.

Dunphy

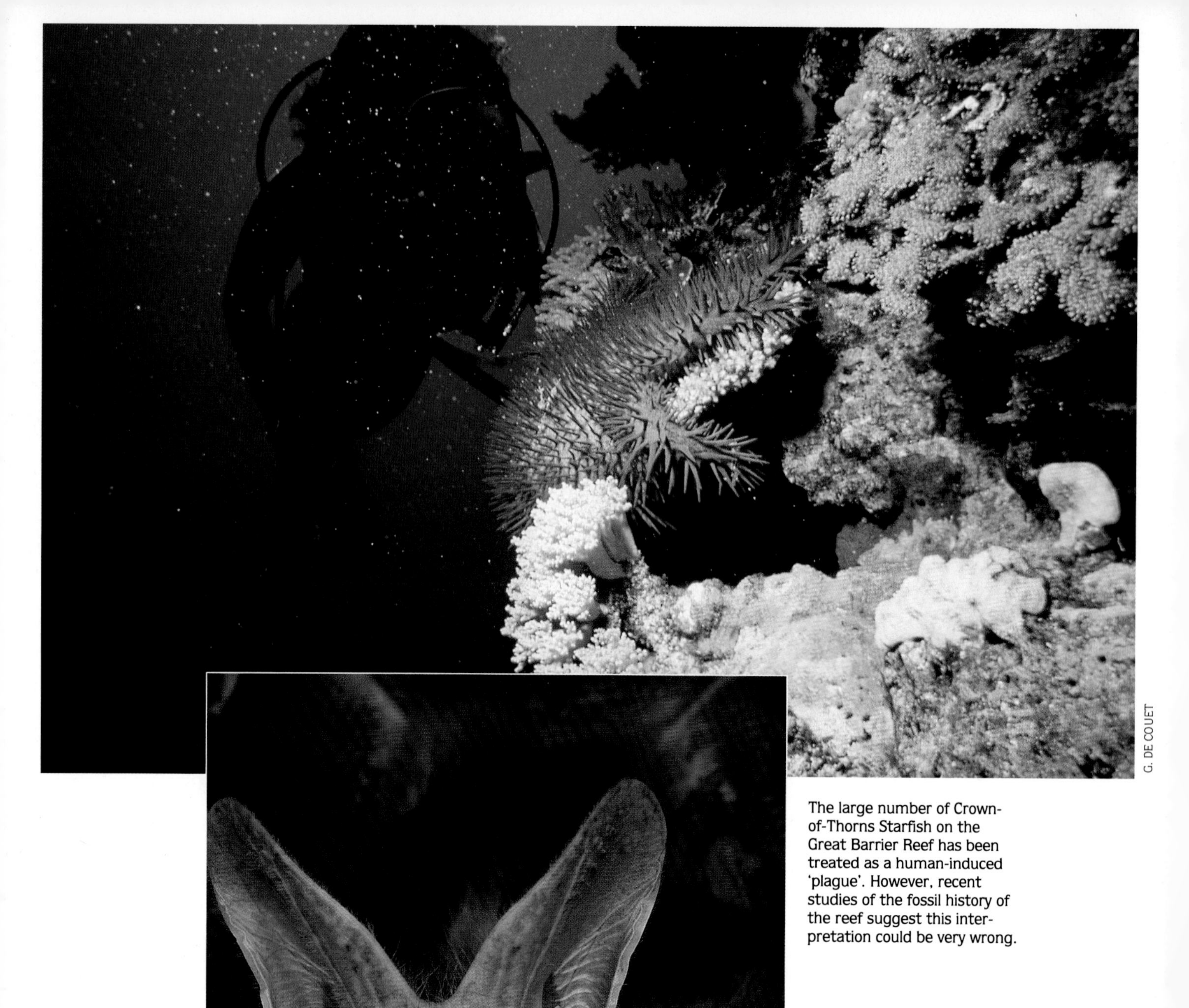

G. DE COUET

The large number of Crown-of-Thorns Starfish on the Great Barrier Reef has been treated as a human-induced 'plague'. However, recent studies of the fossil history of the reef suggest this interpretation could be very wrong.

Ghost Bats, like the living *Macroderma gigas*, have never been diverse or particularly common during their long history in Australia but the Quaternary contraction in range may be cause for concern.

R. WILLIAMS

little Cook could have done to avert what was about to happen—except perhaps to have turned around and left. But this southland was ripe for plunder. Although Europeans had destroyed almost 90% of their own forests by the time Cook sailed, it would be centuries before the long-term implications of this bastardry concerned anyone. Even less thought was given to the viability—let alone the morality—of pillaging a land so far from Europe. So the heels of English boots plunged into the soft heart of Australia.

Not content with harvesting the native herbivores for food, they introduced hard-hoofed ungulates such as goats, cattle, pigs and sheep everywhere they could find a toe-hold on the virgin land. Inevitably, soon after Tasmania was swarming with sheep, the precious remnants of the last population of Thylacines came face to face with these mindless woollies. The outcome was like a classic Greek tragedy—predictable yet unstoppable. From the moment that the first Thylacine followed its instincts for a meal of mutton, the last population was doomed. Europeans have never had the courage or sense to allow a native carnivore to co-exist with their domesticated animals or, for that matter, themselves. Carnivores through the ages of 'humanity' have been the focus of negative human attention through religious rituals, 'manhood' ceremonies or straight-out extermination campaigns. The Thylacine fared no better. By the 1930s, the entire species, last representative of its noble lineage that had withstood more than 25 million years of natural change, was no more than a memory and a few stuffed skins fading in museum display cases.

To Jeanette and the rest of us faced with Riversleigh's warnings which came 200 years too late, loss of the Thylacine and of its magnificently

The Tasmanian Thylacine, rapidly exterminated by Europeans in the 1930s, was the last of a very ancient and distinctive lineage. The fossil record could have provided a warning that the survival of these carnivores was in jeopardy.

TASMANIAN MUSEUM AND ART GALLERY

distinctive potential, steels our determination to have all of the other warnings, many of which are *not* too late, set firmly on the top of the urgent tray of modern conservationists.

The Orange Horseshoe Bat and the Ghost Bat: a Tale of Two Flitties. Accumulating data gathered from the Riversleigh deposits also provide information important for the conservation of two of Australia's most spectacular bats: the carnivorous Ghost Bat (*Macroderma gigas*) and the insect-eating Orange Horseshoe Bat (*Rhinonicteris aurantius*). Today, these cave-dwelling bats are sparsely distributed within their broad northern Australian ranges but the Riversleigh fossil deposits indicate that in the Oligo-Miocene (25-15 million years ago) their ancestors dominated local cave-dwelling bat faunas, outnumbering all other contemporary Riversleigh bats (13 of 23 taxa).

These groups were also once far more diverse. In at least one cave deposit, five Orange Horseshoe like-species (*Brachipposideros* spp.) appear to have coexisted and in other Riversleigh deposits two species of ghost bats (*Macroderma* spp.) are commonly present. By the early Pliocene (5-3 million years ago), diversity had dropped slightly to four Orange Horseshoe types and one ghost bat, including what appear to be early populations of the modern Ghost Bat and Orange Horseshoe Bat. In southern Australia, Pleistocene and Holocene fossil and sub-fossil deposits indicate that the past distribution of the Ghost Bat was once much wider. It has been estimated that there may be less than 6000 individuals surviving today and the species is currently listed as a vulnerable by the International Union for the Conservation of Nature (IUCN).

For the Orange Horseshoe Bat lineage there has been a substantial decline in diversity since the early Pliocene; for ghost bats this has been less dramatic. The decline in the former group paralleled changes in habitat as Australia dried and cooled during the later Tertiary and the Quaternary. In Europe, too, the disappearance of this group coincided with a gradual change in climate from warm, tropical forests to cooler, drier and more open woodlands. Although the group's living Australian descendant, the Orange Horseshoe Bat, exists today in semi-arid areas of northern Australia, in the dry season the species selects roosts in the most humid and warmest parts of suitable caves where temperatures reach up to 32°C and 100% humidity. Such refugia may have enabled the Orange Horseshoe Bat to hang on while its close relatives were pushed to the brink of extinction by the increasing aridity and marked seasonality that accompanied the later Pliocene and Quaternary.

The number of ghost bat species co-existing in northern Australia over the past 25 million years appears to have fluctuated between one and two, and thus the Pleistocene drop in diversity is probably not as significant as in the Orange Horseshoe Bat lineage. However, the modern Ghost Bat's range has contracted markedly during the Quaternary. Its disappearance from south-western Western Australia and New South Wales sometime during the last 10-20 000 years may have been due to broad climatic changes, although this is by no means clear, particularly since its present range includes everything from hot dry mulga to humid rainforest.

The disappearance of the species from areas of southern, central and western Australia as recently as 50 years ago is difficult to interpret. So is the abandonment of many caves within its present range. For example, Carrington's Cave at Riversleigh was evidently used for a very long period before being abandoned, as the enormous guano pile peppered with prey remains testifies. It is likely that at least the more recent disappearances of the Ghost Bat are due to direct and/or indirect human disturbance — human predation, competition for caves, interference and destruction of roost caves, clearing and burning of habitat, and human-induced decline in local vertebrate prey.

Remains of Orange Horseshoe Bats and their relatives have not been found in caves further south than the living species's range, and it is probable that the past distribution of the group in Australia was not much broader than it is today. However, within the very recent past, roost caves within its modern range have been abandoned, in at least one case almost certainly because of human disturbance.

Although presently both may be staring extinction in the face, the Ghost Bat and Orange Horseshoe Bat have reached the brink via two quite different paths. Diversity has nose-dived in the Orange Horseshoe Bat lineage during the last 3-5 million years perhaps heralding its imminent demise. For the ghost bat lineage, diversity appears to have remained relatively constant over at least 25 million years. It is, however, the shrinking range of the modern Ghost Bat that has rung alarm bells among most biologists and conservationists. We believe that the Orange Horseshoe Bat is in need of equally urgent attention.

Re-examination of the timing of disappearances from various parts of Australia will be important in identifying the habitat requirements for both living species and possibly halting their downhill slide. If, as cave deposits in southern South Australia now suggest, the Ghost Bat was present in the area until as recently as 200 years ago, then it is unlikely that climatic change (and associated changes in roost temperatures and humidity) was responsible for the contraction and other factors must be more fully explored.

For both species, the opportunity to move between seasonally suitable caves (either to meet microhabitat requirements and/or to follow seasonally available prey) has probably ensured their survival since the Pliocene. Identification and protection of these caves will be crucial for their continued survival.

Mountain Pygmy-possums: Abandoned in the Alps. The Mountain Pygmy-possum, *Burramys parvus*, has had a most interesting as well as instructive phylogenetic history. From the point of view of helping to understand the conservation status of the living animal, the fossil record has been critical. Without it we would have no idea that the living animal occupies a miniscule fraction of its former range nor that, like the Thylacine, it is the last of a once diverse 'bush' of biodiversity widespread throughout at least northern, central, southern and eastern Australia.

As summarised in Chapter 4, a second species of

Burramys was very common in the lowland rainforests of Riversleigh between 25 and 15 million years ago. A third occurs in sediments of about the same age in central Australia which represent, at least in part, lowland rainforests. Yet a fourth occurs in 4.5 million-year-old lowland rainforest deposits at Hamilton, western Victoria.

The living species, *Burramys parvus*, was first discovered as a Pleistocene fossil in cave breccia near Wombeyan, New South Wales. Although there is some doubt (following the research of Honours student Jenni Brammall) that the fossil and modern populations are the same, they are most similar to each other and probably closely related. Modern populations occur in two areas separated by a deep uninhabited valley (the Mitta Mitta River Valley) between Mt McKay and Mt Higginbotham, Victoria and Kosciusko National Park, New South Wales. If we put this information together with that understood by palaeobotanists, we may hypothesise that by 25 million years ago, ancestral *Burramys* were widespread in lowland rainforests throughout at least the eastern half of the continent. In blind anticipation of the microhabitat preference of the living species, the Oligo-Miocene forms may have already adapted to rocky substrates within the rainforests. Certainly the Riversleigh rainforest floors would have been strewn with limestone blocks. Then, as the climates deteriorated in the late Miocene and rainforests became restricted to eastern Australia, *Burramys* species died out in northern and central Australia but persisted along the eastern margin of the continent. The early Pliocene Hamilton species would have been part of this retreating biota. Subsequently, sometime between middle Pliocene and Pleistocene time, driven by drying conditions in the lowlands, the rainforests of eastern Australia would have contracted up the rocky slopes of the alps dragging with them one population of *Burramys*. This population was the source of the abundant specimens of *B. parvus* in the Broom breccia.

These Pleistocene mountain possums then crossed the last, perhaps behavioural, threshold—they became more dependent on the resources of the rock piles than the rainforest *per se*. So, when the rainforest finally abandoned the tops of the Alps, presumably sometime during the Pleistocene, it left behind two isolated populations. Finally, from the southernmost population, a furry-faced descendant, many generations removed, heralded the survival of its lineage from the middle of that now infamous woodpile in 1966 (see Chapter 4 for details).

It would seem that the solution to understanding the distribution pattern of this snow-bound pygmy-possum must be sought in the decline of Australia's prehistoric rainforests. Further, it is clear that the living species and its populations are a paltry remnant of a once more diverse and far more widespread lineage. Hence, while ecological understanding about the viability of the living populations might suggest they are stable, our longer-term view is much less optimistic.

Like that of the Platypus, the house of *Burramys* is clearly in the twilight of its existence. Looking ahead, what will happen to its already declining populations when greenhouse conditions drive the wintry habitat off the peaks of its last stronghold? Faced with the long-term decline and perhaps imminent threat to its future, we argue that it is in need of much closer conservation attention than other more widespread kinds of animals with which it presently keeps company such as the Common Wombat, the Common Brushtail Possum and the Eastern Grey Kangaroo. Otherwise, the snow burial that the remnant populations endure each year may presage this furry-faced Lazarus's return to the stoney earth from which it mysteriously arose in 1966.

The Musky Rat-kangaroo: Tiny Mystery with a Long Green History. Like many other animals that live on the Atherton Tableland today, the diminutive Musky Rat-kangaroo *(Hypsiprymnodon moshatus)* is cherished for its uniqueness, in this case as Australia's most 'primitive' kangaroo. For all intents and purposes, we presume it to be safe because it occurs within what is now a National Park. Barring narrow-sighted decisions to allow logging in these forests, its immediate future ought to be secure.

But how, through study of the living populations, can we determine the long-term conservation health of this lineage? Does its distribution, population size or morphology provide information about whether it is in decline, on the rise or in a steady state? The honest response is, we cannot tell from study of the living populations. Sound predictions about the future can only arise from constrasting the present and past—and until 1983 we had little idea about the history of this lineage prior to 1876 let alone any idea about its *prehistory*.

Now that Riversleigh's fossil record is unfolding, it is clear that this lineage has been an integral part of Australian rainforests, and as far as we can tell no other habitats, for at least the last 20 million years. This lineage is also present in the early Pliocene Hamilton cool temperate rainforest assemblage of south-western Victoria. So we begin to understand that it has always been tied to rainforests although these have been temperate as well as tropical.

Has it declined in abundance or 'evolutionary fitness' within these rainforest habitats? Certainly in terms of morphology, the Riversleigh animal was larger and more generalised than its living descendant. But the differences are not of the 'degenerative' kind that distinguish, say, the living Platypus from the Oligo-Miocene species of *Obdurodon*. About relative abundance within Australian rainforest communities, we are not yet in a position to provide estimates of change. Once the Riversleigh computerised database has a grip on the majority of specimens so far recovered, we should be able to document trends through time in population parameters, as well as to understand patterns in species associations in Cainozoic communities as a whole.

'Thingodontans' and Other Casualties: When the Bough Breaks. When we consider Riversleigh's value in helping to determine conservation strategies, it is appropriate to note that some of its warnings focus on Oligo-Miocene groups that were even then about to slide beneath the foot of extinction. Extinction is not just a bad side-effect of the twentieth century—it is a natural process that went on long before we became its most successful grim reaper. Although many extinction events were pre-human, they may provide us with an understanding of factors that can stress already declining populations.

One of the most intriguing of Riversleigh's long-exinct groups is Thingodonta or, as it is technically known, *Yalkaparidon* (see Chapter 4). It helps to make the general point about faunal change in Australia. In the last 25 million years, we have lost a great many fundamentally different *kinds* of creatures that were just as distinctively Australian as Koalas and Red Kangaroos. To those of us who wallow in the delights of biological diversity, this is the somewhat regrettable aspect of evolution, the flip-side of diversification. Given that all life struggles to maintain itself in a capriciously changing world, nothing can be more certain than eventual extinction. Ideally, in terms of maintenance of long-term biotic diversity, there are compensations—overall, for every extinction, there is an opposite and equal origination.

Unfortunately for Australia, this does not appear to have been the case for the major *kinds* of mammals lost. Looking at changes in family-level diversity in Australia's rainforests over the last 25 million years (see Appendix 1), it is clear that we have lost many families without equivalent replacement (e.g. yalkaparidontids), although the total number of *species* in the modern rainforests may have increased. The main reason for the decline in major *kinds* has presumably been the progressive decline in distribution and probably diversity of Australia's rainforests precipitated mainly by the severe climatic changes of the late Caínozoic. Progressive drying of the continent occurred as rainfall declined in part because of an overall drop in world temperatures and growth of the polar ice caps as the world headed towards the ice ages of the Pleistocene.

Yingabalanarids, ilariids, wynyardiids, several as yet unnamed families and many genera of other families made their last stand in the early to middle Miocene rainforests at Riversleigh. Humans clearly had nothing to do with these losses although we may well have been involved in the late Pleistocene extinctions of other families such as the diprotodontids, palorchestids and thylacoleonids.

Certainly there is evidence for habitat—if not niche—replacement by invaders from the north for at least some of the ancient families. We have identified a decline in species-level diversity of marsupials and corresponding rise in the diversity of placental mammals. It is also evident that New Guinea's mid-montane rainforests, which contain the most diverse modern animal assemblages in the Australasian region, have suffered greater proportional change in this regard than the rainforests of north-eastern Queensland, although these too were hammered by late Pleistocene environmental changes that, at least once, shrank the wet tropics to areas much smaller than they are at present.

Of the changes in rainforest communities, the most striking are: the loss of species-level diversity among marsupials (including total loss of all groups with representatives more than a metre in length such as diprotodontids, palorchestids and wynyardiids) from 84% to 35% of the faunas; the decline of potoroid ('rat') kangaroos; the rise in diversity of macropodid ('ordinary') kangaroos; the rise of vespertilionid bats; the appearance of rhinolophid, emballonurid and pteropodid ('fruit') bats; and the appearance and rapid increase in diversity in murid rodents.

Clearly extinction in Australian ecosystems has been a significant natural process, one usually followed or accompanied by expansion of species-level diversity in other families to make up the losses. What is not, however, normal is the kinds and rates of extinction that Australia has suffered during the last 200 years. While these losses have taken place, there has been no concomitant expansion of species-level diversity in other groups—only a population explosion of a single African invader, humans, and its commensals.

Conserving Australia's Rainforests: a View from the Past

In general conservationists do not have the time or resources to conduct studies focused on single species. So the alternative approach to securing the world's threatened creatures is to conserve as much as possible of their habitat. In so doing, conservationists trust that the endangered creatures within these habitats will be given the next best chance for survival.

From our point of view, the vegetation type most illuminated by the Riversleigh resource is rainforest. It starts our story in the Oligo-Miocene and ties it inextricably to the modern remnants of rainforest in north-eastern Queensland and more specifically, those of the Wet Tropics. And from what we have learned at Riversleigh, it now seems to us a matter of utmost importance to conserve every leaf and splinter of these forests, for the future of Australia may very well be stillborn without them.

A special resource in need of help. While it is widely understood that half of the world's living species are gathered in tropical rainforests, it is equally clear from our studies that almost every group of Australian terrestrial mammals had ancestors in the lowland rainforests of at least north-western Queensland. How much more widespread these seminal rainforests were, we do not yet know but what we can be sure of today is how much *less* there is now.

This flags the importance of conserving whatever remains of Australia's living rainforests for a vital reason *additional* to those commonly cited. Given our increased understanding about the nature of rainforest as the source for the majority of this continent's terrestrial creatures, from Green Ringtail Possums to Marsupial Moles, it is probable that it is the single most important biological resource we hold in trust for Australia's future. If we continue to exterminate species from the Australian continent at the present rate, which hindsight tells us is far in excess of any in the past, the genetic resource of the depths of our rainforests may not have the capacity to produce the 'seeds' to oncce again populate this continent. Hopefully by the time we are wise enough or perhaps frightened enough to realise how important biological diversity is to our own survival, enough rainforest will have survived to enable the vital processes of repair to begin.

Appendix 1

The Changing Face of Australia's Rainforest Communities

In Chapter 7, we explored the changes in climate and vegetation that took place within the Riversleigh region over the last 25 million years. The primary change evidenced by the vertebrate communities over this period was the late Cainozoic reduction of rainforest, a change that appears to have been well advanced in the Riversleigh region by late Miocene to Pliocene time. What happened to Australia's rainforest mammal lineages after these inland rainforests declined? To follow their fate, we need to consider fossil and living rainforest assemblages outside of the Riversleigh region.

To do this, we have compared here species diversity within mammal families for four Cainozoic rainforest mammal faunas: Upper Site LF, a Riversleigh Oligo-Miocene assemblage; Hamilton LF, an early Pliocene rainforest assemblage from western Victoria; the modern rainforests of the Wet Tropics of north-eastern Queensland (between Cooktown and Townsville); and the mid-montane rainforests of modern Papua New Guinea.

Oligo-Miocene Upper Site Local Fauna

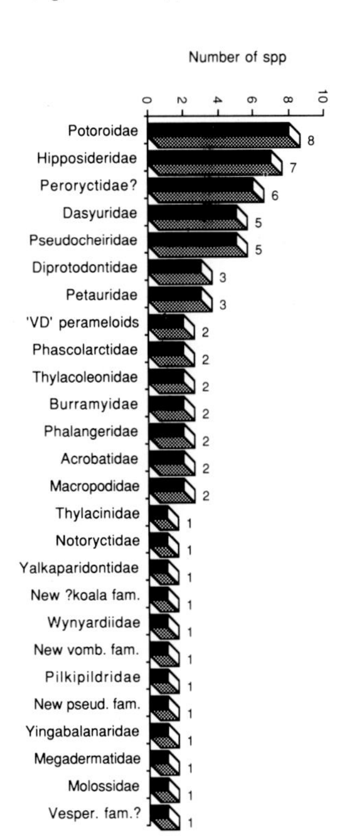

Early Pliocene Hamilton Local Fauna

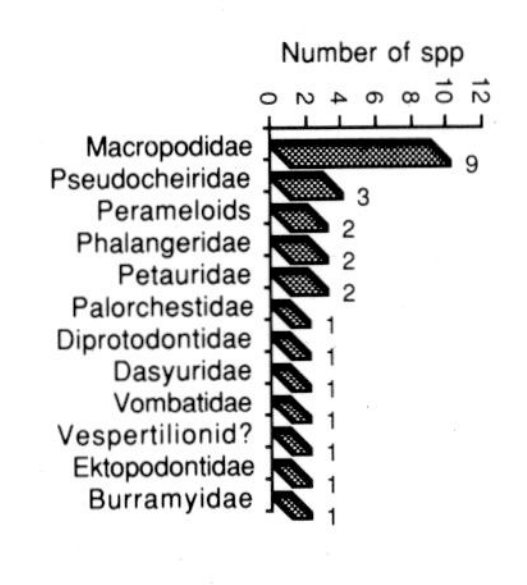

Modern NEQ Wet Tropics diversity

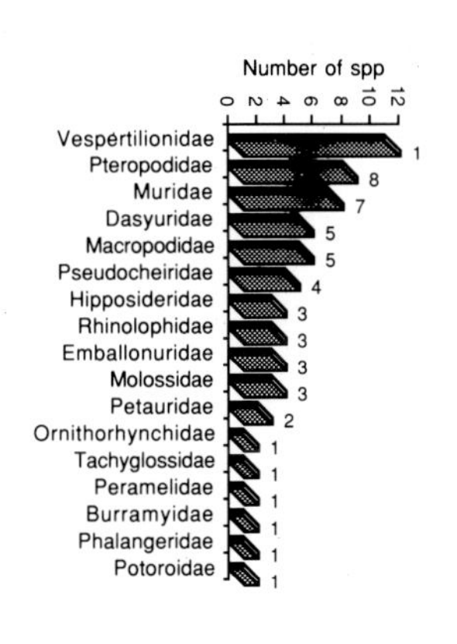

Modern PNG Mid-montane diversity

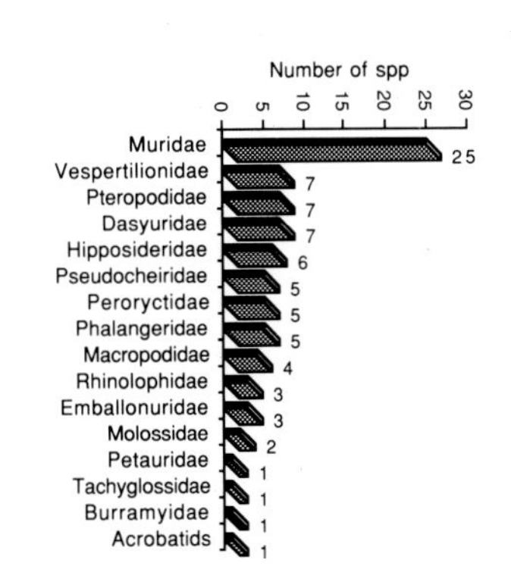

We chose these four for the following reasons. The Upper Site LF is now one of the oldest-known undoubted rainforest mammal assemblages from Australia. The Hamilton LF is the only known Pliocene rainforest mammal assemblage from Australia. (Although Riversleigh's Rackham Roost is Pliocene in age, it appears to represent an open forest or woodland habitat.) There are no undoubted early or mid-Pleistocene Australian rainforest mammal assemblages known. The Wet Tropics rainforests of north-eastern Queensland are among the richest of Australia's remaining rainforest refugia. The Papua New Guinean mid-montane rainforest was chosen because it contains many taxa that are at least distant relatives of taxa in Riversleigh's Oligo-Miocene rainforest assemblages and because it contains New Guinea's richest mammal assemblages.

The diagrams depict species-level diversity within families of mammals in four rainforest assemblages: A, the Oligo-Miocene Upper Site LF of Riversleigh; B, the early Pliocene Hamilton LF of western Victoria; C, the rainforests of the Wet Tropics, north-eastern Queensland; and D, the mid-montane rainforests of modern New Guinea. The data for the Hamilton LF comes from Rich *et al.* 1982 and various papers by Turnbull *et al.* in Archer 1987; that for the modern rainforests of the Wet Tropics of north-eastern Queensland comes from John Winters' 1988 study and Greg Richards (personal communication); that for mid-montane P.N.G. comes from Tim Flannery and Ken Aplin (personal communications).

Trends and changes in these rainforest assemblages invite comment. Looking at the data we see:

1. a progressive extinction of families, particularly among marsupials;
2. an increase in diversity within particular families (e.g. kangaroos, rodents, and some bats) and the rise of placental groups in general; and
3. changes in the size and type of mammals that persisted in rainforests.

1. Extinction of family-level groups from rainforest communities

The Upper Site LF has 27 families (with at least 63 species); the Hamilton LF 13 (with at least 26 species); the Wet Tropics 17 (with 60 species); and the Papuan New Guinea mid-montane rainforest 16 (with 83 species). Clearly, there has been an overall decline since the Oligo-Miocene in family level diversity.

We attribute this trend in part to changes in the plant communities since the Oligo-Miocene and to the arrival of various placental groups, including, in the late Pleistocene, humans. Changes in Tertiary rainforest plant communities may have resulted from increased seasonality and/or deterioration of Australia's rainfall regimes in the late Miocene.

It is possible that the high diversity characteristic of the Oligo-Miocene local faunas of Riversleigh reflects factors no longer evident in modern Australian rainforests. For example, high hipposiderid diversity (7 species) in the older Riversleigh local faunas is best matched not in modern Australian rainforests but in areas of south-eastern Asia (e.g. Borneo), where lowland tropical rainforest still carpets limestone terrains riddled with caves.

If we look at the families that declined to extinction in Australia's rainforests, most were represented by only one or two species of small to medium-sized mammals. Losses of this kind may not have required very significant alterations to the rainforest habitats to push them over the edge. Of these groups, only the marsupial moles survive as descendant populations in desert environments.

Other losses from rainforest communities were of groups that were relatively common in those habitats. These include diprotodontids, palorchestids and marsupial lions, all of which were middle-sized to large mammals. For losses of this kind, we presume that more profound changes affected the rainforests, although they may have been changes that occurred in the Quaternary, rather than the Tertiary. These groups vanished also from mesic environments in the late Pleistocene, their disappearance possibly reflecting the arrival of Australia's first humans at least 50 000 years ago. The times of their disappearances from Australia's rainforests are not clear but *Hulitherium tomasettii*, a zygomaturine diprotodontid, persisted in upland rainforests of New Guinea until as recently as 20 000 years ago.

2. Increase in diversity within particular families in rainforest communities

Among marsupials, one group that shows a striking increase in diversity between Oligo-Miocene and Pliocene time is the 'ordinary' kangaroos. Most other rainforest marsupial groups declined in diversity over this same period. By early Pliocene time there were many macropodid lineages in and outside of rainforests. Their increase appears to correspond with the decline in balungamayine potoroids, a diverse group of Oligo-Miocene 'rat-kangaroos' that had evolved lophodont teeth similar to those of macropodids, presumably as an adaptation for a more herbivorous diet see Chapter 4.

The rise of placental diversity in Australian rainforests is also dramatic. Placentals have come to represent approximately 64% of the modern rainforest mammal faunas of both Australia and New Guinea in contrast to the 16% of the Oligo-Miocene Upper Site LF and 4% of the Pliocene Hamilton LF. However, the changes from the archaic Australian rainforest assemblages, such as that of Upper Site, appear to be different from both the New Guinean mid-montane communities and from those of the Wet Tropics of Australia.

When considering changes in the diversity of placental mammals in Australia's rainforests, it is important to review the relationships of the drifting Australia to adjacent land masses. In contrast to previous views that depict a vast, vacant ocean barrier between drifting Australia and south-eastern Asia, an albeit ephemeral archipelago appears to have linked north-western Australia to south-eastern Asia since at least the late Cretaceous. Bats may have used this archipelago to disperse to and from Australia in the early to middle Tertiary. Marsupials do not appear to have made effective use of this route but it is clear that it was used by the invading rodents in the late Tertiary and by humans and dingoes in the Quaternary.

By 30 million years ago, New Guinea was probably a low, rainforest-covered area that would have encouraged colonists from northern Australia. The middle Miocene uplift of the New Guinean Highlands preserved, in increasingly higher altitudes, part of the formerly widespread lowlands rainforest of the northern area of the continent. By 10 million years ago, Australia's geographic position was much as it is at present. Late Miocene submergence of the New Guinean lowlands need not have endangered the mid-montane biotas but it may have obstructed dispersal of land mammals to and from Australia.

Perhaps the higher diversity of rodents in mid-montane New Guinea has resulted from western New Guinea's relatively longer period of exposure to Asian immigrants. However, considering the comparable antiquity of a more southern archipelago linking south-eastern Asia directly to Australia (via Java and probably Timor, thus by-passing New Guinea), Asian placentals could have been arriving on the shores of north-western Australia for an equally long period of time.

Conversely, perhaps many of the Oligo-Miocene marsupial groups in Australia were unable to disperse to New Guinea. Tim Flannery has argued that by early Miocene time, mammals in the Meganesian region (i.e. Australia plus New Guinea) had separated into two biogeographically isolated subregions: Australia with a blend of archaic and diverse, relatively modern marsupial groups; and New Guinea with a subset of only the more modern groups. Placental invaders into New Guinea may have simply insinuated into ecological niches unoccupied by marsupials but found the same niches occupied in Australia.

It is interesting to speculate about why rat-kangaroos (potoroids) and ghost bats (megadermatids), groups common

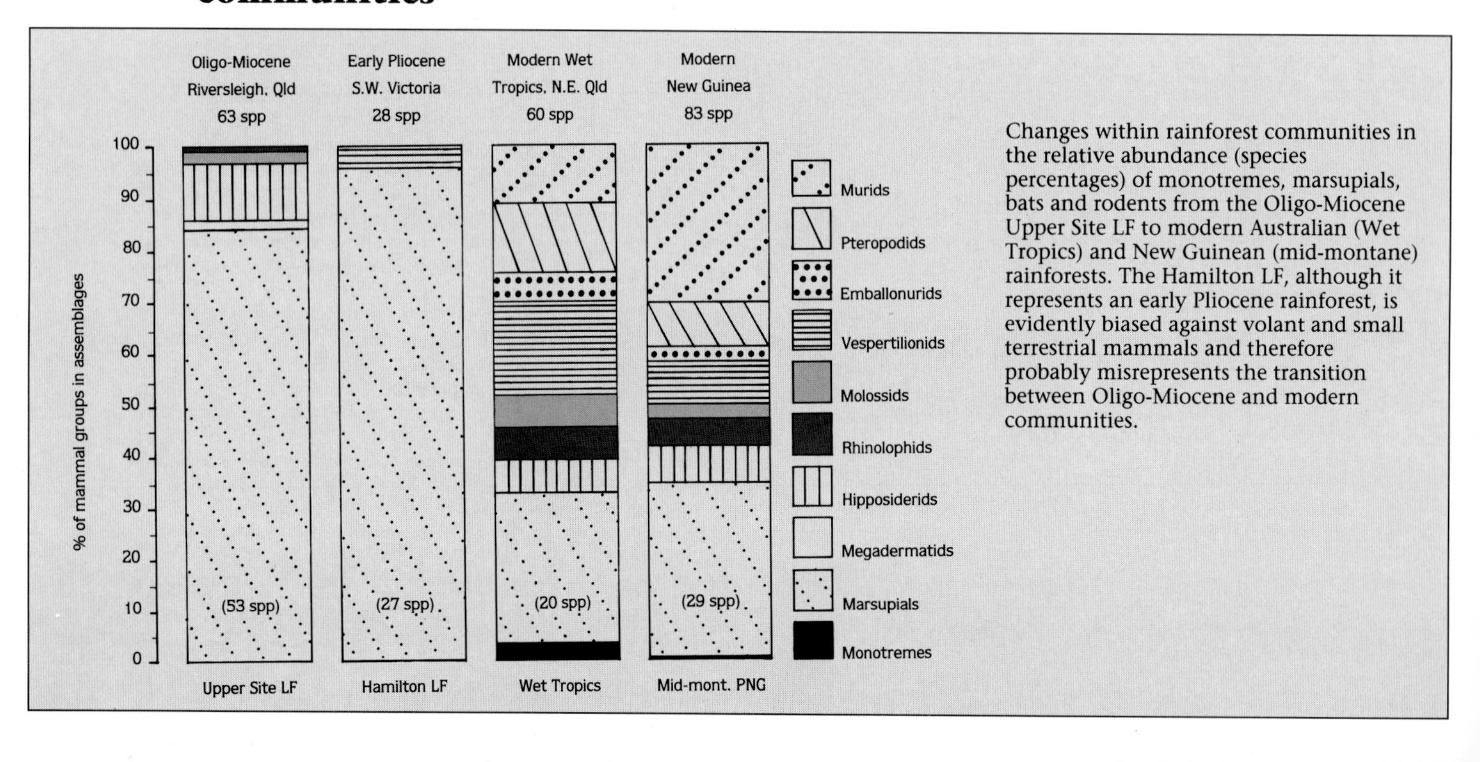

Changes within rainforest communities in the relative abundance (species percentages) of monotremes, marsupials, bats and rodents from the Oligo-Miocene Upper Site LF to modern Australian (Wet Tropics) and New Guinean (mid-montane) rainforests. The Hamilton LF, although it represents an early Pliocene rainforest, is evidently biased against volant and small terrestrial mammals and therefore probably misrepresents the transition between Oligo-Miocene and modern communities.

throughout the Cainozoic in Australian rainforests, didn't manage to establish in New Guinea. Three possible explanations come to mind. First, New Guinea may not have been available for colonisation in the Oligo-Miocene because much of it was under water, (although it has been suggested that New Guinea and Australia were a continuous landmass from at least the Oligocene until the early Miocene, after which time New Guinea was isolated from Australia by water possibly until the Pleistocene Flannery 1989). Second, there may have been filter barriers isolating New Guinea. For example, large murids in New Guinea may have presented a competitive barrier to the late Tertiary dispersal of rat-kangaroos from Australia. Third, the presently 'missing' groups *may* have reached New Guinea but suffered extinction. Submergence of much of lowland New Guinea in the late Miocene could have contributed to the extinction of any potoroids that had previously managed to colonise the area because most Australian potoroids live in lowland environments.

The only known Tertiary mammal fauna from New Guinea is the Pliocene Awe LF of Papua. It contains a subset of modern groups such as macropodine macropodids which are unknown in Australia prior to the late Miocene, diprotodontids which are similar to late Miocene or even Pliocene taxa in Australia, one dasyurid representing a species of *Myoictis* (a derived dasyurine) and an indeterminate number of murids. Hence it is not of much use in clarifying the Tertiary history of New Guinea's, let alone Australia's, rainforest mammals.

In Australia, rodents do not appear in the fossil record of rainforests prior to the Pleistocene. Although this young age may be the result of an inadequate fossil record, the oldest Australian rodents, from the Pliocene of Queensland, appear to represent dry country types (see Chapter 4). It is not clear how soon after their arrival murids invaded Australia's rainforests but they now represent 31% of the Papuan mid-montane mammal fauna and 12% of the mammal fauna of Australia's Wet Tropics.

The increase in vespertilionid bat diversity is almost as striking as that of murids. From 1 or 2 taxa in the Oligo-Miocene rainforest local faunas of Riversleigh, to 4 in the open forest Rackham's Roost LF, to at least 8 in Riversleigh's modern savannah woodlands fauna, 11 in the rainforests of the Wet Tropics and 7 in the mid-montane rainforests of New Guinea, the increase in diversity within and without rainforests is clear.

Sheathtail bats (emballonurids) show a similar, if more recent, increase in rainforest diversity. None are known from the Oligo-Miocene rainforest faunas of Riversleigh; 1 species is known from the Pliocene Rackham's Roost LF; 2 co-exist in the Riversleigh region today; 3 occur today in the rainforests of the Wet Tropics; and 3 live in the mid-montane rainforests of Papua. The slightly higher diversity in modern rainforests suggests that during Rackham's Roost time they may have been present in higher numbers in east coastal rainforests and possibly entered Australia from the north, via rainforest corridors, sometime between the middle Miocene and early Pliocene.

This same scenario may apply to horseshoe bats (rhinolophids). This group does not appear to be in any of the various-aged Riversleigh fossil deposits; nor are they found in the area today. However, two species are known from modern Australia and both are restricted to the east coast. Although four species occur in the mid-montane rainforests of New Guinea, no more than three may be sympatric. Their presence in the rainforests of eastern Australia evidently represents a relatively recent invasion from New Guinea, probably via north-eastern Queensland, perhaps in the Pliocene or Pleistocene.

Fruit bats (pteropodids) appear in the Riversleigh record suddenly, in the late Quaternary, there being not a single tooth to represent this distinctive group of herbivorous bats in the Oligo-Miocene, Pliocene or Pleistocene local faunas. Considering modern diversity in Australia (e.g. 8 spp in the Wet Tropics) and abundance, particularly of the species of *Pteropus*, the absence of this group from the Australian fossil record puzzles us.

Fruit bats may have colonised New Guinea prior to the establishment of arboreal fruit-, blossom- and nectar-feeding possums; conversely, they may have had difficulty invading Australian rainforests because of the abundance there of a variety of arboreal possums. The Quaternary establishment of fruit bats in Australia's northern rainforests may have followed late Tertiary declines of many of Australia's possums. But without a pre-Pleistocene record of fruit bats in New Guinea or Australia or of pre-Pleistocene possums from New Guinea, evidence for or against possible late Cainozoic competition between these groups does not exist.

Overall, the increases in species-level diversity of rodents, vespertilionid bats, fruit bats and macropodid kangaroos more than accounts for the high diversities of the Wet Tropics and mid-montane Papuan rainforest communities compared with that of the Upper Site LF. In the rainforests of the Wet Tropics, 52% of the mammal diversity is now made up by these four groups. In contrast, these groups comprise no more than 5% of the mammal assemblage in the Oligo-Miocene Upper Site LF.

3. Changes in size and type of mammals that persisted in rainforests

One of the most striking contrasts between the older rainforest mammal communities and those that survive today is the presence in the former of many large to medium-sized marsupials. In particular, the diprotodontids, palorchestids, thylacinids, wynyardiids, ilariids and marsupial lions were mostly dog- to cow-sized. In contrast, in the modern New Guinean and Wet Tropics rainforests, although Dingoes and/or Thylacines were present, the rest of the native mammals are no larger than a tree-kangaroo. What happened in these forests to cause the loss of the large mammals? As noted above, at least for diprotodontids and probably palorchestids, thylacinids and marsupial lions, the arrival of humans in Australia may well have contributed to their extinction. However, ilariids and wynyardiids did not survive beyond the middle Miocene so their declines require a different explanation.

Most of the now extinct large rainforest herbivores from Riversleigh represent vombatiform groups. The early Pliocene Hamilton LF retains only three vombatiform groups: palorchestids, diprotodontids and wombats. The decline in diversity of these large groups coincides with the apparent rise in diversity of rainforest possums and kangaroos which, although already diverse in the Upper Site LF, today are the only diprotodont marsupials in rainforests. Koalas, which were not uncommon in the Riversleigh rainforests, had evidently vanished from this habitat by the early Pliocene.

The Oligo-Miocene assemblages of Riversleigh appear to mark an important transition in the composition of rainforest mammal assemblages because they contain the vestiges of an older declining vombatiform radiation as well as the rapidly diversifying possums and macropodid kangaroos.

Although vombatiform marsupials declined during the middle Tertiary, during the late Tertiary, some such as diprotodontids, wombats, palorchestids and marsupial lions, diversified in the open forests and grasslands as did the mesic and arid kangaroos. By the end of the Pleistocene, however, all of the vombatiform groups in these drier habitats, except for wombats and koalas, had vanished.

Among the smaller rainforest mammals that declined or were lost after the Oligo-Miocene, there are several enigmatic groups known only from the older Riversleigh local faunas including thingodontans (yalkaparidontids) and 'weirdodontans' (yingabalanarids). Similarly, pilkipildrids, miralinids and several other odd possum groups vanished after the Oligo-Miocene. Other small possum groups, such as pygmy-possums, feathertail possums, petauroids and ektopodontids persisted in Australian rainforests until at least the early Pliocene and then vanished from rainforests by the Pliocene or, in the case of the ektopodontids, in the early Pleistocene.

Most of Riversleigh's larger Oligo-Miocene rainforest mammals, apart from kangaroos, were quadrupedal, terrestrial browsers similar to the tapirs that survive in the modern rainforests of south-eastern Asia. Consequently, we presume that they fed on an understorey of relatively soft-leaved plants within the forest. The same may have been the case for the gigantic (up to 400 kilograms) dromornithid birds that were abundant in the rainforests of Riversleigh. If the rainforest food resources of these giants vanished, it would explain their demise and that of the carnivores (e.g. marsupial lions) that may have fed on them. However, in other situations, the extinction of the giants may have been caused by things other than vegetation changes. For example, the spines on many wattles (*Acacia* spp.), which seem unnecessarily large to deter living herbivorous marsupials, may be persistent defence mechanisms against browsers that no longer exist.

4. A cautionary tale

Having said all this, it is important to consider any potential weaknesses in the data that forms the basis for these comparisons.

How representative are the communities examined in this comparison?

Because of its richness, particularly in the diversity of marsupials, there might be doubt about the extent to which the Upper Site LF fairly represents biotic diversity of the time. However, the early to middle Miocene Kutjamarpu LF of central Australia is not that much less diverse and could be used to argue that comparably complex rainforests extended at least that far south. In any case, when *species*-level diversity of Upper Site as a whole is considered, it is not significantly higher than that found today in the Wet Tropics. The appearance of higher diversity mainly arises from the greater family level representation of marsupials. Comparable or higher diversity, despite a decline in marsupial diversity, is maintained in Australia's and Papua's modern rainforests mainly because of the addition of rodents and fruit bats.

The early Pliocene Hamilton LF probably represents a rainforest refugium on the south-eastern edge of the continent. There is no evidence of how widespread this fauna was, but it seems likely that it extended at least to the east coast where conditions would probably have been similar. Riversleigh's Upper Site LF and the Hamilton LF share several genera, such as *Hypsiprymnodon*, *Cercartetus*, *Burramys*, *Strigocuscus* and *Trichosurus*, which might then be regarded as 'persistent rainforest genera'.

The rainforests of the Wet Tropics of north-eastern Queensland represent a similar although modern refugium for tropical rainforest. They share several 'persistent rainforest genera' with the Upper Site LF including *Hypsiprymnodon* and *Pseudochirops*. They also share *Hypsiprymnodon*, *Strigocuscus*, *Trichosurus* and possibly *Pseudocheirus* with the Hamilton LF. Studies of pollen preserved in the sediments of Lynch's Crater (on the Atherton Tableland) suggest that during the Pleistocene these rainforests underwent episodes of severe contraction culminating, about 10 000 years ago, in a stress that almost certainly reduced mammal diversity in that region.

John Winters has pointed out that the absence of what he defines as 'rainforest specialist species' (i.e. mammals restricted to rainforest) from the rainforests further south probably reflects the same climatic crises in their history. During the Pleistocene, periods of montane glaciation, cold climates and reduced rainfall probably diminished these southern rainforests below the size necessary to maintain viable populations of the 'rainforest specialists'. Subsequent amelioration of the climate allowed the rainforest to expand again but, unfortunately, without its former 'specialists'. This hypothesis fits well with what we understand about the Cainozoic history of a number of mammal groups. For example, see the discussion of the probable fate of *Burramys* pygmy-possums in Chapters 4 and 14.

The Papua New Guinea mid-montane rainforest mammals may have been isolated from their Australian counterparts since the early Miocene, although ephemeral Pliocene corridors or dispersal routes to and from Australia are indicated by shared genera of kangaroos (*Protemnodon*) and diprotodontids (*Nototherium*) in the Pliocene Awe LF and rodents and tree kangaroos (*Dendrolagus*) in Pliocene/Pleistocene local faunas of eastern Australia and New Guinea but evidently not the Hamilton LF or the older Oligo-Miocene rainforest faunas of Riversleigh.

Limitations in making comparisons. Each of the fossil and modern rainforest assemblages compared above has its own limitations and/or biases. Hence, comparisons between them must be made with caution. Specifically, despite the Upper Site LF's relatively high marsupial diversity, perhaps a better indication of original diversity would involve addition to this list of the rare taxa (e.g. ilariids and ektopodontids) preserved in adjacent sites.

The same is true for the Hamilton assemblage. None of the Hamilton bones or teeth show signs of having been transported. However, diversity for this assemblage is almost certainly under-represented because many rainforest groups uncommon or absent here, such a dasyures, bandicoots, diprotodontids and bats, are diverse before and after the early Pliocene. It is, of course, possible that the lower diversity of the Hamilton LF reflect a less productive rainforest (i.e. Hamilton' cool temperate *Nothofagus* rainforest vs Riversleigh's warm tropical rainforests). Unfortunately, the Hamilton LF is Australia's only known Pliocene rainfores mammal assemblage, so we are stuck with it as our only link between the Oligo-Miocene rainforest local faunas of Riversleigh and those of the present.

Although the Wet Tropics is a broad area of north-eastern Queensland, betwee Townsville and Cooktown, it is relatively uniform in terms of vegetation and mammal diversity and hence comparison of its mammals with those of the geographically much more restricted Upper Site may not be *significantly* misleading.

However, this is probably *not* the cas with the data for the mid-montane rainforests of Papua New Guinea. Although Ken Aplin and Tim Flannery came up with a list of 83 potentially co-existing mammals, they note that it is unlikely that all 83 would occur close enough together to end up in a single fossil deposit like Upper Site or Hamilton. So again, we caution that by comparison, the fossil data from Upper Site probably under-represent regional diversity.

It is also important to remember that there are an enormous number of variable involved in comparing species lists from different sites. To give an example, while i is a relatively simple matter to compare species diversity between faunas, there are at least two quite different factors hidden in such simplistic comparisons: the *frequency* of individuals per species (i.e. species A may be ten times as abundant as species B); and the *biomass* per habitat for each species may differ significantly (i.e. one individual of species A may weigh ten times as much as 100 individuals of species B).

5. In passing

With rapid progress in computerising the Riversleigh data base, we expect to be able to quantify many of the factors and relationships important in comparing biotas and monitoring change through time. But long before this statistical overview is available, the basic message of Riversleigh is not difficult to understand—it is the emergent theme of this book. Where once Australia's 'Green Cradle' was a fertile source that populated the continent, today it is fighting for its life. To the extent that trends through time documented in the fossil record of Riversleigh and the rest of the continent are misunderstood or mismanaged, the future of Australia's rainforest remnants may be catastrophically brief.

Appendix 2

Species lists for selected local faunas from Riversleigh

In Chapter 9 we compared five fossil assemblages and the modern mammal fauna from Riversleigh. While analysis of most of these faunas is still very much in its infancy, we have presented here as much detail as we and our colleagues can muster until more of the systematic studies currently underway are completed.

Riversleigh Local Fauna from Site D

From Sites A-F on the D-Site Plateau and adjacent hills. As currently understood, this assemblage is late Oligocene to early Miocene in age.

Authorities responsible for particular as yet unpublished determinations of Site D Local Faunal taxa, in addition to Archer, Godthelp and Hand include: Walter Boles (Australian Museum); Winston Ponder (Australian Museum); Jeanette Muirhead (University of New South Wales); John Scanlon (University of New South Wales); and Paul Willis (University of New South Wales); Neville Pledge (South Australian Museum); and Peter Murray (Museums and Art Galleries of the Northern Territory).

Mollusca (molluscs)
Gastropoda (snails)
Family?
Physastra rodingae McMichael, 1967
Meracomelon lloydi McMichael, 1967

Pisces (fish)
Teleostii (boney fish)
Ceratodidae
—*Neoceratodus* sp.
Family?
—Genus & sp.

Reptilia (reptiles)
Testudines
?Chelidae
—Genus & sp.

Squamata (lizards and snakes)
Ophidia (snakes)
Family Pythonidae
—*Montypythonoides riversleighensis* Smith & Plane 1985
Family uncertain
—Genus & sp.
Lacertilia (lizards)
Family uncertain
—Genus & sp.
Crocodilia (crocodiles)
Crocodylidae
—*Baru* sp. new

Aves (birds)
Ratites (most large flightless birds)
Dromornithidae (giant mihirung flightless birds)
—*Barawertornis tedfordi* Rich, 1979
Family uncertain
—Genus & sp.

Mammalia (mammals)
Marsupialia (marsupials)
Dasyuromorphia (dasyures)
Dasyuridae (dasyures)
—Genus & sp.
Thylacinidae (thylacines)
—*Nimbacinus dicksoni* Muirhead & Archer, 1989
Diprotodontia (diprotodont marsupials)
Vombatiformes ('wombat-like' marsupials)
Thylacoleonidae (marsupial lions)
—*Wakaleo* sp. cf. *W. oldfieldi* Clemens & Plane, 1974
Wynyardiidae (sheep-sized browsers)
—?*Namilamadeta* n. sp.
Diprotodontidae (giant extinct browsers)
—*Neohelos tirarensis* Stirton, 1967
—Genus & sp. new
—Genus? & sp. new
Palorchestidae
—*Ngapakaldia* sp. new
—*Bematherium angulum* Tedford, 1967
—Genus? & sp. new
Phalangerida ('phalanger'-like marsupials)
Petauroidea ('petaurid-like' marsupials)
Petauridae? (gliders & relatives)
—Genus & sp. new
Macropodoidea (kangaroos)
Potoroidae (rat-kangaroos)
Balungamayinae (lophodont rat-kangaroos)
—*Wabularoo naughtoni* Archer, 1979
—*Balungamaya delicata* Flannery, Archer & Plane, 1983
Potoroinae ('ordinary' rat-kangaroos)
—*Gumardee pascuali* Flannery, Archer & Plane, 1983
Macropodidae ('ordinary' kangaroos)
Balbarinae (primitive 'ordinary' kangaroos)
—*Balbaroo gregoriensis* Flannery, Archer & Plane 1983
Macropodoidea (family uncertain)
—*Galanarla tesselata* Flannery, Archer & Plane 1983
Placentalia (placentals)
Chiroptera (bats)
Microchiroptera ('insect-eating' bats)
Family indet.
—Genus & sp.

Upper Site Local Fauna

From Upper Site on Godthelp Hill the fauna is currently understood to be approximately early Miocene in age.

This System B assemblage, from Upper Site on Godthelp Hill, was published by Archer, Godthelp, Hand and Megirian (1989). It is modified here after consultation with the following authorities: Walter Boles (Australian Museum); Phil Colman (Australian Museum); Bernie Cooke (Queensland University of Technology); Patrick Couper (Queensland Museum); Jeanette Covacevich (Queensland Museum); Miranda Gott (University of New South Wales); Mark Hutchinson (South Australian Museum); Peter Jell (Queensland Museum); Ralph Molnar (Queensland Museum); Jeanette Muirhead (University of New South Wales); John Scanlon (University of New South Wales); Mike Tyler (University of Adelaide); and Paul Willis (University of New South Wales).

Mollusca (molluscs)
Gastropoda (snails)
Family?
Genus & sp.
Family?
Genus & sp.
Family?
Genus & sp.

Arthropoda ('joint-legged' invertebrates)
Diplopoda (millipedes)
Family? (no. of genera undet.)
—Genus & sp. 1
—Genus & sp. 2
—Genus & sp. 3
—Genus & sp. 4

Insecta (insects)
Hymenoptera (bees, ants etc.)
Formicidae (ants; no. of genera undet.)
—Genus & sp. 1
—Genus & sp. 2
Hemiptera (sucking insects)
Cicadidae (cicadas)
—Genus & sp. 1
Coleoptera (beetles)
Curculionidae (weevils)
—Genus & sp. 1
Families? (no. of genera undet.)
—Genus & sp. 1
—Genus & sp. 2

Crustacea
Isopoda (slaters)
Family?
—Genus & sp. 1

Pisces (fish)
Teleostii (boney fish)
Family?
—Genus & sp.

Amphibia
Anura (frogs)
Leptodactylidae (leptodactylid frogs)
—*Crinia* sp.
—*Kyarranus* sp. 1
—*Kyarranus* sp. 2
—*Lechriodus intergerivus* Tyler, 1989
—*Limnodynastes* sp. 1
—*Limnodynastes* sp. 2
Hylidae (hylid frogs)
—*Litoria* sp. 1
—*Litoria* sp. 2
—*Litoria* sp. 3
—*Litoria* sp. 4
—*Litoria* sp. 5
—*Litoria* sp. 6
—*Litoria* sp. 7
—*Litoria* sp. 8

Reptilia (reptiles)
Squamata (lizards and snakes)
Ophidia (snakes)
Pythonidae (pythons)
—*cf. Montypythonoides riversleighensis* Smith & Plane, 1985
Madtsoiidae (large extinct snakes)
—Genus & sp. new
Typhlopidae (blind burrowing snakes)
—?*Ramphotyphlops* sp.
Elapidae (poisonous snakes)
—New genus? & sp.
Lacertilia (lizards)
Agamidae (dragons)
—*Physignathus* sp. *cf. P. leseurii* (Gray, 1831)
—*Physignathus* sp.
—Genus & sp. new
Gekkonidae (geckoes)
—[not yet determined]
Scincidae (skinks)
—[many taxa but not yet determined]
Varanidae (monitor lizards)
—[not yet determined]
Crocodilia (crocodiles)
Crocodylidae
—Genus? & sp. new; *cf. Quinkana*

Aves (birds)
Ratites (most large flightless birds)
Dromornithidae (giant mihirung flightless birds)
—*Barawertornis tedfordi* Rich, 1979
—*Bullockornis* sp. *cf. B. planei* Rich, 1979
Casuariidae (emus and cassowaries)
—*Dromaius gidju* Patterson &Rich, 1987
Passeriformes (songbirds)
Family(ies)?
—Genus & sp. new 1
—Genus & sp. new 2
—Genus & sp. new 3

Mammalia (mammals)
Marsupialia (marsupials)
Dasyuromorphia (dasyures)
Dasyuridae (tiny to medium-sized carnivorous marsupials)
—Genus & sp. new 1
—Genus & sp. new 2
—Genus & sp. new 3
—Genus? & sp. new 4

?Dasyuridae
—Genus & sp. new
Thylacinidae (thylacines)
—*Nimbacinus dicksoni* Muirhead & Archer, 1989
Peramelemorphia (bandicoots)
New family 1 (extinct 'V.D.' bandicoots)
—Genus & sp. new 1
—Genus? & sp. new 2
?Peroryctidae (forest bandicoots)
—Genus & sp. new 1
—Genus? & sp. new 2
—Genus? & sp. new 3
—Genus? & sp. new 4
—Genus? & sp. new 5
—Genus? & sp. new 6
Notoryctemorphia (marsupial moles)
Notoryctidae
—Genus & sp. new
Yalkaparidontia ('thingodontans')
Yalkaparidontidae (extinct family)
—*Yalkaparidon coheni* Archer, Hand & Godthelp, 1988
Diprotodontia (diprotodont marsupials)
Vombatiformes ('wombat-like' marsupials)
Phascolarctomorphia ('koala-like' marsupials)
Phascolarctidae (koalas)
—*Litokoala* n. sp.
—Genus and sp. indet.
?New family 2
—Genus & sp. new
Vombatomorphia ('wombat-shaped' marsupials)
Thylacoleonidae (marsupial lions)
—*Wakaleo* sp. *cf. W. oldfieldi* Clemens & Plane, 1974
—Genus & sp. new cf. *Priscileo*
Wynyardiidae (sheep-sized browsers)
—?*Namilamadeta* n. sp.
Diprotodontidae (giant extinct browsers)
—*Neohelos tirarensis* Stirton, 1967
—Genus & sp. new
—Genus? & sp. new
New family 3
—Genus & sp. new
Phalangerida ('phalanger-like' marsupials)
Burramyidae (pygmy possums)
—*Burramys* n. sp.
—*Cercartetus* n. sp.
Phalangeroidea
Phalangeridae (brushtail possums & cuscuses)
—*Trichosurus* sp. *cf. T. dicksoni* Flannery & Archer, 1987a
—?*Strigocuscus* sp.
Pilkipildridae (small omnivorous possums)
—?*Djilgaringa* sp.
Petauroidea ('petaurid-like' marsupials)
Petauridae? (gliders & relatives)
—Genus & sp. new
New family 4
—Genus & sp. new 1
—Genus & sp. new 2
Pseudocheiridae (ringtail possums)
—*Paljara* n. sp.
—*cf. Pildra* sp.
—*cf. Pseudochirops* 1 (small)
—*cf. Pseudochirops* 2 (middle)
—*cf. Pseudochirops* 3 (large)
?New family 5
—Genus & sp. new
Acrobatidae (feather-tail possums)
—New gen.? *cf. Acrobates*
—Genus? & sp. new
Macropodoidea (kangaroos)
Potoroidae (rat-kangaroos)
Hypsiprymnodontinae (musky-rat kangaroos)
—*Hypsiprymnodon* sp. new
Propleopinae ('carnivorous' kangaroos)
—*Ekaltadeta* sp. *cf. E. ima* Archer & Flannery, 1985
Balungamayinae (lophodont rat-kangaroos)
—*cf. Wabularoo* n. sp.
—*cf. Gumardee* n. sp.
—Genus & sp. new 1
—Genus? & sp. new 2
—Genus? & sp. new 3
Potoroinae ('ordinary' rat-kangaroos)
—*Wakiewakie lawsoni* Woodburne, 1984a (Godthelp, Archer, Hand & Plane 1989)
Macropodidae ('ordinary' kangaroos)
Balbarinae (primitive 'ordinary' kangaroos)
—?*Nambaroo* sp. new 1
—?*Nambaroo* sp. new 2
?Marsupialia
Yingabalanaridae (extinct family of probable marsupials)
—*Yingabalanara richardsoni* Archer *et al.*, 1989
Placentalia (placental mammals)
Chiroptera (bats)
Microchiroptera (mostly 'insect-eating' bats)
Rhinolophoidea ('rhinolophid-like' bats)
Megadermatidae (ghost bats)
—*Macroderma* sp.
Hipposideridae (leaf-nosed bats)
—*Brachipposideros* sp. 1
—*Brachipposideros* sp. 2
—*Brachipposideros* sp. 3
—*Brachipposideros* sp. 4
—*Brachipposideros* sp. 5
—Genus & sp. indet. 1
—Genus & sp. indet. 2
Vespertilionoidea ('ordinary' bats)
Molossidae (free-tail bats)
—Genus & sp. indet.
Family?
—Genus & sp. indet.

Bitesantennary Site Local Fauna

This assemblage, from Bitesantennary Site, comes from the north-eastern margin of the D-Site Plateau. It is interpreted to be an open cave deposit, probably early Miocene in age.

Authorities responsible for particular determinations of Bitesantennary Local Faunal taxa, in addition to Hand, Archer and Godthelp, include: Walter Boles (Australian Museum); Phil Colman (Australian Museum); Bernie Cooke (Queensland University of Technology); Jeanette Muirhead (University of New South Wales); John Scanlon (University of New South Wales); Mike Tyler (University of Adelaide); and Paul Willis (University of New South Wales).

Mollusca (molluscs)
Gastropoda (snails)
Families?
—Genera? & 3 spp.

Arthropoda ('joint-legged' invertebrates)
Families?
—Genera? & 2 spp.

Pisces (fish)
Teleostii (boney fish)
Family?
—Genus & sp.

Amphibia
Anura (frogs)
Leptodactylidae (leptodactylid frogs)
—*Lechriodus intergerivus* Tyler, 1989
Families?
—Genera & 2 spp.

Reptilia (reptiles)
Squamata (lizards and snakes)
Ophidia (snakes)
Pythonidae (pythons)
—Genus & sp.
Lacertilia (lizards)
Agamidae (dragons)
—Genus & sp.

Aves (birds)
Non-Passeriformes
Families?
—Genus & sp. 1
—Genus & sp. 2

Mammalia (mammals)
Marsupialia (marsupials)
Dasyuromorphia (dasyures)
Dasyuridae (tiny to medium-sized carnivorous marsupials)
—Genus & sp. n.
Peramelemorphia (bandicoots)
Families?
—Genera & 2 spp. n.
Diprotodontia (diprotodont marsupials)
Phalangerida ('phalanger-like' marsupials)
Macropodoidea (kangaroos)
Potoroidae (rat-kangaroos)
Balungamayinae (lophodont rat-kangaroos)
—*Balungamaya* sp.
Placentalia (placental mammals)
Chiroptera (bats)
Microchiroptera (mostly 'insect-eating' bats)
Rhinolophoidea ('rhinolophid-like' bats)
Hipposideridae (leaf-nosed bats)
—*Brachipposideros* sp. 1
—*Brachipposideros* sp. 2
—*Brachipposideros* sp. 3
—*Brachipposideros* sp. 4
—*Hipposideros* sp.

Rackham's Roost Local Fauna

This Pliocene assemblage, accumulated on the floor of a cave, is interpreted to be mainly the feeding residues of ancestral ghost bats. Some taxa, such as the python and large kangaroo, are likely to have used the cave for shelter.

Authorities responsible for particular determinations of the Rackham's Roost Local Faunal taxa, in addition to Godthelp, Hand and Archer, include: Walter Boles (Australian Museum); Jeanette Muirhead (University of New South Wales); John Scanlon (University of New South Wales); Mike Tyler (University of Adelaide); and Paul Willis (University of New South Wales).

Crustacea
Family?
—Genus indet. (gastrolith)

Pisces (fish)
Teleostii (boney fish)
Family?
—Genus & sp.

Amphibia
Anura (frogs)
Family?
—Genus & sp.

Reptilia (reptiles)
Squamata (lizards and snakes)
Ophidia (snakes)
Family indet.
—Genus & sp.
Lacertilia (lizards)
Agamidae (dragons)
—Genus & sp.
Gekkonidae (geckoes)
—Genus & sp.
Varanidae (monitor lizards)
—Genus & sp.
Crocodilia (crocodiles)
Crocodylidae
—Genus & sp. (two small teeth & frag. of scute)

Aves (birds)
Family indet.
—Genus & sp.
—Genus & sp.

Mammalia (mammals)
Marsupialia (marsupials)
Dasyuromorphia (dasyures)
Dasyuridae (tiny to medium-sized carnivorous marsupials)
—*Sminthopsis* n. sp.
—*Planigale* n. sp.
—Genus & sp. new

Peramelemorphia (bandicoots)
?Peramelidae (ordinary bandicoots)
—?*Perameles* sp. 1
—?*Perameles* sp. 2
Diprotodontia (diprotodont marsupials)
Phalangerida ('phalanger-like' marsupials)
Petauroidea ('petaurid-like' marsupials)
Pseudocheiridae (ringtail possums)
—Genus & sp.
Macropodoidea (kangaroos)
Potoroidae (rat-kangaroos)
Potoroinae ('ordinary' rat-kangaroos)
—*Bettongia* sp.
—Genus & sp.
Macropodidae ('ordinary' kangaroos)
Macropodinae ('ordinary' kangaroos)
—*Macropus* sp. 1
—*Macropus* sp. 2
—*Protemnodon* sp. *cf. P. snewini* Bartholomai, 1978
Placentalia (placental mammals)
Chiroptera (bats)
Microchiroptera (mostly 'insect-eating' bats)
Rhinolophoidea ('rhinolophid-like' bats)
Megadermatidae (ghost bats)
—*Macroderma gigas* Dobson, 1880
Hipposideridae (leaf-nosed bats)
—?*Rhinonicteris aurantius* (Gray, 1845)
—*cf. Brachipposideros* sp. 1
—*cf. Brachipposideros* sp. 2
—*cf. Brachipposideros* sp. 3
Vespertilionoidea ('ordinary' bats)
Vespertilionidae (plain faced bats)
—*cf. Chalinolobus* sp.
—*cf. Scotorepens* sp.
—Genus & sp. 1
—Genus & sp. 2
Emballonuroidea ('sheathtail' bats)
Emballonuridae (sheathtail bats)
—*Taphozous* n. sp. 1?
—*Taphozous* n. sp. 2?
Microchiroptera
Family uncertain
—Genus & sp.
Rodentia (rodents)
Muridae (rats & mice)
—*Pseudomys* n. sp. 1
—*Pseudomys* n. sp. 2
—*Pseudomys* n. sp. 3
—*Pseudomys* n. sp. 4
—*Pseudomys* n. sp. 5
—*Pseudomys* n. sp. 6
—*Pseudomys* n. sp. 7
—*Leggadina* n. sp. 1
—*Leggadina* n. sp. 2
—*Zyzomys* n. sp.
—Genus & sp. new
—?Genus & sp. 1
—?Genus & sp. 2

Terrace Site Local Fauna

This possibly late Pleistocene assemblage was collected from eroded fluviatile sediments exposed near the western bank of the Gregory River on Riversleigh Station.

Authorities responsible for particular determinations of Terrace Site Local Faunal taxa, besides Archer, Godhelp and Hand, include: Paul Willis (University of New South Wales); Winston Ponder (Australian Museum); Angela Davis (Australian National University); and Arthur White (University of New South Wales).

Arthropoda (joint-legged invertebrates)
Crustacea (crustaceans)
Decapoda (crabs etc.)
Family?
—Genus & sp. (crab gastrolith)

Mollusca
Pelecypoda (bivalves)
Family?
—?*Unio* sp.
Gastropoda (snails)
Families?
—Genera & ?3 spp

Pisces (fish)
Teleostii (boney fish)
Family?
—Genus & sp.
Reptilia (reptiles)
Chelodina (turtles)
Chelidae (chelid turtles)
—*Elseya* n. sp.
—*Emydura* sp.
Squamata (lizards and snakes)
Lacertilia (lizards)
Varanidae (monitor lizards)
—*Varanus* sp. (a goanna)
Crocodilia (crocodiles)
Crocodylidae
—*Crocodilus johnstoni* Krefft, 1873 (Freshwater Crocodile)
—?*Palimnarchus* sp. (giant extinct crocodile)

Mammalia (mammals)
Marsupialia (marsupials)
Diprotodontia (diprotodont marsupials)
Phalangerida ('phalanger-like' marsupials)
Macropodoidea (kangaroos)
Macropodidae ('ordinary' kangaroos)
Macropodinae ('ordinary' kangaroos)
—*Macropus agilis* (Gould, 1842) (Agile Wallaby)
Sthenurinae ('short-faced' kangaroos)
—*Sthenurus* sp.
Vombatiformes ('wombat-shaped' marsupials)
Diprotodontidae (giant extinct marsupials)
—*Diprotodon optatum* Owen, 1838 ('Diprotodon')
Palorchestidae (giant trunked diprotodontoid marsupials)
—*Palorchestes azael* Owen, 1873 ('Marsupial tapir')
Placentalia (placental mammals)
Rodentia (rodents)
Muridae (rats & mice)
—*Hydromys chrysogaster* Geoffroy, 1804 (Water Rat)
—Genus & sp. (indet.)

Riversleigh's modern mammals

The following list of modern mammals from the Riversleigh region has been compiled from several, sometimes limited surveys (see Chapter 9).

Authorities responsible for particular determinations of modern mammal taxa, in addition to Hand, Godthelp and Archer, include: Ray and Anne Williams (University of New South Wales); Susin Churchill; Patricia Woolley and students (La Trobe University); Colin and Cheryl O'Keefe (Queensland National Parks and Wildlife Service).

Mammalia (mammals)
Monotremata (egg-laying mammals)
Tachyglossidae (echidnas)
—*Tachyglossus aculeatus* (Shaw, 1792) (Echidna)
Marsupialia (marsupials)
Dasyuromorphia (dasyures)
Dasyuridae (dasyures)
—*Planigale* sp. (Planigale; species not yet determined)
—*Sminthopsis macroura* (Gould, 1845) (Stripe-faced Dunnart)
Diprotodontia (diprotodont marsupials)
Phalangerida ('phalanger-like' marsupials)
Petauroidea ('petaurid-like' marsupials)
Pseudocheiridae (ringtail possums)
—*Petropseudes dahli* (Collett, 1895) (Rock-Haunting Ringtail; Lawn Hill)
Macropodoidea (kangaroos)
Macropodidae ('ordinary' kangaroos)
—*Macropus agilis* (Gould, 1842) (Agile Wallaby)
—*M. antilopinus* (Gould, 1842) (Antilopine Kangaroo; doubtfully identified)
—*M. rufus* (Desmerest, 1822) (Red Kangaroo)
—*M. robustus* Gould, 1841 (Euro)
—*Onychogalea unguifera* (Gould, 1841) (Northern Nail-tail Wallaby)
—*Petrogale* sp. (Rock Wallaby; species indet. but ?*lateralis*)
Placentalia (placentals)
Chiroptera (bats)
Microchiroptera (flesh & insect-eating bats)
Rhinolophoidea (rhinolophid-like bats)
Megadermatidae (ghost bats)
—*Macroderma gigas* (Dobson, 1880) (Ghost Bat; perhaps only recently gone from the area)
Hipposideridae (horseshoe-bats)
—*Hipposideros ater* Templeton, 1848 (Dusky Horseshoe-bat)
—*Rhinonicteris aurantius* (Gray, 1845) (Orange Horseshoe-bat—Lawn Hill)
Vespertilionoidea ('ordinary' bats)
Vespertilionidae ('ordinary' bats)
—*Nyctophilus arnhemensis* (Johnson, 1959) (Arnhem Land Long-eared Bat)
—*N. walkeri* Thomas, 1892 (Pygmy Long-eared Bat)
—*N.* sp. *cf. N. bifax?* (North Queensland Long-eared Bat)
—*Myotis adversus* (Horsfield, 1824) (Large-footed Mouse-eared Bat)
—*Scotorepens greyi* (Gray, 1843) (Little Broad-nosed Bat)
—*Eptesicus finlaysoni Kitchener*, 1987 (Finlayson's Eptesicus)
—*E.* sp.
—*Chalinolobus nigrogriseus* (Gould, 1856) (Hoary Bat)
Molossidae (free-tail bats)
—*Chaerophon jobensis* (Miller, 1902) (Northern Mastiff-bat)
Emballonuroidea ('sheathtail bats')
Emballonuridae (sheathtail bats)
—*Taphozous georgianus* Thomas, 1915 (Common Shealthtail-bat)
—*Saccolaimus flaviventris* (Peters, 1867) (Yellow-bellied Sheathtail-bat)
Megachiroptera (fruit bats)
Pteropodidae (fruit bats)
—*Pteropus scapulatus* Peters, 1862 (Little Red Flying-fox)
Rodentia (rodents)
Muridae (rats and mice)
—*Pseudomys nanus* (Gould, 1858) (Western Chestnut Mouse)
—*Zyzomys argurus* (Thomas, 1889) (Common Rock-rat)
—*Rattus villosissimus* (Waite, 1898) (Long-haired Rat)
—*Hydromys chrysogaster* Geoffroy, 1804 (Water-rat; Lawn Hill)
Carnivora (carnivores)
Canidae (dogs)
—*Canis familiaris* Linnaeus, 1758 (Dingo)
Primates (monkeys, great apes etc.)
Hominidae (humans & australopithecines)
—*Homo sapiens* (humans)
Introduced species
Canidae (dogs)
—*Vulpes vulpes* (Linnaeus, 1758) (Fox)
Felidae (cats)
—*Felis catus* Linnaeus, 1758 (Feral Cat)
Equidae (horses)
—*Equus caballus* Linnaeus, 1758 (Horse)
Bovidae (cattle)
—*Bos taurus* Linnaeus, 1758 (European Cattle)
Suidae (pigs)
—*Sus scrofa* Linnaeus, 1758 (Feral Pig)
Muridae (rats & mice)
—*Mus musculus* Linnaeus, 1758 (House Mouse)

Glossary

acrobatids: small, insectivorous to omnivorous marsupial possums in the genera *Acrobates* (Feather-tailed Possum of Australia & New Guinea) and *Distoechurus* (Pen-tailed Possum of New Guinea).
acrochordids: aquatic, fish-eating, non-venomous file snakes of the genus *Acrochordus* (e.g. the Javan File Snake of Australia and adjacent tropical islands)
agamids: lizards common in the lands around the Pacific with approx. 50 species living in Australia (e.g. the Eastern Water Dragon *Physignathus lesueurii*).
ammonite: predatory relatives of the Chambered Nautilus that dominated the oceans before a major extinction event 65 million years ago.
anaerobic: able to survive without oxygen.
anhingas: long-necked fishing birds called darters, common around the waterways of Australia; they are close relatives of cormorants.
anterior: towards the front end of something such as a tooth or head.
aquatic: living or spending time in water (e.g. turtles).
arboreal: living or spending time in trees (e.g. the Koala).
archaic: old, ancient or of a primitive form.
arcuate: curved like a bow (e.g. the shape of the cutting crests of the teeth of some animals).
arthropod: joint-legged invertebrate animal (e.g. insects, crustaceans, spiders, centipedes).
articular: bone in the lower jaw of reptiles that hinges with the quadrate of the skull to articulate the lower jaw; in mammals, these bones are still present but transmit sound as the 'hammer' and 'anvil' of the middle ear.
articulate: two things bound together to form a movable joint (such as the adjacent surfaces of the upper arm bone and the shoulder blade).
australopithecines: human relatives and ancestors in the extinct genus *Australopithecus* (e.g. 'Lucy', *A. afarensis*), sometimes called 'ape-men', that lived in Africa between 5 and 1 million years ago.
avifauna: assemblage of birds known from a particular region.

balbarines: Oligo-Miocene browsing kangaroos (e.g. Riversleigh's *Balbaroo riversleighensis*), once common in Australia; evidently ancestral to macropodine kangaroos such as the tree kangaroos and pademelons.
balungamayines: Oligo-Miocene browsing rat-kangaroos (e.g. Riversleigh's *Balungamaya delicata*) that converged in tooth form on more conventional kangaroos such as balbarines and macropodines.
belemnites: extinct, straight-shelled marine molluscs that were distantly related to the living Chambered Nautilus.
bettongine: omnivorous rat-kangaroos with short heads and prehensile tails (e.g. the living Burrowing Bettong and Riversleigh's *Bettongia moyesi*).
biocorrelation: demonstration that the same taxon (usually a species or genus) occurs in isolated deposits thereby leading to the hypothesis that the isolated deposits are probably of similar if not the same age.
biodiversity: the number of species in an area (e.g. the biodiversity of rainforests exceeds that of mangrove swamps).
biomass: the weight of an organism or groups of organisms (e.g. the biomass of a Swamp Wallaby is more than 100 times the biomass of a Brown Antechinus)
biostratigraphy: determination of the relative age of rocks from study of the fossils they contain.
biota: collective term referring to all living things that occupy a particular region or habitat; hence the 'Australian biota' and the 'biota of the Australian Wet Tropics'.
boids: non-venomous, constricting snakes of the family Boidae present in New Guinea and other areas of the world (or a mispronunciation of the word 'birds' by residents of Brooklyn). Australia has no boids but has plenty of pythons (family Pythonidae).
brachiopods: shelled, filter-feeding invertebrates that once dominated the sea floors throughout the world but are now a relatively minor group.
breccia: sedimentary rock type made up of large angular bits of rock cemented together by a matrix (e.g. a limestone breccia consists of pieces of broken limestone joined together by a carbonate cement).
browser: all animals that eat plants are called herbivores. Of these, grazers eat mainly grass while browsers eat mainly other types of leaves such as those of trees, shrubs and ferns. Browsers dominated Australia's Oligo-Miocene forests; grazers dominated the more open, grass-dominated forests of the Pliocene, Pleistocene and Holocene.
burramyids: Australian and New Guinean pygmy-possums in the family Burramyidae—omnivorous, tiny marsupials such as the Mountain Pygmy-possum (*Burramys parvus*) and Long-tailed Pygmy-possum (*Cercartetus caudatus*).

canids: members of the dog family Canidae (e.g. the Australian Dingo *Canis familiaris* and the Fox *Vulpes vulpes*).
carettochelyids: pitted-shelled turtles (*Carettochelys insculpta*) that live in freshwater and estuarine waters of New Guinea and the Northern Territory.
carnivore: any animal that feeds mainly on other animals. Special types of carnivores include active predators (killers of their prey), scavengers (eaters of dead animals), insectivores (eaters of small animals such as insects and spiders) and sanguinivores (eaters of blood such as mosquitoes).
cartilage: non-cellular, semi-rigid matrix that provides flexible support for (e.g.) the ear, nose and ends of limb bones.
casuariids: emus and cassowaries; giant flightless birds of Australia and New Guinea in the ratite family Casuariidae.
chelids: family of aquatic, side-necked freshwater turtles from Australia, New Guinea and South America.
chert: cryptocrystalline sedimentary rock that often forms as siliceous nodules in pre-existing sediments. At Riversleigh, chert nodules are common in the Cambrian Thorntonia Limestone.
chrysochlorids: blind, burrowing 'golden moles' from Africa, placental counterparts of the marsupial notoryctids.
clast: in geological terms, a piece of foreign material embedded in another kind of rock (e.g. a clast of older limestone is sometimes found within a younger limestone that developed around it).
cochlea: coiled, fluid-filled sensory organ forming part of the inner ear in the periotic bone that distinguishes frequencies of sounds transmitted by the bones of the middle ear.
coelacanth: large fleshy-finned, bony fish belonging to a group that was close to the ancestry of the amphibians. Although once thought to be entirely extinct, a single living form (*Latimeria*) was found in the marine waters off Madagascar.
coleopterans: beetles; members of the order Coleoptera.
colubrids: world-wide, mostly non-venomous snakes in the family Colubridae (e.g. the Freshwater Snake *Amphiesma mairii* of Australia).
commensal: an organism that benefits from living within or on another organism without causing the host any benefit or harm.
conglomerate: rock type mostly made up of rounded pebbles or clasts of other materials cemented together. At Riversleigh, beds of rounded chert nodules are frequently cemented together by carbonates into broad pavements.
crocodilians: crocodiles, carnivorous reptiles of the order Crocodilia. In contemporary Australia, two kinds exist (the Freshwater and Saltwater Crocodiles) but many more kinds are represented in the fossil record.
crustaceans: ten-legged, hard-shelled arthropoda invertebrates in the class Crustacea (e.g. yabbies, lobsters, slaters, crabs and barnacles).
cuticle: in animals, the hard, regularly shed, outer layer that protects the soft bodies of many groups of invertebrates such as arthropods from damage and dehydration.
cytoplasm: all of the substances of cells (including the organelles) occurring between the nucleus and the cell membrane.

dasyurids: carnivorous marsupials, such as quolls, antechinuses and phascogales, that live in Australia and New Guinea. They range in siz from the 4-g Long-tailed Planigale to the small-dog-sized Tasmanian Devil.
dentary: the bone of the lower jaw that supports an animal's teeth. In reptiles with many bon in the lower jaw, it is normally the most anterior. In modern mammals, it is the only bone of the lower jaw.
dentition: all of the teeth of an animal (e.g. in humans, the adult dentition consists of two incisors, one canine, two premolars and thre molars in each quadrant of the mouth; the dental formula is thus I1-2, C1, P3-4, M1-3).
depauperate: diminished in some attribute (e.g. a depauperate local fauna may lack species present in another assemblage).
diagenetic: processes (other than weathering or erosion) that have affected a sediment, fossil or other geological object after it was initially deposited.
digitise: in the present context, to convert information about individual fossils into a format that can be used as database in a computer.
dimorphism: having two shapes such as common occurs in populations of mammals in which the males are larger than females or have sexually distinctive structures (e.g. antlers or larger canines).
diorama: in the present context, an artistic or reconstructed portrayal of a prehistoric habitat as it might have looked in life.
diprotodont: having two enlarged, forward projecting front teeth in the lower jaw (one i each dentary), a characteristic of diprotodontian marsupials, such as possums, kangaroos, marsupial lions and wombats, an of the enigmatic Riversleigh marsupials known as yalkaparidontids ('thingodontans').
diprotodontids: herbivorous, commonly large species in the extinct marsupial family Diprotodontidae; diverse in the Cainozoic record of Australia and New Guinea.
diprotodontines: diprotodontids of the subfamily Diprotodontinae (e.g. species of *Diprotodon*, *Euryzygoma* and *Nototherium*).
diprotodontoids: members of the marsupial family Diprotodontoidea which includes all diprotodontids and palorchestids.
disarticulated: having undergone separation; such as the bones of a skeleton after the soft tissue of the animal have decayed.
dromornithids: extinct, giant, flightless, herbivorous birds from Australia, sometimes called 'mihirungs' or 'thunder birds'. While they may be distant relatives of emus and cassowaries, they may also be distantly related to chickens.

echolocation: determination of the nature or position of objects in the environment through analysis of echoes bounced off surfaces. Microchiropteran bats use echolocation at night to home in on prey and navigate in hazard-filled areas.
ecosystem: inorganic environment and associated community of organisms whose members interact with one another.
ektopodontids: extinct, Australian, rodent-like possums in the family Ektopodontidae that survived between the Oligocene and Pleistocene.
elapids: venomous snakes in the family Elapidae; present on most continents and common in Australia (e.g. King Browns and Tiger Snakes).

electrophoresis: technique for determining relatedness or distinction between organisms on the basis of the electric properties of their enzymes.
emballonurids: Old and New World sheath-tailed bats in the family Emballonuridae. Common in Australia (e.g. the Common Sheath-tailed Bat *Taphozous georgianus*).
endemic: restricted to a particular geographic area. For example: Koalas and Emus are endemic to Australia.
entoconid: inner cusp on the heel (talonid) of the lower teeth of 'advanced' mammals (marsupials and placentals).
epiphytes: plants that grow upon (but do not necessarily feed from) the body of another (e.g. epiphytic orchids and ferns attached to the trunks or branches of large rainforest trees).
epoch: interval of geological time that represents a subdivision of a period.
era: interval of geological time that spans two or more periods.
evaporite: mineral or sediment that accumulates as a consequence of water evaporating from a surface (e.g. salt crusts on the edges of salt lakes; carbonate crusts that develop on still waters in the Riversleigh area).

family: group of related genera of organisms (e.g. the family Thylacoleonidae, or marsupial lions, contains the genera *Thylacoleo, Wakaleo* and *Priscileo*).
fauna: animals that characterise a broad region made up of many isolated localities, such as the 'mammal fauna of the Gulf of Carpentaria'. This contrasts with the more restricted 'local fauna'.
felids: members of the carnivorous family of cats, the Felidae (e.g. lions, tigers, sabre-toothed tigers, leopards and domestic cats).
fluviatile: of or belonging to a river or produced by the action of a river, e.g. gravels and sands transported in the bed-load of a river are called fluviatile sediments.
folivores: animals that eat the leaves of plants; normally this refers to eaters of the leaves of trees, such as koalas and ringtail possums.
fossil: any part, trace or indication of a previously alive organism found in the geological record of the earth (e.g. bones, teeth, wood or excreta, insects in amber, footprints, coal, or space in a rock where the organism decayed).
fossilisation: natural processes that result in the exchange of elements between the fossil and its surrounding environment. For example, elements in ground water will gradually exchange with those in a buried bone until the fossil has a chemical composition that is in balance with that of the surrounding rock.

galliforms: large order of birds that includes pheasants, quails, turkeys and chickens; most are poor long-distance fliers.
gastrolith: stone or pebble swallowed and normally retained (e.g. in a crop or gizzard) by an animal to assist pulverising its food. Many fish, reptiles (including dinosaurs) and birds use gastroliths as adjuncts or substitutes for teeth. In some crustaceans, calcium carbonate temporarily stored as a hemispherical concretion is also called a gastrolith.
gastropods: members of the molluscan class Gastropoda (e.g. snails, slugs, limpets and nudibranchs), most but not all of which have obviously spiralled shells.
gavials: long-snouted, commonly fish-eating crocodiles found in tropical waters of the eastern hemisphere.
genera: plural of genus.
genus: cluster of related species (e.g. the genus *Homo* contains the species *sapiens, erectus* and *habilis*).
glaciation: the formation, movement and retreat of generally continental ice sheets. Glaciers 'flow' from a central region of accumulation when their increasing weight makes them unstable.
granivores: animals that eat the seeds or nuts of plants (e.g. many parrots and some possums).
grazers: herbivores that mainly eat grasses, such as wombats and most modern kangaroos (e.g. the Eastern Grey Kangaroo), in contrast to a browser.

gymnosperms: seed-bearing but non-flowering plants belonging to the division Pinophyta (e.g. pines, conifers, ginkoes and cycads). Before the rise of the flowering plants in the Cretaceous, gymnosperms dominated much of the world.

half-life: that period of time required for half of the original mass of a radioactive element to decay to one or more daughter products. For example, ^{14}C (radioactive carbon) has a half-life of 5730 years by which time 50% of its original mass will have decomposed.
herbivores: animals that eat plants. There are many types of herbivores including granivores (eaters of seeds), folivores (eaters of leaves), frugivores (eaters of fruit such as many possums) and lignovores (eaters of wood such as termites).
hipposiderids: leaf-nosed bats of the Old World family Hipposideridae. They are common in tropical south-eastern Asia and Australia (e.g. the Orange Horseshoe Bat *Rhinonicteris aurantius*).
hydromyines: omnivorous to carnivorous water rats of the subfamily Hydromyinae in the rodent family Muridae (e.g. the Common Water Rat of Australia, *Hydromys chrysogaster*).
hypertrophy: extraordinary enlargement or development of some attribute (e.g. the long tails of desert Hopping Mice are hypertrophied relative to the more 'normal' tails of Bush Rats).
hypsiprymnodontines: omnivorous, five-toed, galloping rat-kangaroos in the kangaroo subfamily Hypsiprymnodontinae. All are placed in the genus *Hypsiprymnodon* (e.g. the Musky Rat-kangaroo of the Atherton Tableland).

ilariids: Oligo-Miocene, medium-sized to large herbivorous marsupials in the family Ilariidae.
ilium (plural, ilia): one of three main bones in the pelvis (hip) of four-footed animals. The ilium normally projects towards the head end of the animal and is the site of attachment of many of the powerful muscles that move the hind limb.
illite: datable clay mineral found, for example, in the fossiliferous sediments at Murgon, south-eastern Queensland.
incusive: operating in the manner of a hammer as it strikes a surface. In the present context, some of the cusps of mammalian teeth (e.g. the hypoconid) incus the corresponding surfaces (e.g. the protocone) of the opposite teeth.
indurated: a surface or material secondarily hardened, e.g. by fossilisation, mineralisation or dehydration.
insectivorous: the habit of eating mainly insects and other small animals such as spiders or centipedes.
invertebrates: animals lacking backbones (e.g. worms, crinoids, brachiopods, sea squirts, corals etc.).

labile: in the present context, being able to respond or adapt to changing conditions.
labyrinthodonts: group of extinct amphibians that retained many features of more primitive amphibians as well as a range of features otherwise restricted to reptiles. In this sense, they are found 'missing links' between amphibians and reptiles.
lacustrine: of or belonging to a lake or the habit of living in a lake (e.g. turtles can be lacustrine).
laterite: sediment secondarily enriched with concentrations of aluminium and/or iron oxides.
leptodactylids: large group of frogs found in Australia, the Americas and southern Africa. In Australia, most are burrowing or terrestrial but some are entirely aquatic.
lithic: pertaining to or made of stone.
local faunas: assemblages of animals obtained from a single locality (e.g. the total fauna from Riversleigh's Upper Site is the Upper Site Local Fauna). Assemblages of contemporaneous local faunas from the same region may be called a fauna.
lophodont: teeth whose crowns are dominated by wide, transverse ridges or blades (e.g. the molars of kangaroos and diprotodontoids). Normally, this tooth morphology characterises herbivores.

lycopods: relatively simple plants with long branching stems (e.g. club mosses and selaginellas among still-living groups; one extinct group, the lepidodendrons, became tree-like and reached heights of 30 metres).

macropodids: 'ordinary' kangaroos in the family Macropodidae. The family contains the subfamilies: Macropodinae (most modern kangaroos); Sthenurinae (mostly extinct, commonly short-faced Pliocene to Holocene kangaroos); Balbarinae (extinct Oligo-Miocene browsers).
macropodines: 'ordinary' kangaroos in the subfamily Macropodinae (e.g. tree kangaroos, forest wallabies, pademelons, grey kangaroos, most hare wallabies etc.).
madtsoiids: extinct family of constricting snakes known as fossils from South America, Madagascar and Australia.
marsupials: live-bearing mammals today found chiefly in the Australian region but also in South, Central and North America. Fossils are known from all continents, the oldest being 90 million years old from Texas. Young are born in a little-developed state and commonly shelter in a pouch which contains the teats of the milk glands.
matrix: interstitial material of a sediment lying between larger fragments, crystals, particles or fossils.
maxilla: bone of the upper jaw that bears the canine, premolar and molar teeth (the incisors are in the premaxilla).
megachiropterans: suborder of bats containing flying foxes and fruit bats; 11 living species in Australia.
megadermatids: tropical, eastern hemisphere family of false vampire bats; in Australia, the only living species is the Ghost Bat (*Macroderma gigas*).
megafauna: term applied to the gigantic animals that developed during the later part of the Cainozoic era culminating with the giants of the late Pleistocene (e.g. *Diprotodon optatum, Megalania prisca* etc.). Most of the megafauna died out during the late Pleistocene for reasons that are controversial, or was succeeded by dwarfed forms.
megapodes: turkey-like birds that build mounds of dirt and vegetation in which they incubate their eggs.
meiolaniids: extinct, giant, horned turtles in the family Meiolaniidae, known as fossils from South America, Madagascar and Australia (including Lord Howe Island).
mesa: flat-topped hill of more limited extent than a plateau.
metaconid: postero-internal cusp of the anterior half (trigonid) of the lower molars of most mammals.
microchiropterans: suborder of echolocating, mostly insect-eating bats; approximately 54 living species in Australia.
microcontinent: small, often isolated landmass that behaves geologically like an independent continent.
microenvironment: in contrast to the overall environment of a region, the microenvironment is the combination of immediate, local conditions that affect a single species or cluster of species. It may include, for example, the humidity and temperature of a hollow space within a tree.
microhabitat: very small and commonly isolated habitat such as a pool of water accumulated in a tree stump.
microstriations: fine striations or scratches such as may be seen on the enamel surfaces of teeth. These often provide clues about the types of foods being eaten and about how the teeth are actually used and resharpened (thegosed).
mihirungs: extinct, large, flightless birds of the family Dromornithidae (e.g. the gigantic *Dromornis stirtoni* from the late Miocene Alcoota LF of the Northern Territory).
molossids: cosmopolitan family of free-tailed bats, commonly called mastiff bats.
monotremes: egg-laying mammals (echidnas, platypuses and the extinct *Steropodon galmani*) of the subclass Monotremata so far known only from Australia. They appear on balance to be more 'primitive' than any other living mammals and diverged from the mammal family tree sometime between 120 and 200 million years ago.

montane: of or pertaining to mountains (e.g. mid-montane rainforests of New Guinea).
morganucodontids: family of small, primitive (reptile-like) mammals from the early Mesozoic.
morphology: of or pertaining to the shape of something.
mosasaurids: Giant, extinct, carnivorous marine lizards that lived during the late Mesozoic.
multituberculates: rodent-like, primitive mammals that were common in the Cretaceous and early Tertiary of the northern continents. None are known from Australia.
murids: rats and mice in the family Muridae that comprise a quarter of Australia's modern, native, non-marine mammals.
murines: subfamily of murids that includes true rats of the genera *Rattus* and *Mus*. In Australia these include native bush rats as well as the introduced Black Rat and House Mouse.

notoryctids: small, blind, earless, burrowing marsupials in the family Notoryctidae. The single living species, the Marsupial Mole *Notoryctes typhlops*, is confined to the central Australian deserts. A second type lived in the Oligo-Miocene rainforests of Riversleigh.

obligate: restricted to a particular condition of life, such as eating leaves.
omnivores: organisms that eat both animals and plants.
oncolite: rock made up of concentrically laminated, calcareous structures formed by blue-green algae. Normally these form in water exposed to sunlight.
order: taxonomic category between class and family. Orders of placental mammals include, for example, Primates (apes, humans, etc.), Carnivora (bears, cats, dogs, etc.) and Cetacea (whales). Orders of marsupials include the Dasyuromorphia (dasyurids, thylacinids, etc.), Diprotodontia (possums, kangaroos, wombats, etc.) and Yalkaparidontia (thingodontans).
ostracods: marine and freshwater crustaceans enclosed by two shell-like structures (Cambrian to modern times). Some are useful for determining the age of a sediment.

palaelodids: family of flamingo-like birds that were common in the Oligo-Miocene sediments of central Australia as well as other areas of the world.
palaeoanthropology: study of fossil humans, their origin, development, variety and cultural development.
palaeoaustral: pertaining to regions that once comprised the ancient southern supercontinent of Gondwana, including Australia, India, New Zealand, Africa, Madagascar, South America, Antarctica and possibly continental fragments in south-eastern Asia.
palaeobiogeography: the search for patterns in the distribution of extinct organisms. Sometimes, for example, these indicate that once-joined landmasses subsequently split.
palaeobotany: study of fossil plants.
palaeoclimates: prehistoric climates.
palaeocommunity: the original community that contained an extinct assemblage of organisms.
palaeodiversity: prehistoric diversity.
palaeoecology: study of the relationship between ancient organisms and their environment.
palaeoenvironment: prehistoric environment.
palaeogeography: geography of the past.
palaeontology: study of prehistoric organisms and trace fossils, the relationships of prehistoric creatures, their evolutionary history and environments.
palaeotopography: reconstruction of a prehistoric terrain.
palorchestids: large to giant, herbivorous marsupials in the family Palorchestidae (e.g. species of *Ngapakaldi, Pitikantia, Palorchestes* and *Propalorchestes*), all of which became extinct before the end of the Pleistocene. Species of *Palorchestes* had tapir-like trunks and huge claws on their feet.
palynology: study of fossil pollens.
paracone: main anterior and original cusp of the mammalian upper molar.
passerines: small to medium-sized, widely varying birds in the order Passeriformes. There are in excess of 5000 species in the world (60% of all birds).

peramelids: small to medium-sized, insectivorous to omnivorous bandicoots in the family Peramelidae (*Perameles, Isoodon, Chaeropus*), all named species except one, the early Pliocene *P. allinghamensis*, being extant or recently extinct).
perameloids: omnivorous to carnivorous marsupials in the superfamily Perameloidea. Four families are known: the 'ordinary' bandicoots or peramelids; forest bandicoots or peroryctids (Australia and New Guinea); the bilbies or thylacomyids (Australia); and the extinct 'VD' bandicoots (the Tertiary of Australia).
period: interval of geological time smaller than that of era and larger than that of epoch (e.g. the Cambrian and Tertiary Periods). The beginnings and ends of periods are commonly marked by world-wide extinction events.
periotic bone: a thick bone at the base of the skull, behind the jaw joint, that surrounds and protects the inner ear of vertebrates.
permafrost: ground that remains frozen throughout the year such as occurs in the arctic and antarctic regions of the world.
peroryctids: members of the 'forest bandicoot' family Peroryctidae most of which occur today in New Guinea (e.g. species of *Peroryctes* and *Microperoryctes*) but one of which occurs in Australia (*Echymipera rufescens*). Fossil peroryctids, however, are common in the Riversleigh deposits.
petaurids: small arboreal possums of the family Petauridae including gliders and striped possums (Australia and New Guinea) and non-gliders (Australia).
petaurines: members of the petaurid subfamily Petaurinae which includes the gliders (*Petaurus* species) and the non-gliding Leadbeater's Possum (*Gymnobelideus leadbeateri*).
petrify: to turn animal or plant tissue to stone; i.e. fossilise.
phalangerids: arboreal herbivorous to omnivorous possums in the family Phalangeridae including cuscuses (Australia and New Guinea), brushtail possums and scaly-tailed possums (Australia).
phascolarctids: arboreal, Australian tree-leaf-eaters in the family Phascolarctidae, commonly called koalas.
phororhacoids: extinct, predatory, flightless birds, some of which had heads a metre in length, that once stalked South America.
phylogeny: the evolutionary relationships of organisms as determined from morphological and molecular studies of living as well as fossil species.
phyllostomids: bats in the New World family Phyllostomidae; includes fruit- , flower- and nectar-feeders, as well as insect- and flesh-eaters.
pilkipildrids: Oligo-Miocene, omnivorous Australian possums in the family Pilkipildridae (e.g. *Djilgaringa gillespieae* from Riversleigh).
placentals: majority of living, live-bearing mammals, in the subclass Placentalia (e.g. cats, bats, rats, elephants and humans), in which the embryo develops entirely in the uterus (rather than partly in a pouch as it does in most marsupials and some monotremes).
planigales: very small, commonly flat-headed, Australian and New Guinean dasyurids of the genus *Planigale*. The earliest known member comes from the Pliocene Bluff Downs and Riversleigh's Rackham's Roost LFs.
plesiosaurs: extinct, long-necked, carnivorous aquatic reptiles that lived from the late Triassic to Cretaceous; some grew to 16 metres in length.
polydolopids: extinct, small, omnivorous South American marsupials that had converged on some of Australia's diprotodont marsupials (e.g. rat-kangaroos) in having enlarged, procumbent lower incisors as well as blade-like, sectorial premolars.
post-cranial: parts of an animal's body that occur behind the head.
posterior: situated towards the hind end of a structure or whole animal (as opposed to anterior).
potoroids: small to medium-sized, omnivorous Australian and New Guinean rat-kangaroos in the family Potoroidae. There are four subfamilies: hypsiprymnodontines (the musky rat-kangaroos); propleopines (extinct carnivorous rat-kangaroos); potoroines ('ordinary' rat-kangaroos); and balungamayines (extinct, Oligo-Miocene lophodont rat-kangaroos).
potoroines: members of the omnivorous, Australian potoroid subfamily Potoroinae (e.g. species of *Bettongia, Aepyprymnus, Potorous* and *Caloprymnus*).
primordial: primitive, initial, first.
procumbent: reclined and jutting forward. In the present context refers to the shape and posture of the lower first incisors of diprotodont marsupials.
propleopines: extinct, carnivorous potoroids in the subfamily Propleopinae (Australian Pliocene and Pleistocene species of *Propleopus*; Oligo-Miocene species of *Ekaltadeta* so far only known from Riversleigh).
protoplasm: living matter in all plant and animal cells.
pseudocheirids: diverse, leaf-eating, arboreal marsupials in the family Pseudocheiridae known as ringtail possums and greater gliders. The fossil record for ringtails extends back to the late Oligocene; that for greater gliders may extend back to the Pliocene.
pteropodids: tropical, fruit- and leaf-eating bats of the eastern hemisphere in the family Pteropodidae, commonly called flying foxes and fruit bats.
pythonids: small to giant, non-venomous, constricting snakes in the family Pythonidae. They are common in Australia and much of the Old World where some reach lengths of 11 metres.

quinkanines: extinct (Oligo-Miocene to Pleistocene) flesh-eating, semi-terrestrial Australian crocodiles in the subfamily Quinkaninae.

radiation: all of the creatures of a particular kind that have evolved from a common ancestor (e.g. the human radiation would include all of the species of *Homo* and *Australopithecus*; the ektopodontid radiation would include all species of *Ektopodon, Chunia* and *Darcius*).
radiogenic: of or pertaining to radioactive decay.
ranids: insectivorous frogs of the family Ranidae. Although abundant on most continents, there is only one species in Australia.
raptors: birds of prey in the order Raptores (e.g. eagles, hawks and falcons).
ratites: large, flightless, running birds with no keel on their sternum and a 'primitive' palate. They are found on all southern continents except Antarctica and, in Australia, include emus and cassowaries (Casuariidae) and the extinct mihirungs (Dromornithidae).
recurved: curved, bent backwards. Term commonly used, for example, to describe the shape of the teeth of crocodiles and snakes.
relict: organism that survived while other related ones became extinct.
rhinolophids: small, insectivorous bats of the eastern hemisphere, in the family Rhinolophidae, commonly known as horseshoe bats.
rhinolophoid: bats in the superfamily Rhinolophoidea including the horseshoe bats (Rhinolophidae), ghost bats (Megadermatidae) and leaf-nosed bats (Hipposideridae).
rostral: pertaining to a rostrum, beak or beak-like projection; also, towards the nose end of an animal.

scincid: small to medium-sized lizards in the family Scincidae. Most are smooth-scaled, terrestrial, burrowing lizards with wedge-shaped heads, elongate, streamlined bodies and small or absent limbs. They are common in Australia (e.g. the Blue-tongued Lizard *Tiliqua scincoides*).
sclerophyll: of or pertaining to thick-leaved plants (which help to reduce water loss) common in low rainfall areas. In Australia, these include gum trees, acacias and other mainly arid-adapted groups.
sectorial: adapted for cutting. In the present context, the horizontal sectorial blades on the molar and premolar teeth of many marsupials aid in cutting up tough foods.
selenodont: in the present context, sets of crescent-shaped, parallel blades on the molar teeth of some marsupials (such as koalas and ringtail possums) which adapt the tooth for cutting the leaves of trees.

serological: of or pertaining to the serum of blood. In the present context, serological studies of blood proteins have been important as indicators of animal relationships.
sites: individual fossil localities or quarries (e.g. Upper Site, Dredge's Ledge Site) from which collections have been obtained.
speciation: the development of new species and the processes that lead to the production of new species. Allopatric speciation occurs when an isolated part of the ancestral population evolves reproductive isolation. Parapatric speciation occurs when a new species develops on the periphery of the range of the pre-existing species. Sympatric speciation occurs when the new species arises from within the range of the pre-existing species.
species: group of organisms that are capable of interbreeding with each other to produce fertile offspring. Fossil species are normally defined on the basis of morphology (shape) rather than reproductive boundaries. The amount of shape variation within living species is often used as a guide to recognise comparable boundaries between extinct species.
spicules: tiny, calcareous or siliceous bodies that support the otherwise soft bodies of various organisms such as sponges.
sthenurines: mostly large to giant, browsing or grazing kangaroos in the macropodid subfamily Sthenurinae (species of *Sthenurus, Simosthenurus, Procoptodon, Troposodon* and *Lagostrophus*). All are extinct (ranging in time from the Pliocene to Pleistocene) except the central Australian Banded Hare-wallaby, *Lagostrophus fasciatus.*
stochastic: in the present sense, occuring at unpredictable intervals. The term has been used to describe the way in which some speciation events may occur.
stratigraphy: study of the disposition, relationships and origins of sedimentary rocks.
stromatolites: reef-like structures built up over time by layers or mats of cyanobacteria (blue-green algae) which deposit, as well as trap, sedimentary material. Although stromatolites include the oldest evidence of life on Earth (at 3.5 billion years), some still survive (e.g. in Shark Bay, Western Australia).
strontium: metallic element whose chemical reactivity causes organisms to treat it as if it were calcium; hence it ends up in the bones of animals in proportion to the extent to which it is encountered. For this reason, the ecological role of an extinct animal may be roughly indicated by the amount of strontium in its bones.
subduction: during 'continental drift', the converging of plates normally involves the overriding of one edge by the other. Where the edge of one plate nose-dives under the other, the area is called a zone of subduction.
subfamily: a taxonomic rank between that of family and genus. For example, the family Diprotodontidae contains two subfamilies (Diprotodontinae and Zygomaturinae) each of which contain a number of genera (e.g. *Neohelos* , *Kolopsis* and *Zygomaturus* are zygomaturines).
substrate: underlying layer or material upon which an organism grows. In the present context, the substrate of a Riversleigh rainforest was soil and older limestone.
superfamily: a taxonomic rank between family and order. For example, the superfamily Diprotodontoidea which contains the families Diprotodontidae and Palorchestidae is contained with other superfamilies in the order Diprotodontia.
superposition: in sequences of sedimentary rocks, an upper layer is presumed to be younger than a lower one unless earth movements or erosion followed by deposition (e.g. as in cave formation) have subsequently reversed the order of strata.
suture: where two bones in a skull, pelvis, sacrum or other complex set of bones meet as they develop, the boundary may persist without fusion often leaving an interdigitated junction called a suture. As an animal ages, more of its sutures tend to fuse.
sympatric: in the present context, two species that occupy the same region such that they might encounter each other.
tarsipedids: tiny nectivorous to insectivorous possums in the family Tarsipedidae. There is only a single species known, the living Australian Honey Possum, *Tarsipes rostratum,* of Western Australia.
taxon (plural, taxa): a member of a taxonomic category. The primary categories are (in ascending order): species, genus, family, order, class, phylum, kingdom.
tectonics: geological study of the broad architecture of the outer part of the earth including major structural features, their origin and development.
teleost: fish that have a skeleton at least in part made of bone rather than cartilage; includes majority of living fish (e.g. barramundi, catfish etc.).
terrestrial: living on the ground; i.e. not aquatic. However, as used in the present context, it includes arboreal as well as aerial organisms.
thanatocoenosis: an assemblage of fossils found in one place but not all representative of the original biota of the immediate area (which would, in contrast, have been a bioceonosis). Sometimes, secondary enrichment processes such as streams and carnivores, can bring together organic remains gathered from distant habitats thereby creating a fossil assemblage whose diversity is misleadingly high.
Thylacine: the marsupial wolf or Tasmanian Tiger, *Thylacinus cynocephalus,* of the Pliocene to Holocene in New Guinea and Australia. It became extinct in the 1930s in its last stronghold, Tasmania, following systematic persecution by Europeans.
thylacinids: small to large carnivorous marsupials of the family Thylacinidae (e.g. the Oligo-Miocene *Nimbacinus dicksoni* from Riversleigh and the Pliocene to Holocene *Thylacinus cynocephalus*).
thylacoleonids: extinct (Oligo-Miocene to Pleistocene), cat- to leopard-sized flesh-eating marsupial lions of the family Thylacoleonidae; includes species of *Thylacoleo, Wakaleo* and *Priscileo.*
topography: the relief features or surface configuration of an area.
travertine: crystalline but frequently massive deposit of calcium carbonate precipitated from water. It commonly develops in limestone caves as shawls, rims, flowstones, straws, stalactites and stalagmites.
triconodont: tooth form typical of the earliest mammals with three main cusps lined up in a longitudinal row (e.g. late Triassic morganucodontids).
trilobites: extinct arthropods belonging to the class Trilobita that lived in Palaeozoic seas. Their remains are widely distributed throughout the world. They became extinct at the end of the Permian.
trionychids: highly aquatic, soft-shelled turtles of the family Trionychidae, with a proboscis, fleshy lips and a reduced bony shell. Although Australia has no living representatives, fossil species are known from most of the Cainozoic record.
trophic: of or pertaining to the level in an ecosystem or food web occupied by an organism. Trophic levels within an ecosystem commonly include producers (autotrophs), primary consumers (herbivores), secondary consumers (carnivores) and decomposers (saprophages).
tufa: porous mass of calcium carbonate (limestone) deposited around a springhead or as the walls of rim pools ('tufa dams') developed within calcium-rich streams (e.g. the Gregory River).
typhlopids: primitive, blind, burrowing worm-like snakes in the family Typhlopidae; 200 living species are known, from the New and Old Worlds including Australia.

uromyins: murid rodents of the tribe Uromyini (a subgroup of the subfamily Hydromyinae), includes species of *Uromys* and *Melomys*, both primarily rainforest groups of Australian and New Guinea.

varanids: large, agile, predatory lizards of the family Varanidae, commonly known as monitors. There are 30 living species in tropical and subtropical regions of the Old World and Australia. The group includes the gigantic, Pleistocene *Megalania prisca* of Australia and the living Komodo Dragon of Indonesia.
vespertilionids: cosmopolitan, 'ordinary' or plain-faced bats of the family Vespertilionidae.
vertebrates: animals with backbones as well as notochords: fish, amphibians, reptiles, birds and mammals.
vespertilionoids: bats of the superfamily Vespertilionoidea, including plain-faced bats (vespertilionids) and free-tailed bats (molossids), both groups being well represented in Australia.
vombatiforms: wombat-like marsupials in the suborder Vombatiformes (a subgroup of the order Diprotodontia) including koalas and wombats as well as the extinct marsupial lions, diprotodontoids, ilariids and wynyardiids.

wynyardiids: extinct (Oligo-Miocene), browsing vombatiform marsupials of the family Wynyardiidae (species of *Wynyardia, Murramura* and *Namilamadeta*).

zygomaturines: extinct (Oligo-Miocene to Pleistocene), sheep- to bullock-sized, browsing diprotodontids of the subfamily Zygomaturinae (e.g. *Raemeotherium, Neohelos, Kolopsis, Plasiodon* and *Zygomaturus*).

Bibliography

Preface

Archer, M., 1989. The four-dimensional 'bioblob' called life. *Aust. Nat. Hist.* **22**: 512-13.

Chapter 1

Archer, M., Godthelp, H., Hand, S.J., Megirian, D., 1989. Fossil mammals of Riversleigh, northwestern Queensland: preliminary overview of biostratigraphy, correlation and environmental change. *Aust. Zool.* **25**: 29-65.

Chapter 2

Archer, M., 1979. *Wabularoo naughtoni* gen. et sp. nov., an enigmatic kangaroo (Marsupialia) from the middle Tertiary Carl Creek Limestone of northwestern Queensland. *Mem. Qld Mus.* **19**: 299-307.

Archer, M., 1988. Riversleigh: window into our ancient past. *Aust. Geog.* **9**: 40-57.

Archer, M., 1991. Additional publications concerning the research work at Riversleigh: part 7. *Riversleigh Notes* **13**: 7-8.

Archer, M., Bartholomai, A., 1978. Tertiary mammals of Australia: a synoptic review. *Alcheringa* **2**: 1-19.

Archer, M., Clayton, G. (eds), 1984. 'Vertebrate zoogeography and evolution in Australasia'. Hesperian Press: Perth.

Archer, M., Hand, S.J., Godthelp, H., 1986. 'Uncovering Australia's dreamtime'. Surrey Beatty & Sons Pty Ltd: Sydney.

Beale, B., 1990. The fabulous fossils of Riversleigh. *Reader's Digest* **132 (813)**: 15-21.

Cameron, W.E., 1901. Geological observations in north-western Queensland. *Ann. Rep. Geol. Surv. Qld* **1900**: 10-15.

Fairall, J., 1991. Montypythonoides before the Dreaming. *Geo* **13(2)**: 24-33.

Hand, S.J., Archer, M. (eds), 1987. 'The Antipodean ark'. Angus & Robertson: Sydney.

Harland, W.B., Armstrong, R.L., Cox, A.V., Craig, L.E., Smith, A.G., Smith, D.G., 1989. 'A geological time scale'. Cambridge University Press: Cambridge.

Rich, P.V., Thompson, E.M. (eds), 1982. 'The fossil vertebrate record of Australasia'. Monash University Offset Printing Unit: Melbourne.

Savage, D.E., Russell, D.E., 1983. 'Mammalian paleofaunas of the world'. Addison-Wesley Publishing Co.: London.

Sigé, B., Hand, S.J., Archer, M., 1982. An Australian Miocene *Brachipposideros* (Mammalia, Chiroptera) related to Miocene representatives from France. *Palaeovertebrata* **12**: 149-72.

Stirton, R.A., Tedford, R.H., Miller, A.H., 1961. Cenozoic stratigraphy and vertebrate paleontology of the Tirari Desert, South Australia. *Rec. S. Aust. Mus.* **14**: 19-61.

Stirton, R.A., Tedford, R.H., Woodburne, M.O., 1967. A new Tertiary formation and fauna from the Tirari Desert, South Australia. *Rec. S. Aust . Mus.* **15**: 427-62.

Tedford, R.H., 1967. Fossil mammal remains from the Tertiary Carl Creek Limestone, north-western Queensland. *Bur. Miner. Res. Aust., Bull.* **92**: 217-36.

Chapter 3

Archer, M., 1988. Riversleigh: window into our ancient past. *Aust. Geog.* **9**: 40-57.

Archer, M., Godthelp, H., Hand, S.J., Megirian, D., 1989. Fossil mammals of Riversleigh, northwestern Queensland: preliminary overview of biostratigraphy, correlation and environmental change. *Aust. Zool.* **25**: 29-65.

Fenton, C.L., Fenton, M.A., Rich, P.V., Rich, T.H., 1990. 'The fossil book: a record of prehistoric life'. Doubleday: New York.

Godthelp, H., 1987. Riversleigh scene 4: Rackham's Roost—the beginnings of the modern world. Pp. 81-83 *in* "The Antipodean ark" ed S. Hand, M. Archer. Angus & Robertson Publishers: Sydney.

Hand, S.J., Archer, M., Godthelp, H., 1989. A bat-rich Oligo-Miocene cave-fill from Riversleigh. *Conference on Australasian Vertebrate Evolution, Palaeontology & Systematics, Sydney 1989.* Abstracts: p. 7.

Chapter 4

Anon. [Archer, M.], 1987. The koala and the gum tree. *Riversleigh Notes* **1**: 4-5.

Aplin, K., Archer, M., 1987. Recent advances in marsupial systematics with a new syncretic classification. Pp. xv-lxxii in "Possums and opossums: studies in evolution" ed by M. Archer. Surrey Beatty & Sons Pty Ltd and the Royal Zoological Society of New South Wales: Sydney.

Archer, M., 1974. New information about the Quaternary distribution of the Thylacine (Marsupialia, Thylacinidae) in Australia. *J. Proc. Roy. Soc. West. Aust.* **57**: 43-50.

Archer, M., 1979. *Wabularoo naughtoni* gen. et sp. nov., an enigmatic kangaroo (Marsupialia) from the middle Tertiary Carl Creek Limestone of northwestern Queensland. *Mem. Qld Mus.* **19**: 299-307.

Archer, M., 1982. A review of Miocene thylacinids (Thylacinidae, Marsupialia), the phylogenetic position of the Thylacinidae and the problem of apriorisms in character analysis. Pp. 445-76 *in* "Carnivorous marsupials" ed by M. Archer. The Royal Zoological Society of New South Wales: Sydney.

Archer, M., 1982. Review of the dasyurid (Marsupialia) fossil record, integration of data bearing on phylogenetic interpretation, and suprageneric classification. Pp. 397-43 *in* "Carnivorous marsupials" ed M. Archer. The Royal Zoological Society of New South Wales: Sydney.

Archer, M., 1984. The Australian marsupial radiation. Pp. 633-808 *in* "Vertebrate zoogeography & evolution in Australasia" ed M. Archer & G. Clayton. Hesperian Press: Perth.

Archer, M., 1984. The Australian marsupial radiation. Pp. 633-808 in "Vertebrate zoogeography & evolution in Australasia" ed M. Archer, G. Clayton. Hesperian Press: Perth.

Archer, M., 1987. "Thingodonta": last remnant of an ancient group. Pp. 56-58 *in* "The Antipodean ark" ed S. Hand, M. Archer. Angus & Robertson Publishers: Sydney.

Archer, M., 1988. Riversleigh: window into our ancient past. *Aust. Geographic* **9**: 40-57.

Archer, M., 1988. The logo of the Riversleigh Society. *Riversleigh Notes* **3**: 5.

Archer, M., 1989. The science of being wrong. *Aust. Nat. Hist.* **23**: 170-71.

Archer, M., 1990. Distribution patterns in space and time for Australian 'possums' and 'possum-like' diprotodontians. *Aust. Mamm. Soc. Abstracts AGM, Canberra, 1990*: 10.

Archer, M., 1991. Tharalkoo's child: an ugly duckling story. *Aust. Nat. Hist.* **23**: 574-75.

Archer, M., Bartholomai, A., 1978. Tertiary mammals of Australia: a synoptic review. *Alcheringa* **2**: 1-19.

Archer, M., Dawson, L., 1982. Revision of marsupial lions of the genus *Thylacoleo* Gervais (Thylacoleonidae, Marsupialia) and thylacoleonid evolution in the late Cainozoic. Pp. 477-94 *in* 'Carnivorous marsupials' ed M. Archer. Royal Zoological Society of New South Wales: Sydney.

Archer, M., Every, R.G., Godthelp, H., Hand, S., Scally, K., 1990. Yingabalanaridae, a new family of enigmatic mammals from Tertiary deposits of Riversleigh, northwestern Queensland. *Mem. Qld Mus.* **28**: 193-202.

Archer, M., Flannery, T.F., 1985. Revision of the extinct gigantic rat kangaroos (Potoroidae: Marsupialia), with description of a new Miocene genus and species and a new Pleistocene species of *Propleopus*. *J. Paleont.* **59**: 1131-49.

Archer, M., Flannery, T.F., 1985. 'The kangaroo'. Weldon Pty Ltd: Sydney.

Archer, M., Flannery, T.F., Ritchie, A., Molnar, R.E., 1985. First Mesozoic mammal from Australia—an early Cretaceous monotreme. *Nature* **318**: 363-66.

Archer, M., Godthelp, H., Hand, S.J., Megirian, D., 1989. Fossil mammals of Riversleigh, northwestern Queensland: preliminary overview of biostratigraphy, correlation and environmental change. *Aust. Zool.* **25**: 29-65.

Archer, M., Hand, S.J., 1987. Evolutionary considerations. Pp. 79-106 *in* "The Koala: Australia's endearing marsupial" ed L. Cronin. Reed Books: Sydney.

Archer, M., Hand, S.J., Godthelp, H,. 1988. Cainozoic changes in Australian vertebrate diversity: growth of a pre-Quaternary data base. *Abstract. The University of Sydney Extinction Conference.*

Archer, M., Hand, S.J., Godthelp, H., 1986. 'Uncovering Australia's dreamtime'. Surrey Beatty & Sons Pty Ltd: Sydney. [Extended introduction to Riversleigh discoveries and their significance]

Archer, M., Hand, S.J., Godthelp, H., 1988. A new order of Tertiary zalambdodont marsupials. *Science* **239**: 1528-31.

Archer, M., Hand, S.J., Godthelp, H., 1989. Dentition of the Oligocene/Miocene ornithorhynchid genus *Obdurodon* and the phylogenetic relationships of monotremes. *Conference on Australasian Vertebrate Evolution, Palaeontology and Systematics, Sydney 1989.* Abstracts: p. 1.

Archer, M., Hand, S.J., Godthelp, H., 1989. Ghosts from green gardens: preliminary hypotheses about changes in Australia's rainforest mammals through time based on evidence from Riversleigh. *Riversleigh Notes* **7**: 4-7.

Archer, M., Muirhead, J., 1990. Odontology. Part 3: The basic marsupial molar pattern and variation within the dasyurid pattern. *Riv. Notes* **11**: 4-5.

Archer, M., Plane, M.D., Pledge, N.S., 1978. Additional evidence for interpreting the Miocene *Obdurodon insignis* Woodburne and Tedford, 1975, to be a fossil platypus (Ornithorhynchidae: Monotremata) and a reconsideration of the status of *Ornithorhynchus agilis* De Vis, 1885. *Aust. Zool.* **20**: 9-27.

Archer, M., Rich, T.H., 1982. Results of the Ray E. Lemley Expeditions. *Wakaleo alcootaensis* n. sp. (Thylacoleonidae, Marsupialia), a new marsupial lion from the Miocene of the Northern Territory with a consideration of early radiation in the family. Pp. 495-502 *in* "Carnivorous marsupials" ed M. Archer. Royal Zoological Society of New South Wales: Sydney.

Archer, M., Tedford, R.H., Rich, T.H., 1987. The Pilkipildridae, a new family and four new species of ?petauroid possums (Marsupialia: Phalangerida) from the Australian Miocene. Pp. 607-27 *in* "Possums and opossums: studies in evolution" ed M. Archer. Surrey Beatty & Sons Pty Ltd and The Royal Zoological Society of New South Wales: Sydney.

Bergdolt, S., Muirhead, J., 1989. A Late Oligocene/Early Miocene perameloid skull: a look at the bandicoot bauplan. *Conference on Australasian Vertebrate Evolution, Palaeontology*

Boles, W., 1987. Riversleigh's fossil feathered fliers. *Riversleigh Notes* **2**: 4-5.

Boles, W., 1989. Preliminary analysis of fossil avifaunas from Riversleigh, Qld. *Conference on Australasian Vertebrate Evolution, Palaeontology and Systematics, Sydney 1989.* Abstracts: p. 2.

Boles, W., 1991. The origin and radiation of Australasian birds: perspectives from the fossil record. *Proceedings of the XXth International Ornithological Congress* (in press).

Boles, W., 1991. Work on the Riversleigh birds. *Riversleigh Notes* **13**: 2-4.

Case, J.A., 1984. A new genus of Potoroinae (Marsupialia: Macropodidae) from the

Miocene Ngapakaldi Local Fauna, South Australia, and a definition of the Potoroinae. *J. Paleont.* **58**: 1074-86.
Clemens, W.A., Plane, M.D., 1974. Mid-Tertiary Thylacoleonidae (Marsupialia, Mammalia). *J. Paleont.* **48**: 652-60.
Cooke, B., 1989. A taxonomically tantalising new 'roo from Riversleigh, Qld. *Conference on Australasian Vertebrate Evolution, Palaeontology and Systematics, Sydney 1989.* Abstracts: p. 3.
Cooke, B., 1989. Ancient 'roos from Riversleigh. *Riversleigh Notes* **8**: 2-3.
Cooke, B., 1991. Primitive macropodids from the Oligo-Miocene freshwater deposits of Riversleigh, northwestern Queensland. *Conference on Australasian Vertebrate Evolution, Palaeontology and Systematics, Alice Springs: Abstracts.*
Covacevich, J., Couper, P., R.E. Molnar, G. Witten, Young, W., 1990. Miocene dragons from Riversleigh: new data on the history of the family Agamidae (Reptilia: Squamata) in Australia. *Mem. Qld Mus.* **29**: 339-360.
Dawson, L., 1983. The taxonomic status of small fossil wombats (Vombatidae: Marsupialia) from Quaternary deposits, and related modern wombats. *Proc. Linn. Soc. N.S.W.* **107**: 101-23.
Estes, R., 1984. Fish, amphibians and reptiles from the Etadunna Formation, Miocene of South Australia. *Aust. Zool.* **21**: 335-43.
Flannery, T.F., 1989. Phylogeny of the Macropodoidea: a study in convergence. Pp. 1-46 *in* "Kangaroos, wallabies and rat-kangaroos" ed G. Grigg, P. Jarman, I. Hume. Surrey Beatty & Sons Pty Ltd: Sydney.
Flannery, T.F., Archer, M., 1987. *Bettongia moyesi*, a new and plesiomorphic kangaroo (Marsupialia: Potoroidae) from Miocene sediments of northwestern Queensland. Pp. 759-67 *in* "Possums and opossums: studies in evolution" ed M. Archer. Surrey Beatty & Sons Pty Ltd and the Royal Zoological Society of New South Wales: Sydney.
Flannery, T.F., Archer, M., 1987. *Hypsiprymnodon bartholomaii* (Potoroidae: Marsupialia), a new species from the Miocene Dwornamor Local Fauna and a reassessment of the phylogenetic position of *H. moschatus* . Pp. 749-58 *in* "Possums and opossums: studies in evolution" ed M. Archer. Surrey Beatty & Sons Pty Ltd and the Royal Zoological Society of New South Wales: Sydney.
Flannery, T.F., Archer, M., 1987. *Strigocuscus reidi* and *Trichosurus dicksoni* , two new fossil phalangerids (Marsupialia: Phalangeridae) from the Miocene of northwestern Queensland. Pp. 527-36 *in* "Possums and opossums: studies in evolution" ed M. Archer. Surrey Beatty & Sons Pty Ltd and the Royal Zoological Society of New South Wales: Sydney.
Flannery, T.F., Archer, M., Maynes, G., 1987. The phylogenetic relationships of living phalangerids (Phalangeroidea: Marsupialia) with a suggested new taxonomy. Pp. 477-506 *in* "Possums and opossums: studies in evolution" ed M. Archer. Surrey Beatty & Sons Pty Ltd and the Royal Zoological Society of New South Wales: Sydney.
Flannery, T.F., Archer, M., Plane, M.D., 1983. Middle Miocene kangaroos (Macropodoidea: Marsupialia) from three localities in northern Australia, with a description of two new subfamilies. *B.M.R.J. Aust. Geol., Geophys.* **7**: 287-302.
Flannery, T.F., Archer, M., Plane, M.D., 1984. Phylogenetic relationships and a reconsideration of higher level systematics within the Potoroidae (Marsupialia). *J. Paleo.* **58**: 1087-97.
Gaffney, E.S., Archer, M., White, A., 1989. Chelid turtles from the Miocene freshwater limestones of Riversleigh Station, northwestern Queensland, Australia. *Novitates* **2959**: 1-10.
Godthelp, H., 1987. The beginnings of the modern world. Pp. 81-83 *in* "The Antipodean ark" ed S.J. Hand, M. Archer. Angus and Robertson: Sydney.
Godthelp, H., 1989. What's that in my vat? *Riversleigh Notes* **4**: 3.
Godthelp, H., 1989. What's that in my vat? *Riversleigh Notes* **6**: 8.
Godthelp, H., 1990. Origins of the modern Australian murid fauna. *Aust. Mamm. Soc. Abstracts AGM, Canberra, 1990*: 21.
Godthelp, H., 1990. *Pseudomys vandycki*, a Tertiary murid from Australia. *Mem. Qld Mus.* **28**: 171-73.
Godthelp, H., Archer, M., Hand, S., Plane, M., 1989. New potoroine from Tertiary Kangaroo Well Local Fauna, N.T. and description of upper dentition of potoroine *Wakiewakie lawsoni* from Upper Site Local Fauna, Riversleigh. *Conference on Australasian Vertebrate Evolution, Palaeontology and Systematics, Sydney 1989.* Abstracts: p. 6.
Gott, M., 1989. Oligocene/Miocene notoryctids (Notoryctidae, Marsupialia) from Riversleigh: first fossil record for the family. *Conference on Australasian Vertebrate Evolution, Palaeontology and Systematics, Sydney 1989.* Abstracts: p. 5.
Gott, M., 1989. There's no mole like an old mole. *Riversleigh Notes* **6**: 2-3.
Hand, S., 1989. On the winds of fortune. *Aust. Nat. Hist.* **23**: 130-38.
Hand, S.J., 1983. An ancient Australian bat. Pp. 76-77 *in* "Prehistoric animals of Australia" ed S. Quirk, M. Archer. Australian Museum: Sydney.
Hand, S.J., 1984. Australia's oldest rodents: master mariners from Malaysia. Pp 905-19 *in* "Vertebrate zoogeography and evolution in Australasia" ed M. Archer, G. Clayton. Hesperian Press: Perth.
Hand, S.J., 1984. Bat beginnings and biogeography: a southern perspective. Pp. 853-904 *in* "Vertebrate zoogeography and evolution in Australasia" ed M. Archer, G. Clayton. Hesperian Press: Perth.
Hand, S.J., 1984. Old bat bones and their discovery down under. *Brit. Mamm. Soc. Newsl.* June, 1984.
Hand, S.J., 1984. The radiation of the Megadermatidae (Mammalia, Chiroptera): evidence from the Australian fossil record. *Aust. Mammal. Soc. Bull. Abstr.* , December, 1984.
Hand, S.J., 1985. New Miocene megadermatids (Megadermatidae, Chiroptera) from Australia with comments on megadermatid phylogenetics. *Aust. Mammal.* **8**: 5-43.
Hand, S.J., 1985. The Nooraleeba bats. Pp. 253-58 *in* "Kadimakara—extinct backboned animals of Australia" ed P.V. Rich, G.F. Van Tets. Pioneer Design Studios: Melbourne.
Hand, S.J., 1986. 'Phylogenetic studies of Australian Tertiary bats'. Unpublished PhD thesis. Macquarie University: Sydney.
Hand, S.J., 1987. Phylogenetic studies of Australian Tertiary bats: summary of PhD thesis. *Macroderma* **3**: 9-12.
Hand, S.J., 1989. The status of two vulnerable Australian bats: view from the 4th dimension. *Abstracts: 8th International Bat Research Conference, Sydney 1989*: 10-11.
Hand, S.J., 1989. On the wings of fortune. The origin of Australia's bat fauna. *Aust. Nat. Hist.* **24**: 130-38.
Hand, S.J., 1990. First Tertiary molossid (Microchiroptera: Molossidae) from Australia: its phylogenetic and biogeographic implications. *Mem. Qld Mus.* **28**: 175-92.
Hand, S.J., 1990. Origins and radiation of the Australian bat fauna: evidence from the fossil record. *Aust. Mamm. Soc. Abstracts AGM, Canberra, 1990*: 23.
Hand, S.J., Archer, M., 1985. Correlative value and biogeographic significance of Australian fossil bats. *Abstract, Proc. 7th Internat. Bat Res. Conf., Aberdeen; August, 1985* .
Hand, S.J., Archer, M., Godthelp, H., 1989. A fossil bat-rich Oligo-Miocene cave-fill from Riversleigh Station, north-western Queensland. *Conference on Australasian Vertebrate Evolution, Palaeontology and Systematics, Sydney 1989.* Abstracts: p. 7.
Hand, S.J., Archer, M., Godthelp, H., 1989. Australian fossil hipposiderids: new evidence about the evolutionary history and radiation of old world leaf-nosed bats. *Abstracts: 8th International Bat Research Conference, Sydney 1989*: 10.
Hand, S.J., Archer, M., Godthelp, H., 1989. The Australian bat fauna: a 30 million year history? *Scientific Meeting and A.G.M. of the Australian Mammal Society, Alice Springs, April 24-25, 1989. Abstracts*: p. 26.
Hope, J., Wilkinson, H.E., 1982. *Warendja wakefieldi*, a new genus of wombat (Marsupialia, Vombatidae) from Pleistocene sediments in McEacherns Cave, western Victoria. *Mem. Nat. Mus. Vic.* **43**: 109-20.
Hutchinson, M.N., 1991. Preliminary report on the Tertiary skinks of Riversleigh, northwestern Queensland. *Conference on Australasian Vertebrate Evolution, Palaeontology and Systematics, Alice Springs: Abstracts.*
Kemp, A., 1991. Australian Mesozoic and Cainozoic lungfish. Pp. 465-98 *in* "Vertebrate palaeontology of Australasia" ed P.V. Rich, J.M. Monaghan, R.F. Baird, T.H. Rich. Pioneer Design: Melbourne.
Lester, K.S., Archer, M., 1986. A description of the molar enamel of a middle Miocene monotreme (*Obdurodon* , Ornithorhynchidae). *Anat. Embryol.* **174**: 145-51.
Lester, K.S., Archer, M., Gilkeson, C.F., Rich, T., 1988. Enamel of *Yalkaparidon coheni* : representative of a distinctive order of Tertiary zalambdodont marsupials. *Scanning Microscopy* **2**: 1491-1501.
Lester, K.S., Boyde, A., Gilkeson, C., Archer, M., 1987. Marsupial and monotreme enamel structure. *Scanning Microscopy* **1**: 401-20.
Muirhead, J., 1987. The extinct Thylacine or Tasmanian Tiger. *Riversleigh Notes* **2**: 8.
Muirhead, J., 1989. The Tertiary radiation of the Family Thylacinidae. *Conference on Australasian Vertebrate Evolution, Palaeontology and Systematics, Sydney 1989. Abstracts*: p. 11.
Muirhead, J., 1990. The decline in diversity of the family Thylacinidae since the Oligo-Miocene. *Aust. Mamm. Soc. Abstracts AGM, Canberra, 1990*: 34.
Muirhead, J., 1991. Tertiary bandicoots. *Conference on Australasian Vertebrate Evolution, Palaeontology and Systematics; Abstracts.*
Muirhead, J., Archer, M., 1990. *Nimbacinus dicksoni*, a plesiomorphic thylacine (Marsupialia: Thylacinidae) from Tertiary deposits of Queensland and the Northern Territory. *Mem. Qld Mus.* **28**: 203-21.
Muirhead, J., Archer, M., 1991. Odontology. Part 4: Dasyurid relatives the numbats and thylacines. *Riversleigh Notes* **12**: 4-6.
Muirhead, J., Archer, M., 1991. Odontology. Part 5: Bandicoots. *Riversleigh Notes* **13**: 4-7.
Murray, P., Wells, R.T., Plane, M.D., 1987. The cranium of the Miocene thylacoleonid *Wakaleo vanderleuri*: click go the shears—a fresh bite at thylacoleonid systematics. Pp. 433-66 *in* "Possums and opossums: studies in evolution" ed M. Archer. Surrey Beatty & Sons Pty Ltd and the Royal Zoological Society of New South Wales: Sydney.
Murray, P., 1990. Primitive marsupial tapirs (*Propalorchestes novaculacephalus* Murray and *P. ponticulus* sp. nov.) from the mid-Miocene of North Australia (Marsupialia: Palorchestidae). *The Beagle* **7**: 39-51.
Pledge, N., 1990. *Wynyardia*: the Riversleigh connection. *Riversleigh Notes* **9**: 2-4.
Pledge, N., Archer, M., Hand, S., Godthelp, H., 1989. Additions to knowledge about ektopodontids (Marsupialia: Ektopodontidae) with description of a new species, *Ektopodon simplicidens. Conference on Australasian Vertebrate Evolution, Palaeontology and Systematics, Sydney 1989.* Abstracts: p. 14.
Pledge, N.S., 1987. A new species of *Burramys* Broom (Marsupialia: Burramyidae) from the middle Miocene of South Australia. Pp. 725-28 *in* "Possums and opossums: studies in evolution" ed M. Archer. Surrey Beatty & Sons Pty Ltd: Sydney.
Pledge, N.S., 1987. *Muramura williamsi*, a new genus and species of ?wynyardiid (Marsupialia: Vombatoidea) form the middle Miocene Etadunna Formation of South Australia. Pp. 393-400 *in* "Possums and opossums: studies in evolution" ed M. Archer. Surrey Beatty & Sons Pty Ltd: Sydney.
Rauscher, B., 1987. *Priscileo pitikantensis*, a new genus and species of thylacoleonid marsupial (Marsupialia: Thylacoleonidae) from the Miocene Etadunna Formation, South Australia. Pp. 423-32 *in* "Possums and opossums: studies in evolution" ed M. Archer. Surrey Beatty & Sons Pty Ltd: Sydney.
Rich, P.V., 1979. The Dromornithidae, an extinct family of large ground birds endemic to Australia. *Bull. Bur. Miner. Res. Aust.* **184**: 1-196.

Rich, P.V., van Tets, G.F. (eds), 1985. 'Kadimakara: extinct vertebrates of Australia'. Pioneer Design Studio: Melbourne.

Rich, P.V., van Tets, G.F., 1982. Fossil birds of Australia and New Guinea: their biogeographic, phylogenetic and biostratigraphic input. Pp. 235-384 *in* "The fossil vertebrate record of Australasia" ed P.V. Rich, E.M. Thompson. Monash University Offset Printing Unit: Melbourne.

Rich, P.V., van Tets, G.F., 1984. What fossil birds contribute towards an understanding of origin and development of the Australian avifauna. Pp. 421-446.

Rich, T.H., Archer, M., Plane, M.D., Flannery, T.F., Pledge, N.S., Hand, S.J., Rich, P.V., 1982. Australian Tertiary mammal localities. Pp. 526-72 *in* "The fossil vertebrate record of Australia" ed P.V. Rich, B. Thompson. Monash University Offset Printing Unit: Melbourne.

Scanlon, J.D., 1988. The snakes of Riversleigh. *Riversleigh Notes* **3**: 7-8.

Scanlon, J.D., 1989. The fossil snakes of Riversleigh: a preliminary survey. *Conference on Australasian Vertebrate Evolution, Palaeontology and Systematics: Abstracts*: p. 15.

Scanlon, J.D., 1991. A new genus of large madtsoiid snakes from Miocene deposits at Bullock Creek, Northern Territory and Riversleigh, Queensland. *Conference on Australasian Vertebrate Evolution, Palaeontology and Systematics; Abstracts.*

Scanlon, J.D. 1990. 'Rainbow Serpents': the family Madtsoiidae in Australia. *Australian Society of Herpetologists, Gemini Downs Meeting; Abstracts.*

Scanlon, J.D., 1991. Origins and radiations of snakes: an Australian slant. In "Animals in space and time: vertebrate zoogeography and evolution in Australasia" ed M. Archer, G. Clayton, S. Hand and J. Long. 2nd ed. Hesperian Press, Perth (in press).

Shea, G.M., Hutchinson, M.N., 1991. A new species of *Tiligua* (Lacertilia, Scincidae) from the Oligo-Miocene deposits of Riversleigh, northwestern Queensland. *Mem. Qld Mus.* (in press).

Sigé, B., Hand, S.J., Archer, M., 1982. An Australian Miocene *Brachipposideros* (Mammalia, Chiroptera) related to Miocene representatives from France. *Palaeovertebrata* **12**: 149-72.

Smith, M.J., Plane, M., 1985. Pythonine snakes (Boidae) from the Miocene of Australia. *Bur. Miner. Res. Aust., J. Geol. & Geophys.* **9**: 191-95.

Tedford, R.H., 1967. Fossil mammal remains from the Tertiary Carl Creek Limestone, northwestern Queensland. *Bur. Miner. Res. Aust., Bull.* **92**: 217-36.

Tedford, R.H., Archer, M., Bartholomai, A., Plane, M.D., Pledge, N.S., Rich, T.H., Rich, P.V., Wells, R.T., 1977. The discovery of Miocene vertebrates, Lake Frome area, South Australia. *Bur. Miner. Res. Aust., J. Geol. & Geophys.* **2**: 53-57.

Turnbull, W.D., Lundelius, E., 1987. Burramyids (Marsupialia: Burramyidae) from the early Pliocene Hamilton Local Fauna, southwestern Victoria. Pp. 729-39 *in* "Possums and opossums: studies in evolution" ed M. Archer. Surrey Beatty & Sons Pty Ltd and the Royal Zoological Society of New South Wales: Sydney.

Tyler, M.J., 1989. A new species of *Lechriodus* (Anura: Leptodactylidae) from the Tertiary of Queensland, with a redefinition of the ilial characteristics of the genus. *Trans. R. Soc. S. Aust.* **113**: 15-21.

Tyler, M.J., 1990. *Lymnodynastes* Fitzinger (Anura: Leptodactylidae) from the Cainozoic of Queensland. *Mem. Qld Mus.* (in press).

Tyler, M.J., 1991. A large new species of *Litoria* (Anura: Hylidae) from the Tertiary of Queensland. *Trans. R. Soc. S. Aust.* (in press).

Tyler, M.J., 1991. *Crinia tschudi* (Anura: Leptodactylidae) from the Cainozoic of Queensland, with a description of a new species. *Trans. R. Soc. S. Aust.* **115** (in press).

Tyler, M.J., Hand, S.J., Ward, V.J., 1990. Analysis of the frequency of *Lechriodus intergerivus* Tyler (Anura: Leptodactylidae) in Oligo-Miocene local faunas of Riversleigh Station, Queensland. *Proc. Linn. Soc. N.S.W.* **112**: 105-109.

Van Dyck, S., 1989. Biting remarks on the Riversleigh Antechinus. *Riversleigh Notes* **7**: 2-3.

Van Dyck, S., Archer, M., 1989. A new *Antechinus*-like dasyurid from the Oligo-Miocene deposits of Riversleigh. Poster paper at Scientific Meeting and A.G.M. of the Australian Mammal Society, Alice Springs, April 24-25.

Watts, C.H.S., Aslin, H.J., 1981. 'The rodents of Australia'. Angus and Robertson: Sydney.

Watts, C.H.S., Kemper, C.M., 1989. The Muridae. Pp 939-56 *in* "The fauna of Australia. Vol. 1B. Mammalia" ed D.W. Walton, B.J. Richardson. Australian Government Publishing Service: Canberra.

Wells, R.T., Horton, D.R., Rogers, P., 1982. *Thylacoleo carnifex* Owen (Thylacoleonidae: Marsupialia): marsupial carnivore? Pp. 573-86 *in* "Carnivorous marsupials" ed M. Archer. Royal Zoological Society of New South Wales: Sydney.

Wells, R.T., Nichol, B., 1977. On the manus and pes of *Thylacoleo carnifex* Owen (Marsupialia). *Trans. R. Soc. S. Aust.* **101**: 139-46.

White, A., 1989. Riversleigh turtles: all shell be revealed. *Riversleigh Notes* **4**: 4-5.

White, A., Archer, M., 1989. Riversleigh's turtles—so far. *Conference on Australasian Vertebrate Evolution, Palaeontology and Systematics, Sydney 1989.* Abstracts: p. 17.

Willis, P.M.A., Murray, P., Megirian, D., 1990. *Baru darrowi gen. et sp. nov.*, a large, broad-snouted crocodiline (Eusuchia, Crocodylidae) from mid-Tertiary freshwater limestones in northern Australia. *Mem. Qld Mus.* **29**: 521-40.

Woodburne, M.O., 1967. The Alcoota Fauna, central Australia: an integrated palaeontological and geological study. *Bull. Bur. Min. Res. Geol. Geophys. Aust.* **87**: 1-187.

Woodburne, M.O., Clemens, W.A. (eds), 1986. Revision of the Ektopodontidae (Mammalia; Marsupialia; Phalangeroidea) of the Australian Neogene. *Univ. Calif. Publ. Geol. Sci.* **131**.

Woodburne, M.O., Tedford, R.H., 1975. The first Tertiary monotreme from Australia. *Amer. Mus. Novit.* **2588**.

Woodburne, M.O., Tedford, R.H., Archer, M., Pledge, N.S., 1987. *Madakoala.* a new genus and two new species of Miocene koalas (Marsupialia: Phascolarctidae) from South Australia, and a new species of *Perikoala*. Pp. 293-317 *in* "Possums and opossums: studies in evolution" ed M. Archer. Surrey Beatty & Sons Pty Ltd and the Royal Zoological Society of New South Wales: Sydney.

Woodburne, M.O., Tedford, R.H., Archer, M., Turnbull, W.D., Plane, M.D., Lundelius, E.D., 1985. Biochronology of the continental mammal record of Australia and New Guinea. *Spec. Publ., S. Aust. Dept Mines and Energy* **5**: 347-63.

Chapter 5

Archer, M., Godthelp, H., Hand, S.J., Megirian, D., 1989. Fossil mammals from Riversleigh, northwestern Queensland: preliminary overview of biostratigraphy, correlation and environmental change. *Aust. Zool.* **25**: 29-65.

Chapter 6

Archer, M., Clayton, G. (eds), 1984. 'Vertebrate zoogeography and evolution in Australasia'. Hesperian Press: Perth.

Glaessner, M., 1984. 'The dawn of animal life: a biohistorical study'. Cambridge University Press: Cambridge.

Hand, S.J., Archer, M. (eds), 1987. 'The Antipodean Ark'. Angus & Robertson: Sydney.

Mackness, B., 1987. 'Prehistoric Australia: 4000 million years of evolution in Australia.' Golden Press: Sydney.

Morrison, R., Morrison, M., 1988. 'Australia: the four billion year journey of a continent'. Weldon Publishing: Sydney.

Quirk, S., Archer, M. (eds), 1983. 'Prehistoric animals of Australia'. Australian Museum: Sydney.

Rich, P.V., Rich, T.H., Wagstaff, B.E., Mason, J.M., Douthitt, C.B., Gregory, R.T., Felton, E.A., 1988. Evidence for low temperatures and biologic diversity in Cretaceous high latitudes of Australia. *Science* **242**: 1403-06.

Rich, P.V., Thompson, B. (eds), 1982. 'The fossil vertebrate record of Australia'. Monash University Offset Printing Unit: Melbourne.

Rich, P.V., Van Tets, G. (eds), 1985. 'Kadimakara: extinct vertebrates of Australia'. Pioneer Design Studio: Lilydale, Victoria.

White, M.E., 1986. 'The greening of Gondwana'. Reed Books Pty Ltd: Sydney.

White, M.E., 1990. 'Hidden worlds'. Reed Books Pty Ltd: Sydney.

Woodburne, M.O., Tedford, R.H., Archer, M., Turnbull, W.D., Plane, M.D., Lundelius, E.D., 1985. Biochronology of the continental mammal record of Australia and New Guinea. *Spec. Publ., S. Aust. Dept Mines and Energy* **5**: 347-63.

Chapter 7

Archer, M., 1984. Earth-shattering concepts for historical zoogeography. Pp. 45-59 *in* "Vertebrate zoogeography and evolution in Australasia" ed M. Archer, G. Clayton. Hesperian Press: Perth.

Barlow, B.A., Hyland, B.P.M., 1988. The origins of the flora of Australia's wet tropics. *Proc. Ecol. Soc. Aust.* **15**: 1-17.

Egerton, L. (ed), 1990. 'Do-it-yourself Earth repair'. Reed Books: Sydney.

Figgis, P., 1985. 'Rainforests of Australia'. Weldon: Sydney.

Filewood, W., 1984. The Torres connection: zoogeography of New Guinea. Pp. 1121-31 *in* "Vertebrate zoogeography and evolution in Australasia" ed M. Archer & G. Clayton. Hesperian Press: Perth.

Flannery, T.F., 1989. Origins of the Australo-Pacific land mammal fauna. *Aust. Zool. Rev.* **1**: 15-24.

Hope, G., Kirkpatrick, J., 1988. The ecological history of Australian forests. Pp. 3-22 *in* "Australia's ever changing forests" ed by K.J. Frawley, N.M. Semple. Dept. Geog. Oceanogr. Aust. Def. Force Acad.: Special Paper No. 1.

Kershaw, A.P., 1988. Australasia. Pp. 237-306 *in* "Vegetation history" ed by B. Huntley, T. Webb. Kluwer Academic Publishers: London.

Martin, H.A., 1990. Tertiary climate and phytogeography in southeastern Australia. *Rev. Palaeobotany & Palynology* **65**: 47-55.

Martin, H.A., 1990. The palynology of the Namba Formation in the Wooltana-1 bore, Callabonna Basin (Lake Frome), South Australia, and its relevance to Miocene grasslands in central Australia. *Alcheringa* **14**: 247-55.

Molnar, R.E., O'Reagan, M., 1989. Dinosaur extinctions. *Aust. Nat. Hist.* 22: 560-70.

Plane, M.D., 1967. Stratigraphy and vertebrate fauna of the Otibanda Formation, New Guinea. *Bull. Bur. Miner. Resour. Geol. Geophys. Aust.* **86**: i-64.

Prance, G.T., Caldecott, J., Lovejoy, T., Bellamy, D. (chapters in) Silcock, L. (ed), 1990. "The rainforests — a celebration". Doubleday: Sydney.

Truswell, E.M., 1990. Australian rainforests: the 100 million year record. Pp. 7-22 in "Australian tropical rainforests: science, values, meaning" ed L.J. Webb, J. Kikkawa. CSIRO Australia: Melbourne.

Veevers, J.J., 1984. 'Phanerozoic earth history of Australia'. Clarendon Press: Oxford.

Webb, L.J., Keto, A. *et al.* in Figgis, P. (ed.), 1985. 'Rainforests of Australia'. Weldons: Sydney.

Webb, L.J., Tracey, J.G., 1981. Australian rainforests: pattern and change. Pp. 605-94 *in* "Ecological biogeography of Australia" ed A. Keast. Junk: The Hague.

White, M.E., 1986. 'The greening of Gondwana'. Reed Books Pty Ltd: Sydney.

White, M.E., 1991 (in press). Environments of the geological past *in* "Animals in space and time: vertebrate zoogeography and evolution in Australasia" ed M. Archer, G. Hickey, S.J. Hand, J. Long. Hesperian Press: Perth

Chapter 8

Anon., 1989. Selected publications involving Riversleigh fossils and their significance. *Riversleigh Notes* **5**: 3-6.

Archer, M., 1991. Additional publications concerning the research work at Riversleigh. *Riversleigh Notes* **13**: 7-8.
Archer, M., 1988. Riversleigh: window into our ancient past. *Aust. Geog.* **9**: 40-57.
Archer, M., Godthelp, H., Hand, S.J., Megirian, D., 1989. Fossil mammals from Riversleigh, northwestern Queensland: preliminary overview of biostratigraphy, correlation and environmental change. *Aust. Zool.* **25**: 29-65.
Archer, M., Hand, S.J., Godthelp, H., 1986. 'Uncovering Australia's Dreamtime'. Surrey Beatty
Bates, R.L., Jackson, J.A. (eds), 1984. 'Dictionary of geological terms'. Anchor Press/Doubleday: New York.
Godthelp, H., 1987. Riversleigh scene 4: Rackham's Roost—the beginnings of the modern world. Pp. 81-83 *in* "The Antipodean ark" ed S.J. Hand, M. Archer. Angus & Robertson Publishers: Sydney.
Hand, S.J., Archer, M., Godthelp, H., 1989a. A fossil bat-rich, Oligo-Miocene cave-fill from Riversleigh Station, north-western Queensland. *Conf. Australasian Vert. Evol., Palaeont. and System., Abstracts*: p. 7.
Tedford, R.H., 1967. Fossil mammals from the Carl Creek Limestone, northwestern Queensland. *Bull. Bur. Miner. Resour. Geol. Geophys. Aust.* **92**: 217-36.
Woodburne, M.O., Tedford, R.H., Archer, M., Turnbull, W.D., Plane, M.D., Lundelius, E.L., 1985. Biochronology of the continental mammal record of Australia and New Guinea. *Spec. Publ. S. Aust. Dept Mines and Energy* **5**: 347-63.

Chapter 9

Archer, M., Godthelp, H., Hand, S.J., Megirian, D., 1989. Fossil mammals of Riversleigh, northwestern Queensland: preliminary overview of biostratigraphy, correlation and environmental change. *Aust. Zool.* **25**: 29-65.
Archer, M., Hand, S.J., Godthelp, H., 1988b. Green cradle: the rainforest origins of Australia's marsupials. *7th Internat. Palyn. Cong., Brisbane. Abstracts*: p. 2.
Archer, M., Hand, S.J., Godthelp, H., 1989. Ghosts from green gardens: preliminary hypotheses about changes in Australia's rainforest mammals through time based on evidence from Riversleigh. *Riversleigh Notes* **7**: 4-7.
Boles, W., 1989. Preliminary analysis of fossil avifaunas from Riversleigh, Qld. *Conf. Australasian Vert. Evol., Palaeont. and System., Abstracts*: p. 2.
Godthelp, H., 1988. Riversleigh scene 4: Rackham's Roost—the beginnings of the modern world. Pp. 81-83 *in* "The Antipodean ark" ed S.J. Hand, M. Archer. Angus & Robertson Publishers: Sydney.
Hand, S.J., 1987. Phylogenetic studies of Australian Tertiary bats. Summary of PhD thesis. *Macroderma* **3**: 9-12.
Hand, S.J., 1990. First Tertiary molossid (Microchiroptera: Molossidae) from Australia: its phylogenetic and biogeographic implications. *Mem. Qld Mus.* **28**: 175-92.
Hand, S.J., 1989. On the wings of fortune. The origin of Australia's bat fauna. *Aust. Nat. Hist.* **24**: 130-38.
Hand, S.J., Archer, M., Godthelp, H., 1989. A fossil bat-rich, Oligo-Miocene cave-fill from Riversleigh Station, north-western Queensland. *Conf. Australasian Vert. Evol., Palaeont. and System., Abstracts*: p. 7.
Hand, S.J., Archer, M., Godthelp, H., 1989b. The Australian bat fauna: a 30 million year history? *Aust. Mamm. Soc. Conf. April, 1989, Abstracts*: p. 26.
Scanlon, J., 1988. The snakes of Riversleigh. *Riversleigh Notes* **3**: 7-8.
Woodburne, M.O., Tedford, R.H., Archer, M., Turnbull, W.D., Plane, M.D., Lundelius, E.L., 1985. Biochronology of the continental mammal record of Australia and New Guinea. *Spec. Publ. S. Aust. Dept Mines and Energy* **5**: 347-63.

Chapter 10

Archer, M., Hand, S.J., 1984. Background to the search for Australia's oldest mammals. Pp. 517-65 in "Vertebrate zoogeography & evolution in Australasia" ed M. Archer, G. Clayton. Hesperian Press: Perth.
Fenton, C.L., Fenton, M.A., Rich, P.V., Rich, T.H., 1990. 'The fossil book: a record of prehistoric life'. Doubleday: New York.

Chapter 11

Archer, M., Hand, S.J., 1984. Background to the search for Australia's oldest mammals. Pp. 517-65 in "Vertebrate zoogeography & evolution in Australasia" ed M. Archer, G. Clayton. Hesperian Press: Perth.
Archer, M., Hand, S.J., Godthelp, H., 1988. A new order of Tertiary zalambdodont marsupials. *Science* **239**: 1528-31.
Converse Jr., H.H., 1984. 'Handbook of paleo-preparation techniques'. Florida State Museum: Gainseville, Florida.
Croucher, R., Woolley, A.R., 1982. 'Fossils, minerals and rocks: collection and preservation'. British Museum (Natural History): London.

Chapter 12

Archer, M., 1983. A lion in possum's clothes. *Aust. Nat. Hist.* **20**: 373-79.
Archer, M., Godthelp, H., Hand, S.J., Megirian, D., 1989. Fossil mammals of Riversleigh, northwestern Queensland: preliminary overview of biostratigraphy, correlation and environmental change. *Aust. Zool.* **25**: 29-65.
Horner, J.R., Gorman, J., 1988. 'Digging dinosaurs'. Workman Publishing: New York.
Murray, P., Wells, R., Plane, M.D., 1987. The cranium of the Miocene thylacoleonid *Wakaleo vanderleuri*: click go the shears—a fresh bite at thylacoleonid systematics. Pp. 433-66 *in* "Possums and opossums: studies in evolution" ed M. Archer. Surrey Beatty & Sons Pty Ltd and the Roy. Zool. Soc. N.S.W.: Sydney.
Quirk, S, Archer, M., 1983. 'Prehistoric animals of Australia'. Australian Museum: Sydney.
Wells, R.T., Horton, D.R., Rogers, P., 1982. *Thylacoleo carnifex* Owen (Thylacoleonidae, Marsupialia): marsupial carnivore? Pp. 573-86 *in* "Carnivorous marsupials" ed M. Archer. *Roy. Zool. Soc. N.S.W.*: Sydney.
Willis, P.M.A., Murray, P., Megirian, D., 1990. *Baru darrowi gen. et sp. nov.*, a large, broad-snouted crocodiline (Eusuchia, Crocodylidae) from mid-Tertiary freshwater limestones in northern Australia. *Mem. Qld Mus.* **29**: 521-40.

Chapter 13

Ambrose, W., Duerdon, P., 1982 'Archaeometry: an Australian perspective'. Australian National University: Canberra.
Archer, M., Godthelp, H., Hand, S.J., Megirian, D., 1989. Fossil mammals of Riversleigh, northwestern Queensland: preliminary overview of biostratigraphy, correlation and environmental change. *Aust. Zool.* **25**: 29-65.
Geyh, M.A., Schleicher, H., 1990. 'Absolute age determination: physical and chemical dating methods and their application'. Springer-Verlag: Berlin.
Godthelp, H., Archer, M., Hand, S.J., Plane, M.D., 1989. New potoroine from Tertiary Kangaroo Well Local Fauna, N.T. and description of upper dentition of potoroine *Wakiewakie lawsoni* from Upper Site Local Fauna, Riversleigh. *Conf. Australasian Vert. Evol., Palaeont. and System., Abstracts*: p. 6.
Lindsay, J.M., 1987. Age and habitat of a monospecific foraminiferal fauna from near-type Etadunna Formation, Lake Palankarinna, Lake Eyre Basin. *Dept Mines and Energy South Australia* **Rept Bk**. 87/93.
Sigé, B., Hand, S.J., Archer, M., 1982. An Australian Miocene *Brachipposideros* (Mammalia, Chiroptera) related to Miocene representatives from France. *Palaeovertebrata* **12**: 149-72.
Woodburne, M.O., Clemens, W.A. (eds), 1986. Revision of the Ektopodontidae (Mammalia; Marsupialia; Phalangeroidea) of the Australian Neogene. *Univ. Calif. Publ. Geol. Sci.* **131**.
Woodburne, M.O., Tedford, R.H., Archer, M., Turnbull, W.D., Plane, M.D., Lundelius, E.L., 1985. Biochronology of the continental mammal record of Australia and New Guinea. *Spec. Publ. S. Aust. Dept Mines and Energy* **5**: 347-63.

Chapter 14

Archer, M., 1991. Life's scroll of prophecy: conservation and the fossil record. *Aust. Nat. Hist.* **23**: 654-55.
Archer, M., 1974. New information about the Quaternary distribution of the Thylacine (Marsupialia, Thylacinidae) in Australia. *J. Proc. R. Soc. W. Aust.* **57**: 43-50.
Archer, M., 1990. Why are Australia's animals so different? Pp. 32-40 *in* "The Australian environment: taking stock and looking ahead" ed by S. Neville. Australian Conservation Foundation: Melbourne.
Archer, M., Flannery, T.F., Ritchie, A., Molnar, R.E., 1985. First Mesozoic mammal from Australia—an early Cretaceous monotreme. *Nature* **318**: 363-66.
Archer, M., Godthelp, H., Hand, S.J., Megirian, D., 1989. Fossil mammals of Riversleigh, northwestern Queensland: Preliminary overview of biostratigraphy, correlation and environmental change. *Aust. Zool.* **25**: 29-65.
Archer, M., Hand, S.J., 1987. Evolutionary considerations. Pp. 79-106 *in* "The Koala: Australia's endearing marsupial" ed L. Cronin. Reeds Books: Sydney.
Archer, M., Hand, S.J., Godthelp, H., 1988. A new order of Tertiary zalambdodont marsupials. *Science* **239**: 1528-31.
Archer, M., Hand, S.J., Godthelp, H., 1989. Ghosts from green gardens. *Riversleigh Notes* **7**: 4-7.
Archer, M., Plane, M.D., Pledge, N.S., 1978. Additional evidence for interpreting the Miocene *Obdurodon insignis* Woodburne and Tedford, 1975, to be a fossil platypus (Ornithorhynchidae: Monotremata) and a reconsideration of the status of *Ornithorhynchus agilis* De Vis, 1885. *Australian Zoologist* **20**: 9-27.
Baynes, A., 1990. Biogeography of non-volant mammals in south-western Australia: past and future. *Aust. Mammal Soc. Newsl.* **Spring 1990.**
Flannery, T. F., 1989. Historic extinctions in Australia: aftershock of megafaunal loss? *Aust. Mammal Soc. Newsl.* Spring 1989.
Flannery, T.F., 1990. Pleistocene faunal loss: implications of the aftershock for Australia's past and future. *Archaeol. Oceania* **25**: 45-67.
Hand, S.J., 1990. The conservation status of two vulnerable Australian bats: view from the 4th dimension. *Macroderma* **5**: 11-12.
Kershaw, A.P., 1983. A Holocene pollen diagram from Lynch's Crater, north-eastern Queensland, Australia. *New Phytologist* **94**: 669-82.
Kershaw, A.P., 1986. Climatic change and Aboriginal burning in north-eastern Australia during the last two glacial-interglacial cycles. *Nature* **322**: 47-49.
Muirhead, J., 1987. The extinct Thylacine or Tasmanian Tiger. *Riversleigh Notes* **2**: 8.
Muirhead, J., Archer, M., 1990. *Nimbacinus dicksoni*, a plesiomorphic thylacine (Marsupialia: Thylacinidae) from Tertiary deposits of Queensland and the Northern Territory. *Mem. Qld Mus.* **28**: 203-22.
Rich, T.H., Archer, M., Plane, M., Flannery, T.F., Pledge, N.S., Hand, S.J., Rich, P.V., 1982. Australian Tertiary mammal localities. Pp. 526-72 *in* "The fossil vertebrate record of Australasia" ed by P.V. Rich, E.M. Thompson. Monash University Offset Printing Unit: Clayton.
Strahan, R. (ed.), 1983. 'The Australian Museum complete book of Australian mammals'. Angus & Robertson Publishers: Sydney.
Walbran, P.D., Henderson, R.A., Jull, A.J.T. & Head, M.J., 1989. Evidence from sediments of long-term *Acanthaster planci* predation on corals of the Great Barrier Reef. *Science* **245**: 847-850.

Walker, D., 1970. Conservation: a need for new strategems. *Search* **1**: 96-97.
Walker, D., 1990. Directions and rates of tropical rainforest processes. Pp. 23-32 *in*"Australian tropical rainforests: science, values, meaning" ed L.J. Webb, J. Kikkawa. CSIRO Australia: Melbourne.
Woodburne, M.O., Tedford, R.H., 1975. The first Tertiary monotreme from Australia. *American Museum Novitates* **2588**.

Appendix 1

Archer, M., Godthelp, H., Hand, S.J., Megirian, D., 1989. Fossil mammals of Riversleigh, northwestern Queensland: preliminary overview of biostratigraphy, correlation and environmental change. *Aust. Zool.* **25**: 29-65.
Archer, M., Hand, S.J., Godthelp, H., 1988b. Green cradle: the rainforest origins of Australia's marsupials. *7th Internat. Palyn. Cong., Brisbane. Abstracts*: p. 2.
Archer, M., Hand, S.J., Godthelp, H., 1989. Ghosts from green gardens: preliminary hypotheses about changes in Australia's rainforest mammals through time based on evidence from Riversleigh. *Riversleigh Notes* **7**: 4-7.
Barlow, B.A., Hyland, B.P.M., 1988. The origins of the flora of Australia's wet tropics. *Proc. Ecol. Soc. Aust.* **15**: 1-17.
Filewood, W., 1984. The Torres connection: zoogeography of New Guinea. Pp. 1121-31 *in* "Vertebrate zoogeography and evolution in Australasia" ed M. Archer & G. Clayton. Hesperian Press: Perth.
Flannery, T.F., 1989. Origins of the Australo-Pacific land mammal fauna. *Aust. Zool. Rev.* **1**: 15-24.
Flannery, T.F., Plane, M.D., 1986. A new late Pleistocene diprotodontid (Marsupialia) from Pureni, Southern Highlands Province, Papua New Guinea. *B.M.R. J. Aust. Geol. Geophys* **10**: 65-76.
Hand, S.J., Archer, M., Godthelp, H., 1989b. The Australian bat fauna: a 30 million year history? *Aust. Mamm. Soc. Conf. April, 1989, Abstracts*: p. 26.
Kershaw, A.P., 1976. A late Pleistocene and Holocene pollen diagram from Lynch's Crater, north-eastern Queensland. *New Phytol.* **77**: 469-98.
Plane, M.D., 1967. Stratigraphy and vertebrate fauna of the Otibanda Formation, New Guinea. *Bull. Bur. Miner. Resour. Geol. Geophys. Aust.* **86**: i-64.
Rich, T.H., Archer, M., Plane, M., Flannery, T., Pledge, N., Hand, S., Rich, P., 1982. Australian Tertiary mammal localities. Pp. 525-94 *in* "The fossil vertebrate record of Australasia" ed P.V. Rich, E.M. Thompson. Monash University Offset Printing Unit: Clayton.
Specht, R.L., 1988. Origin and evolution of terrestrial plant communities in the wet-dry tropics of Australia. *Proc. Ecol. Soc. Aust.* **15**: 19-30.
Turnbull, W.D., Lundelius Jr, E.L., 1970. The Hamilton Fauna a late Pliocene mammalian fauna from the Grange Burn, Victoria, Australia. *Fieldiana: Geology* **19**.
Turnbull, W.D., Rich, T.H.V., Lundelius Jr, E.L., 1987. The petaurids (Marsupialia: Petauridae) of the early Pliocene Hamilton Local Fauna, southwestern Victoria. Pp. 629-38 *in* "Possums and opossums: studies in evolution" ed M. Archer. Surrey Beatty & Sons Pty Ltd and The Royal Zoological Society of New South Wales: Sydney.
Turnbull, W.D., Rich, T.H.V., Lundelius Jr, E.L., 1987. Burramyids (Marsupialia: Burramyidae) of the early Pliocene Hamilton Local Fauna, southwestern Victoria. Pp. 729-39 *in* "Possums and opossums: studies in evolution" ed M. Archer. Surrey Beatty & Sons Pty Ltd and The Royal Zoological Society of New South Wales: Sydney.
Webb, L.J., Tracey, J.G., Williams, W.T., 1984. A floristic framework of Australian rainforests. *Aust. J. Ecol.* **9**: 169-98.
Webb, L.J., Tracey, J.G., Jessup, L.W., 1986. Recent evidence for autochthony of Australian tropical and subtropical rainforest floristic elements. *Telopea* **2**: 575-89.
Winter, J.W., 1988. Ecological specialization of mammals in Australian tropical and subtropical rainforest: refugial or ecological determinism. *Proc. Ecol. Soc. Aust.* **15**: 127-38.
Specht, R.L., 1988. Origin and evolution of terrestrial plant communities in the wet-dry tropics of Australia. *Proc. Ecol. Soc. Aust.* **15**: 19-30.
Woodburne, M.O., Tedford, R.H., Archer, M., Turnbull, W.D., Plane, M.D., Lundelius, E.L., 1985. Biochronology of the continental mammal record of Australia and New Guinea. *Spec. Publ. S. Aust. Dept Mines and Energy* **5**: 347-63.

Additional publications between 1991 and 1993 relevant to Riversleigh

Archer, M., 1991. Life's scroll of prophecy: conservation & the fossil record. *Aust. Nat. Hist.* **23**: 654-55.
Archer, M., 1991. Tharalkoo's child: an ugly duckling story. *Aust. Nat. Hist.* **23**: 574-75.
Archer, M., 1992. Ringtail possums (Pseudo cheiridae, Marsupialia) from the Tertiary deposits of Riversleigh. *The Beagle* **9**: 257.
Archer, M., 1993. Evolution of the Australian mammal fauna. *Abstracts Sixth International Theriological Congress* University of New South Wales, Sydney, Australia: 8-9.
Archer, M., 1993. Patagonian platypus. *Aust. Nat. Hist.* **24**: 60-61.
Archer, M., 1993. The origin and diversification of Australia's mammals. *Abstracts Sixth International Theriological Congress* University of New South Wales, Sydney, Australia: 8.
Archer, M., Godthelp, H., 1992. Wombat-like marsupials from the Oligo-Miocene faunal assemblages of Riversleigh. *The Beagle* **9**: 256.
Archer, M., Godthelp, H., Hand, S.J., 1993 (in press). Early Eocene marsupial from Australia. *Kaupia. Darmstadter Beitrage zur Naturgeschichte* **2** (in press).
Archer, M., Hand, S.J., Godthelp, H., 1992. Back to the future: the contribution of palaeontology to the conservation of Australian forest faunas. Pp. 67-80 *in* "Conservation of Australia's forest fauna" ed D. Lunney. Royal Zoological Society of New South Wales: Sydney.
Archer, M., Hand, S.J., Godthelp, H., 1993. Environmental and biotic change in the Tertiary of Australia. Pp. 1-9 *in* Abstracts for 'Conference on Palaeoclimate and Evolution with Emphasis on Human Origins', Airlie Conference Center, Virginia, 17-21 May, 1993.
Archer, M., Hand, S.J., Godthelp, H., 1993 (in press). Patterns in the history of Australia's mammals and inferences about palaeohabitats. *In* 'Origins of the Australian vegetation' ed. R. Hill. Cambridge University Press: London.
Archer, M., Jenkins Jr, F.A., Hand, S.J., Murray, P., Godthelp, H., 1992. Description of the skull and non-vestigial dentition of a Miocene platypus (*Obdurodon dicksoni* n. sp.) from Riversleigh, Australia, and the problem of monotreme origins. Pp. 15-27 *in* 'Platypus and echidnas' ed M.L. Augee. Royal Zoological Society of New South Wales: Sydney.
Archer, M., Murray, P., Hand, S.J., Godthelp, H., 1993 (in press). *In* 'Mammal phylogeny, Vol. 1. Mesozoic differentiation, monotremes, early therians, and marsupials' ed. F.S. Szalay, M.J. Novacek, M.C. McKenna. Springer Verlag: New York.
Boles, W.E., 1991. The origin and radiation of Australasian birds: perspectives from the fossil record. *ACTA XX Congressus Internationalis Ornithologici*: 383-91.
Boles, W.E., 1992. Revision of *Dromaius gidju* Patterson and Rich 1987 from Riversleigh, northwestern Queensland, Australia, with a reassessment of its generic position. *Nat. Hist. Mus. Los Angeles County, Science Series* **36**: 195-208.
Boles, W.E., 1993. A logrunner *Orthonyx* (Passeriformes: Orthonychidae) from the Miocene of Riversleigh, north-western Queensland. *Emu* **93**: 44-49.
Boles, W.E., 1993. A new cockatoo (Psittaci formes: Cacatuidae) from the Tertiary of Riversleigh, northwestern Queensland, and an evaluation of rostral characters in the systematics of parrots. *Ibis* **135**: 8-18.
Boles, W.E., 1993. *Pengana robertbolesi*, a peculiar bird of prey from the Tertiary of Riversleigh, northwestern Queensland, Australia. *Alcheringa* **17**: 19-25.
Boles, W.E., in press. A preliminary analysis of the Passeriformes from Riversleigh, northwestern Queensland, Australia, with the description of a new species of lyrebird. *Senckenbergiana.*
Boles, W.E., Godthelp, H., Hand, S.J., Archer, M., in press. Earliest Australian non-marine bird assemblage from the early Eocene Tingamarra Local Fauna, Murgon, southeastern Queensland, Australia. *Alcheringa.*
Brammall, J., 1993. A new Oligo-Miocene species of *Burramys* and a new look at burramyid diversity and phylogenetics. *Abstracts Sixth International Theriological Congress* University of New South Wales, Sydney, Australia: 34-35.
Cooke, B.N., 1992. Primitive macropodids from Riversleigh, northwestern Queensland. *Alcheringa* **16**: 201-17.
Cooke, B.N., 1992. Primitive macropodids from the Miocene freshwater limestone deposits of Riversleigh, northwestern Queensland. *The Beagle* **9**: 255.
Cooke, B.N., 1993. Bulungamayines, balbarines and evolutionary convergence among Miocene macropodoids. *Abstracts Sixth International Theriological Congress* University of New South Wales, Sydney, Australia: 59.
Cooke, B.N., 1993. New species of *Nambaroo* Flannery & Rich from Riversleigh, northwestern Queensland. *Abstracts CAVEPS, Adelaide, 19-21 April, 1993.*
Gaffney, E.S., Archer, M., White, A., 1992. *Warkalania*, a new meiolaniid turtle from the Tertiary Riversleigh deposits of Queensland. *The Beagle* **9**: 35-47.
Godthelp, H., 1993. *Zyzomys rackhami*, a new species of murid from the Pliocene Rackham's Roost deposit, northwestern Queensland. *Abstracts CAVEPS, Adelaide, 19-21 April, 1993.*
Godthelp, H., Archer, M., Cifelli, R., Hand, S.J., Gilkeson, C.F., 1992. Earliest known Australian Tertiary mammal fauna. *Nature* **356**: 514-16.
Godthelp, H., Archer, M., Hand, S., Sutherland, L., 1992. The Tingamarra Local Fauna, early Tertiary vertebrates from southeastern Queensland. *The Beagle* **9**: 268.
Hand, S.J., 1993. *A Cloud of Bats.* Bookshelf Publishing Australia Pty Ltd: Sydney.
Hand, S.J., 1993. A new and distinctive Oligo-Miocene hipposiderid (Microchiroptera: Hipposideridae) from Riversleigh Station, Queensland. *Abstracts CAVEPS, Adelaide, 19-21 April, 1993.*
Hand, S.J., 1993. Morphological changes in the Australian *Macroderma* lineage (Microchiroptera: Megadermatidae) from the Oligo-Miocene to the present. *Abstracts CAVEPS, Adelaide, 19-21 April, 1993.*
Hand, S.J., 1993. First skull of a species of *Hipposideros (Brachipposideros)* (Microchiroptera: Hipposideridae), from Australian Miocene sediments. *Mem. Qd Mus.* **33** (in press).
Hand, S.J., Archer, M., Godthelp, H., Rich, T.H., Pledge, N.S., 1993. *Nimbadon*, a new genus and three new species of Tertiary zygomaturines (Marsupialia: Diprotodontidae) from northern Australia, with a revision of *Neohelos. Mem. Qd Mus.* **33** (in press).
Hand, S.J., Godthelp, H., Novacek, M.J., Archer, M., 1992. An early Tertiary bat from the Tingamarra Local Fauna of southeastern Queensland. *The Beagle* **9**: 257.

Hand, S.J., Godthelp, H., Novacek, M., Archer, M., 1991. An early Tertiary bat from the Tingamarra Local Fauna of southeastern Queensland, Australia. Abstracts *International Conference Monument Grube Messel—Perspectives and Relationships 6-9 November, 1991* [Hessiches Landesmuseum Darmstadt, Germany].

Hand, S.J., Novacek, M., Godthelp, H., Archer, M., 1993? First Eocene bat from Australia. *J. Vert. Paleont.* (in press).

Hutchinson, M.N., 1992. Origins of the Australian scincid lizards: a preliminary report on the skinks of Riversleigh. *The Beagle* **9**: 61-70.

Hutchinson, M.N., 1993. Scaly-foots and pointy teeth: pygopod mandibular variation and the first record from the Miocene of Queensland. *Abstracts CAVEPS , Adelaide, 19-21 April, 1993.*

Megirian, D., 1992. Interpretation of the Miocene Carl Creek Limestone, northwestern Queensland. *The Beagle* **9**: 219-48.

Muirhead, J., 1992. Tertiary bandicoots. *The Beagle* **9**: 258.

Muirhead, J., 1993. The systematics and evolution of recent and fossil bandicoots (Marsupialia; Peramelemorphia). *Abstracts Sixth International Theriological Congress* University of New South Wales, Sydney, Australia: 215.

Muirhead, J., 1993. *Thylacinus macknessi*, a specialised thylacinid (Marsupialia: Thylacinidae) from Miocene deposits of Riversleigh, northwestern Queensland. *Aust. Mammal.* **15**: 67-76.

Muirhead, J., 1993. *Thylacinus macknessi*, a specialised thylacinid (Marsupialia: Thylainidae) from Miocene deposits of Riversleigh, northwestern Queensland. *Aust. Mammal.* **15**: 67-76.

Pascual, R., Archer, M., Ortiz Jaureguizar, E., Prado, J.L., Godthelp, H., Hand, S.J., 1992. First discovery of monotremes in South America. *Nature* **356**: 704-706.

Pascual, R., Archer, M., Ortiz Jaureguizar, E., Prado, J.L., Godthelp, H., Hand, S.J., 1992. The first non-Australian monotreme: an early Paleocene South American platypus (Monotremata, Ornithorhynchidae). Pp. 1-14 *in* 'Platypus and echidnas' ed M.L. Augee. Royal Zoological Society of New South Wales: Sydney.

Pledge, N.S., 1993. The Wynyardiidae—generalised diprotodont marsupials. *Abstracts Sixth International Theriological Congress* University of New South Wales, Sydney, Australia: 242.

Rich, T.H., Archer, M., Hand, S.J., Godthelp, H., Muirhead, J., Pledge, N.S., Flannery, T.F., Woodburne, M.O., Case, J.A., Tedford, R., Turnbull, W.D., Lundelius Jr, E.L., Rich, L.S.V., Whitelaw, M.J., Kemp, A., Rich, P.V., 1991. Appendix 1, Australian Mesozoic and Tertiary terrestrial mammal localities. Pp. 1005-1058 *in* Vickers-Rich *et al.* (1991).

Scanlon, J.D., 1992. A new large madtsoiid snake from the Miocene of the Northern Territory. *The Beagle* **9**: 49-60.

Scanlon, J.D., 1993. A small madtsoiid snake from the early Eocene Tingamarra Local Fauna, southeastern Queensland. *Darmstadter Beitrage zur Naturgeschichte* **2** (in press).

Shea, G.M., Hutchinson, M.N., 1992. A new species of lizard (*Tiligua*) from the Miocene of Riversleigh, Queensland. *Mem. Qd. Mus.* **32**: 303-10.

Tyler, M.J., 1991. A large new species of *Litoria* (Anura: Hylidae) from the Tertiary of Queensland. *Trans. Roy. Soc. S. Aust.* **115**: 103-105.

Tyler, M.J., 1991. Australian fossil frogs. Pp 592-604 in Vickers-Rich, Monaghan, Baird & Rich (1991).

Tyler, M.J., 1991. *Crinia* Tschudi (Anura: Leptodactylidae) from the Cainozoic of Queensland, with a description of a new species. *Trans. Roy. Soc. S. Aust.* **115**: 99-101.

Vickers-Rich, P., 1991. The Mesozoic and Tertiary history of birds on the Australian Plate. Pp. 722-808 in Vickers-Rich, Monaghan, Baird & Rich (1991).

Vickers-Rich, P., Monaghan, J.M., Baird, R.F., Rich, T.H. (eds), 1991. *Vertebrate palaeontology of Australasia.* Pioneer Design Studio: Melbourne.

White, A., Archer, M., 1993. A new Pleistocene turtle from fluviatile deposits at Riversleigh. *Abstracts CAVEPS , Adelaide, 19-21 April, 1993.*

Willis, P.M.A., 1992. Four new crocodilians from early Miocene sites at Riversleigh Station, northwestern Queensland. *The Beagle* **9**: 26.

Willis, P.M.A., 1993. *Trilophosuchus rackhami*, gen. et sp. nov., a new crocodilian from the early Miocene limestones of Riversleigh, northwestern Queensland. *J. Vert. Paleo.* **13**: 90-98.

Willis, P.M.A., Archer, M., 1990. A Pleistocene longirostrine crocodilian from Riversleigh: first fossil occurrence of *Crocodilus johnstoni* Krefft. *Mem. Qd Mus.* **28**: 159-64.

Willis, P.M.A., Molnar, R.E., Scanlon, J.D., 1993. An early Eocene crocodilian from Murgon, southeastern Queensland. *Darmstadter Beitrage zur Naturgeschichte* **2** (in press).

Wroe, S., 1992. Unique marsupial tooth replacement/function in *Ekaltadeta ima*, an Oligo-Miocene potoroid kangaroo from Riversleigh, northwestern Queensland. *The Beagle* **9**: 256.

Acknowledgements

We wish to acknowledge the vital financial support the Riversleigh Project has had from: the Australian Research Grant Scheme (Grant PG A3 851506P); the National Estate Grants Scheme (Queensland); the Department of Arts, Sport, the Environment, Tourism and Territories; IBM Australia Pty Ltd; ICI Australia Pty Ltd; the University of New South Wales; the Australian Geographic Society; the Queensland Museum; the Australian Museum; Queensland National Parks and Wildlife Service; the Australian Heritage Commission; Mount Isa Mines Pty Ltd; Surrey Beatty & Sons Pty Ltd; Wang Australia Pty Ltd; Ansett/Wridgway Pty Ltd; the Australian Broadcasting Commission; the Linnean Society of New South Wales; the Royal Zoological Society of New South Wales; Prospector's Supplies Pty Ltd; Bridges Consultants Pty Ltd; Reader's Digest Pty Ltd; and private supporters including Elaine Clark, Martin Dickson, Sue & Don Scott-Orr and Margaret Beavis. Critical logistical support in the field and laboratory has been received from: the Riversleigh Society Inc.; the University of New South Wales; the Friends of Riversleigh; the Royal Australian Air Force; the Australian Defence Force; the Queensland Geological Survey; the Queensland Museum; the Riversleigh Consortium (Riversleigh being a privately owned station); Campbell's Coaches Pty Ltd; the Mount Isa Shire; the Northwest Queensland Tourism and Development Board; the Burke Shire; the Gulf Local Authorities Development Association; PROBE; and many able volunteer workers, perceptive graduate students and enthusiastic colleagues.

Reed Books wish to thank *Australian Geographic* magazine for permission to reproduce artwork on pages 24/25 and 48 which are of an original concept from *Australian Geographic* No. 9 Jan–March 1988 — Riversleigh. Window into Our Ancient Past.

Index

Page numbers in italics indicate illustrations